Functional Nanomaterials Based on Self-Assembly

Functional Nanomaterials Based on Self-Assembly

Pavel Padnya

Basel • Beijing • Wuhan • Barcelona • Belgrade • Novi Sad • Cluj • Manchester

Pavel Padnya
A.M. Butlerov Chemistry Institute
Kazan Federal University
Kazan
Russia

Editorial Office
MDPI AG
Grosspeteranlage 5
4052 Basel, Switzerland

This is a reprint of articles from the Special Issue published online in the open access journal *Nanomaterials* (ISSN 2079-4991) (available at: www.mdpi.com/journal/nanomaterials/special_issues/nano_self_assembly).

For citation purposes, cite each article independently as indicated on the article page online and using the guide below:

Lastname, A.A.; Lastname, B.B. Article Title. *Journal Name* **Year**, *Volume Number*, Page Range.

ISBN 978-3-7258-2130-3 (Hbk)
ISBN 978-3-7258-2129-7 (PDF)
https://doi.org/10.3390/books978-3-7258-2129-7

© 2024 by the authors. Articles in this book are Open Access and distributed under the Creative Commons Attribution (CC BY) license. The book as a whole is distributed by MDPI under the terms and conditions of the Creative Commons Attribution-NonCommercial-NoDerivs (CC BY-NC-ND) license

Contents

About the Editor vii

Preface ix

Pavel Padnya
Editorial of Special Issue "Functional Nanomaterials Based on Self-Assembly"
Reprinted from: *Nanomaterials* **2023**, *13*, 3062, doi:10.3390/nano13233062 1

Anastasiya Malanina, Yurii Kuzin, Alena Khadieva, Kseniya Shibaeva, Pavel Padnya and Ivan Stoikov et al.
Voltammetric Sensor for Doxorubicin Determination Based on Self-Assembled DNA-Polyphenothiazine Composite
Reprinted from: *Nanomaterials* **2023**, *13*, 2369, doi:10.3390/nano13162369 4

Nadjib Kihal, Ali Nazemi and Steve Bourgault
Supramolecular Nanostructures Based on Perylene Diimide Bioconjugates: From Self-Assembly to Applications
Reprinted from: *Nanomaterials* **2022**, *12*, 1223, doi:10.3390/nano12071223 24

Dmitriy N. Shurpik, Yulia I. Aleksandrova, Olga A. Mostovaya, Viktoriya A. Nazmutdinova, Regina E. Tazieva and Fadis F. Murzakhanov et al.
Self-Healing Thiolated Pillar[5]arene Films Containing Moxifloxacin Suppress the Development of Bacterial Biofilms
Reprinted from: *Nanomaterials* **2022**, *12*, 1604, doi:10.3390/nano12091604 46

Darya Filimonova, Anastasia Nazarova, Luidmila Yakimova and Ivan Stoikov
Solid Lipid Nanoparticles Based on Monosubstituted Pillar[5]arenes: Chemoselective Synthesis of Macrocycles and Their Supramolecular Self-Assembly
Reprinted from: *Nanomaterials* **2022**, *12*, 4266, doi:10.3390/nano12234266 66

Anna N. Gabashvili, Maria V. Efremova, Stepan S. Vodopyanov, Nelly S. Chmelyuk, Vera V. Oda and Viktoria A. Sarkisova et al.
New Approach to Non-Invasive Tumor Model Monitoring via Self-Assemble Iron Containing Protein Nanocompartments
Reprinted from: *Nanomaterials* **2022**, *12*, 1657, doi:10.3390/nano12101657 76

Marisa Hoffmann, Christine Alexandra Schedel, Martin Mayer, Christian Rossner, Marcus Scheele and Andreas Fery
Heading toward Miniature Sensors: Electrical Conductance of Linearly Assembled Gold Nanorods
Reprinted from: *Nanomaterials* **2023**, *13*, 1466, doi:10.3390/nano13091466 90

Angela Candreva, Renata De Rose, Ida Daniela Perrotta, Alexa Guglielmelli and Massimo La Deda
Light-Induced Clusterization of Gold Nanoparticles: A New Photo-Triggered Antibacterial against *E. coli* Proliferation
Reprinted from: *Nanomaterials* **2023**, *13*, 746, doi:10.3390/nano13040746 101

Cheng-En Yang, Selvaraj Nagarajan, Widyantari Rahmayanti, Chean-Cheng Su and Eamor M. Woo
From Nano-Crystals to Periodically Aggregated Assembly in Arylate Polyesters—Continuous Helicoid or Discrete Cross-Hatch Grating?
Reprinted from: *Nanomaterials* **2023**, *13*, 1016, doi:10.3390/nano13061016 113

Elmira A. Vasilieva, Darya A. Kuznetsova, Farida G. Valeeva, Denis M. Kuznetsov and Lucia Ya. Zakharova
Role of Polyanions and Surfactant Head Group in the Formation of Polymer–Colloid Nanocontainers
Reprinted from: *Nanomaterials* **2023**, *13*, 1072, doi:10.3390/nano13061072 . . . . **131**

Yi-Shen Huang, Dula Daksa Ejeta, Kun-Yi (Andrew) Lin, Shiao-Wei Kuo, Tongsai Jamnongkan and Chih-Feng Huang
Synthesis of PDMS--PCL Miktoarm Star Copolymers by Combinations (*russian*) of Styrenics-Assisted Atom Transfer Radical Coupling and Ring-Opening Polymerization and Study of the Self-Assembled Nanostructures
Reprinted from: *Nanomaterials* **2023**, *13*, 2355, doi:10.3390/nano13162355 . . . . **156**

Anastasia Gileva, Daria Trushina, Anne Yagolovich, Marine Gasparian, Leyli Kurbanova and Ivan Smirnov et al.
Doxorubicin-Loaded Polyelectrolyte Multilayer Capsules Modified with Antitumor DR5-Specific TRAIL Variant for Targeted Drug Delivery to Tumor Cells
Reprinted from: *Nanomaterials* **2023**, *13*, 902, doi:10.3390/nano13050902 . . . . **171**

About the Editor

Pavel Padnya

Dr. Pavel Padnya graduated from the A.M. Butlerov Chemistry Institute of Kazan Federal University in 2010 and defended his PhD thesis in 2015 at Kazan Federal University. After that, he worked as an engineer, junior researcher, researcher, and senior researcher at the Organic and Medicinal Chemistry Department of Kazan Federal University, Russia. Since 2022, he is the leading researcher in the above department. He is the author of 83 publications in peer-reviewed international journals (h-index 18, Scopus). He is the editor of *Chimica Techno Acta* (ISSN 2411–1414) and a topical advisory panel member of *Nanomaterials* (MDPI) and *Journal of Functional Biomaterials* (MDPI). His research activity covers the design, synthesis, and study of applications of macrocyclic compounds and nanomaterials.

Preface

In recent years, the design and creation of new functional nanosystems and nanomaterials similar in properties to biological systems have been actively developing as an interdisciplinary field of research at the intersection of chemistry, biology, and physics. The base for the creation of such nanomaterials can involve both organic and inorganic components, including synthetic molecules (small and macrocyclic molecules), nanostructures (nanoparticles and organic–inorganic complexes), and natural polymers (nucleic acids, proteins, and carbohydrates). This reprint presents the most recent developments in the synthesis, modification, properties, and applications in various areas related to functional nanosystems and nanomaterials. This reprint will be useful to scientists of diverse backgrounds, i.e., material science, organic and inorganic chemistry, biochemistry, nanotechnology, and nanomedicine.

Pavel Padnya
Editor

Editorial of Special Issue "Functional Nanomaterials Based on Self-Assembly"

Pavel Padnya

A.M. Butlerov' Chemistry Institute, Kazan Federal University, 18 Kremlevskaya Street, 420008 Kazan, Russia; padnya.ksu@gmail.com

In recent years, the design and creation of new functional nanosystems and nanomaterials similar in their properties to biological systems showed remarkable progress as an interdisciplinary field of research combining chemistry, biology, and physics [1]. The creation of such nanomaterials can be based on both organic and inorganic components, e.g., synthetic molecules (small ones and macrocyclic molecules), nanostructures (nanoparticles and organic-inorganic complexes), and natural polymers (nucleic acids, proteins, and carbohydrates). This Special Issue covers a wide range of research topics related to the design and application of soft and hard functional nanomaterials.

It is well known that self-assembly is often used to create nanomaterials and composites to modify the sensor interface. This often leads to improvements in their properties, e.g., efficiency and selectivity of the detection of various analytes. The work by Prof. Evtugyn et al. [2] is an example of the creation of a voltammetric sensor for the detection of anticancer drug doxorubicin (DOX). The authors used a DNA-polyphenothiazine composite as a modifier able to recognize, accumulate and detect target species. Such hybrid materials exert unique characteristics to achieve the DOX determination in the concentration range from 10 pM to 0.2 mM (limit of detection 5 pM).

A review article by Prof. Bourgault and colleagues [3] is devoted to the latest achievements in development of functional systems to be obtained by self-assembly of π-conjugated systems on the perylene diimide platform (PDI). These systems show unique optical and electronic properties and demonstrate chemical, thermal and photo-stability. The synthesis, self-assembly, and applications of the PDI bioconjugates with biological molecules such as oligonucleotides and peptides are discussed in detail. The presented relationships between the structure of the PDI derivatives and their properties contribute to the tunable design and creation of innovative functional biomaterials demanded in various biomedical and nanotechnological applications.

In recent decades, the chemistry of synthetic macrocyclic compounds such as 2 (thia)calixarenes and pillararenes has been actively developed. The group of Prof. Stoikov is one of the leaders in this field in Russia. In recent work presented by this research team, Shurpik et al. [4] showed that decasubstituted thiolated pillar[5]arenes were able to form polymer self-healing films via thiol/disulfide redox cross-linking. The addition of the antimicrobial drug moxifloxacin to these films resulted in a marked inhibition of biofilm formation of *Staphylococcus aureus* and *Klebsiella pneumonia* bacteria. In another work by Prof. Stoikov [5], monosubstituted pillar[5]arene derivatives containing a diethylenetriamine spacer with one or two terminal carboxyl groups were synthesized. Solid lipid nanoparticles (SLNs) were assembled based on these macrocyclic compounds. The authors showed that the number of carboxyl groups in the macrocycle dramatically affected the size and shape of the yielding SLNs. In the case of the macrocycles with one carboxyl group, stick-like structures with the diameter of 50–80 nm and the length of 700–1000 nm were formed. The obtained functional systems based on the pillar[5]arenes can be further used to create promising materials.

A new approach to noninvasive tumor cell therapy was proposed by the scientific team of Prof. Abakumov [6]. The authors created a novel genetically encoded material based on *Quasibacillus thermotolerans* encapsulin stably expressed in mouse 4T1 breast carcinoma cells. The material structure contained the enzyme ferroxidase that oxidized Fe(II) ions to Fe(III) to form iron oxide nanoparticles. The authors showed that the expression of transgenic sequences did not affect cell viability, and the presence of magnetic nanoparticles resulted in an increase in T2 relaxivity.

Prof. Fery's group [7] synthesized gold nanorods via bottom-up template-assisted self-assembly. The authors studied the conductivity of the obtained nanomaterials and found that multiple parallel AuNR lines (>11) were required to achieve predictable conductivity properties. The results can be used to create resistance-based sensor wires or as anisotropically conducting surfaces in devices.

A new photoactivatable antibacterial drug based on gold nanoparticles (~50 nm) coated with polyethylene glycol was proposed and obtained by Prof. La Deda and coworkers [8]. The resulting supramolecular systems showed a 46% growth inhibition in the dark and a 99% growth inhibition in the light. The photothermal effect was shown to be due to the formation of a light-induced cluster of gold nanoparticles (at low concentrations). The authors suggested that the bacterium wall catalyzed the formation of these clusters that were increased in the measured temperature and caused bactericidal effect.

Prof. Woo and colleagues [9] studied the self-assembly of model arylate polyester polymers with different numbers of methylene segments (n = 3, 9, 10, and 12) in solid state. These polymers crystallized depending on the conditions as similar types of periodically banded spherulites. The observed phenomena were explained on the basis of crystal-by-crystal self-assembly with periodic branching and discontinuous intersection. The obtained results facilitate better understanding of mechanisms of periodic phenomena in crystal assembly in the scientific community.

Vasilieva et al. [10] studied the self-assembly of surface-active amphiphilic compounds containing different cationic head groups (pyrrolidinium and imidazolium) with anionic structures, i.e., polyacrylic acid and human serum albumin. The addition of polyanions led to a two-order decrease in the critical concentration of surfactant aggregation (from 1000 μM to 10 μM). As a result, nanocontainers capable of efficient drug loading (meloxicam, warfarin, and amphotericin B) were created based on the obtained supramolecular systems.

Another example of the use of self-assembly to create nanomaterials is the work of Huang and colleagues [11] devoted to the creation of star copolymers with a defined number of necessary monomer links. The authors obtained nanostructures of about 15 nm in size with low polydispersity. Thus, this work is an example of successful post-modification of linear polymers to obtain materials with new properties.

Gileva et al. [12] proposed and developed a novel system for targeted delivery of DOX based on polyelectrolyte multicellular capsules modified with the DR5-B protein. The developed supramolecular systems showed improved antitumor activity due to high tumor targeting and the synergistic effect of DR5-B and DOX.

Overall, this Special Issue includes 11 excellent papers of the contributors from eight countries on the design, synthesis and application of self-assembled functional nanomaterials and covers various fields of nanotechnology, chemistry, biology and material science. As guest editor of Special Issue "Functional Nanomaterials Based on Self-Assembly", I thank all authors, reviewers, and assistant editors for their valuable contributions. I hope that the publications presented in this Special Issue will be interesting and useful to readers.

Funding: This work was funded by the subsidy allocated to Kazan Federal University for the state assignment in the sphere of scientific activities no. FZSM-2023-0018 (Ministry of Science and Higher Education of the Russian Federation).

Conflicts of Interest: The author declares no conflict of interest. The funders had no role in the design of the study; in the collection, analyses, or interpretation of data; in the writing of the manuscript, or in the decision to publish the results.

References

1. Antipin, I.S.; Alfimov, M.V.; Arslanov, V.V.; Burilov, V.A.; Vatsadze, S.Z.; Voloshin, Y.Z.; Volcho, K.P.; Gorbatchuk, V.V.; Gorbunova, Y.G.; Gromov, S.P.; et al. Functional Supramolecular Systems: Design and Applications. *Russ. Chem. Rev.* **2021**, *90*, 895–1107. [CrossRef]
2. Malanina, A.; Kuzin, Y.; Khadieva, A.; Shibaeva, K.; Padnya, P.; Stoikov, I.; Evtugyn, G. Voltammetric Sensor for Doxorubicin Determination Based on Self-Assembled DNA-Polyphenothiazine Composite. *Nanomaterials* **2023**, *13*, 2369. [CrossRef] [PubMed]
3. Kihal, N.; Nazemi, A.; Bourgault, S. Supramolecular Nanostructures Based on Perylene Diimide Bioconjugates: From Self-Assembly to Applications. *Nanomaterials* **2022**, *12*, 1223. [CrossRef] [PubMed]
4. Shurpik, D.N.; Aleksandrova, Y.I.; Mostovaya, O.A.; Nazmutdinova, V.A.; Tazieva, R.E.; Murzakhanov, F.F.; Gafurov, M.R.; Zelenikhin, P.V.; Subakaeva, E.V.; Sokolova, E.A.; et al. Self-Healing Thiolated Pillar[5]Arene Films Containing Moxifloxacin Suppress the Development of Bacterial Biofilms. *Nanomaterials* **2022**, *12*, 1604. [CrossRef] [PubMed]
5. Filimonova, D.; Nazarova, A.; Yakimova, L.; Stoikov, I. Solid Lipid Nanoparticles Based on Monosubstituted Pillar[5]arenes: Chemoselective Synthesis of Macrocycles and Their Supramolecular Self-Assembly. *Nanomaterials* **2022**, *12*, 4266. [CrossRef] [PubMed]
6. Gabashvili, A.N.; Efremova, M.V.; Vodopyanov, S.S.; Chmelyuk, N.S.; Oda, V.V.; Sarkisova, V.A.; Leonova, M.K.; Semkina, A.S.; Ivanova, A.V.; Abakumov, M.A. New Approach to Non-Invasive Tumor Model Monitoring via Self-Assemble Iron Containing Protein Nanocompartments. *Nanomaterials* **2022**, *12*, 1657. [CrossRef] [PubMed]
7. Hoffmann, M.; Schedel, C.A.; Mayer, M.; Rossner, C.; Scheele, M.; Fery, A. Heading toward Miniature Sensors: Electrical Conductance of Linearly Assembled Gold Nanorods. *Nanomaterials* **2023**, *13*, 1466. [CrossRef] [PubMed]
8. Candreva, A.; De Rose, R.; Perrotta, I.D.; Guglielmelli, A.; La Deda, M. Light-Induced Clusterization of Gold Nanoparticles: A New Photo-Triggered Antibacterial against E. Coli Proliferation. *Nanomaterials* **2023**, *13*, 746. [CrossRef] [PubMed]
9. Yang, C.-E.; Nagarajan, S.; Rahmayanti, W.; Su, C.-C.; Woo, E.M. From Nano-Crystals to Periodically Aggregated Assembly in Arylate Polyesters—Continuous Helicoid or Discrete Cross-Hatch Grating? *Nanomaterials* **2023**, *13*, 1016. [CrossRef]
10. Vasilieva, E.A.; Kuznetsova, D.A.; Valeeva, F.G.; Kuznetsov, D.M.; Zakharova, L.Y. Role of Polyanions and Surfactant Head Group in the Formation of Polymer–Colloid Nanocontainers. *Nanomaterials* **2023**, *13*, 1072. [CrossRef] [PubMed]
11. Huang, Y.-S.; Ejeta, D.D.; Lin, K.-Y.; Kuo, S.-W.; Jamnongkan, T.; Huang, C.-F. Synthesis of PDMS-µ-PCL Miktoarm Star Copolymers by Combinations ( ) of Styrenics-Assisted Atom Transfer Radical Coupling and Ring-Opening Polymerization and Study of the Self-Assembled Nanostructures. *Nanomaterials* **2023**, *13*, 2355. [CrossRef] [PubMed]
12. Gileva, A.; Trushina, D.; Yagolovich, A.; Gasparian, M.; Kurbanova, L.; Smirnov, I.; Burov, S.; Markvicheva, E. Doxorubicin-Loaded Polyelectrolyte Multilayer Capsules Modified with Antitumor DR5-Specific TRAIL Variant for Targeted Drug Delivery to Tumor Cells. *Nanomaterials* **2023**, *13*, 902. [CrossRef]

Disclaimer/Publisher's Note: The statements, opinions and data contained in all publications are solely those of the individual author(s) and contributor(s) and not of MDPI and/or the editor(s). MDPI and/or the editor(s) disclaim responsibility for any injury to people or property resulting from any ideas, methods, instructions or products referred to in the content.

Article

Voltammetric Sensor for Doxorubicin Determination Based on Self-Assembled DNA-Polyphenothiazine Composite

Anastasiya Malanina [1], Yurii Kuzin [1], Alena Khadieva [1], Kseniya Shibaeva [1], Pavel Padnya [1,*], Ivan Stoikov [1] and Gennady Evtugyn [1,2,*]

[1] A.M. Butlerov Chemistry Institute, Kazan Federal University, 18 Kremlevskaya Street, Kazan 420008, Russia
[2] Analytical Chemistry Department, Chemical Technology Institute, Ural Federal University, 19 Mira Street, Ekaterinburg 620002, Russia
* Correspondence: padnya.ksu@gmail.com (P.P.); gennady.evtugyn@kpfu.ru (G.E.)

Citation: Malanina, A.; Kuzin, Y.; Khadieva, A.; Shibaeva, K.; Padnya, P.; Stoikov, I.; Evtugyn, G. Voltammetric Sensor for Doxorubicin Determination Based on Self-Assembled DNA-Polyphenothiazine Composite. *Nanomaterials* **2023**, *13*, 2369. https://doi.org/10.3390/nano13162369

Academic Editor: Cosimino Malitesta

Received: 21 July 2023
Revised: 11 August 2023
Accepted: 12 August 2023
Published: 18 August 2023

Copyright: © 2023 by the authors. Licensee MDPI, Basel, Switzerland. This article is an open access article distributed under the terms and conditions of the Creative Commons Attribution (CC BY) license (https://creativecommons.org/licenses/by/4.0/).

Abstract: A novel voltammetric sensor based on a self-assembled composite formed by native DNA and electropolymerized N-phenyl-3-(phenylimino)-3H-phenothiazin-7-amine has been developed and applied for sensitive determination of doxorubicin, an anthracycline drug applied for cancer therapy. For this purpose, a monomeric phenothiazine derivative has been deposited on the glassy carbon electrode from the 0.4 M H_2SO_4-acetone mixture (1:1 v/v) by multiple potential cycling. The DNA aliquot was either on the electrode modified with electropolymerized film or added to the reaction medium prior to electropolymerization. The DNA entrapment and its influence on the redox behavior of the underlying layer were studied by scanning electron microscopy and electrochemical impedance spectroscopy. The DNA–doxorubicin interactions affected the charge distribution in the surface layer and, hence, altered the redox equilibrium of the polyphenothiazine coating. The voltametric signal was successfully applied for the determination of doxorubicin in the concentration range from 10 pM to 0.2 mM (limit of detection 5 pM). The DNA sensor was tested on spiked artificial plasma samples and two commercial medications (recovery of 90–95%). After further testing on real clinical samples, the electrochemical DNA sensor developed can find application in monitoring drug release and screening new antitumor drugs able to intercalate DNA.

Keywords: DNA sensor; electropolymerization; self-assembling; phenothiazine electropolymerization; DNA intercalation

1. Introduction

Doxorubicin is an anthracycline drug commonly applied in solid tumor therapy. It was approved for medical application and involved in the WHO List of Essential Medicines [1]. Doxorubicin's effect involves its intercalation in the double-stranded DNA helix followed by suppression of biochemical functions, e.g., transcription and replication required for the DNA biosynthesis [2]. It can also cause the generation of reactive oxygen species in mitochondria by the reaction with dissolved oxygen [3]. The efficiency of doxorubicin is limited by frequently observed side effects, of which delayed cardiomyopathy, acute life-threatening inflammation of the bowel, and dermatological disorders are most commonly mentioned [4–6]. This calls for the effective monitoring of the doxorubicin levels in biological fluids and establishing personal doses of medication depending on its release from the organism. Traditionally, optical [7,8] and chromatographic [9,10] methods of analysis are used for this purpose.

Many efforts have been made in the development and application of new nanomaterials for extending optic detection systems. Thus, specially designed nanophotonic structures have been synthesized and ascribed with complex mathematical functions applicable for sensing elements [11]. Core@shell nanoparticles with an Ag core covered with a silica layer showed absorption in the UV–Vis spectrum that depended on the thickness of the shell [12]. The size and optical properties of the metamaterials obtained can be directly varied by the choice of precursor using the microwave polyol method [13]. The effect can be used for sensitive detection of many species affecting the optic phenomena on the nanoparticles surface. Epsilon-near-zero (ENZ) metamaterials with a narrow metallic waveguide channel are able to detect changes in the permittivity and refractive index of the dielectric microenvironment as a universal approach to the characterization of homogeneous materials [14]. Furthermore, graphene materials offer both optic and electrochemical properties related to the influence of analytes on electromagnetic properties of the materials able to detect DNA, recombinant proteins, and antibodies [15].

Being universal and sensitive, optical methods based on the detection of electromagnetic properties near the sensor interface need sophisticated data treatment and complicated protocols for material synthesis. Electrochemical approaches are more adaptive to the point-of-care testing conditions, including drug residue detection [16–18].

Electrochemical detection techniques are considered as a fast, inexpensive, and reliable alternative to more sophisticated universal instrumentation, Nevertheless, to reach the sensitivity of analysis comparable with chromatography, conventional electrodes should be modified with effective mediators of electron transfer that both amplify the currents recorded and diminish the working potential. The latter condition is important for the direct determination of analytes in complex media containing oxidizable organic species, e.g., biological fluids. Searching for new mediators and characterization of their implementation in the assembly of electrochemical sensors is one of the modern trends of electroanalytical chemistry. In the case of biosensors, such mediators can also provide effective immobilization of the biorecognition elements in the surface layer.

Phenothiazine derivatives have received a wide application in biochemistry and analytical chemistry as specific dyes exert reversible redox properties and are able to concentrate in the cells and organelles. They also exert antipsychotic, antimalarial, insecticidal, antifungal, antibacterial, and anthelmintic properties [19–21]. The redox activity of phenothiazines used as drugs has been applied for their direct electrochemical determination [22–24]. Monomeric dyes and the products of their electropolymerization were also tested as mediators and showed high electroactivity and efficiency as mediators and electrocatalysts [25–32].

Although many phenothiazine derivatives are easily adsorbed on the bare electrodes, their electropolymerization has additional advantages in sensor assembling, e.g., easy control of the reaction by the potential applied, variation in the quantities of the accumulated products on the electrode interface, no auxiliary reagents required in chemical polymerization (chemical oxidants, etc.), and the possibility of combining the deposition of redox active components and of the biopolymer entrapped in the growing polymer film [33]. Analytical application of electropolymerized materials started from polyaniline [34] and polypyrrole [35]; recently the attention of researchers was turned to the electropolymerization of phenazines, phenothiazines, and phenoxazines that can be performed in the absence of organic solvents and in milder conditions [36–40]. As examples, electropolymerization of thionine [41], methylene blue [42], methylene green [43], azure A [44], and azure B [45] can be mentioned. Appropriate polymeric products improved the conditions of the electron exchange and demonstrated catalytic activity in the conversion of hydrogen peroxide [46] and NADH [47]. Meanwhile, spontaneous aggregation and limited solubility of the monomeric dyes affects the regularity of the polymer films obtained and the reproducibility of the modifier properties. Further improvement of the analytical and operational characteristics of electrochemical sensors calls for searching for new derivatives and their testing in the assembly of biosensors for drugs selection, in solar batteries, and in charge storage devices [48–52].

Recently, we have studied the electrochemical behavior of N-phenyl-3-(phenylimino)-3H-phenothiazin-7-amine (PhTz) with electropolymerization, alone and in the assembly of the DNA sensor [53]. Phenyl imine fragments present in its structure allow two mechanisms of polymerization, including the one proposed for polyaniline in strong acidic media [54–56] and another one attributed to the phenothiazine core. In this work, we have for the first time polymerized the PhTz from the acidic media and showed intrinsic redox activity of the product (polyPhTz) in the assembly of the DNA sensor and in the presence of doxorubicin as a model intercalator.

2. Materials and Methods

Low-molecular DNA from salmon testes (DNA1, <5% protein, $A_{260/280}$ = 1.4), from fish sperm (DNA2, mol. weight 40–1000 kDa), and from chicken erythrocytes (DNA3, "Reanal", Budapest, Hungary, average mol. mass 1.2 MDa), doxorubicin hydrochloride (98–102%) (Figure 1a), idarubicin hydrochloride (>98%), daunorubicin hydrochloride (>90%), valrubicin (<100%), cyclophosphamide (EP reference standard), prednisone (dehydrocortisone, >98%), dacarbazine (EP reference standard), potassium hexacyanoferrate (III) (99%), potassium hexacyanoferrate (II) trihydrate (98.5–102%), and HEPES (4-(2-hydroxyethyl)-1-piperazineethanesulfonic acid) were purchased from Sigma-Aldrich, Dortmund, Germany. All solutions were prepared with Millipore Q® water (Simplicity® water purification system, Merck-Millipore, Mosheim, France). Doxorubicin-TEVA® and Doxorubicin-LANS® (lyophilizates for intravascular injection solutions) were purchased at a local pharmacy market.

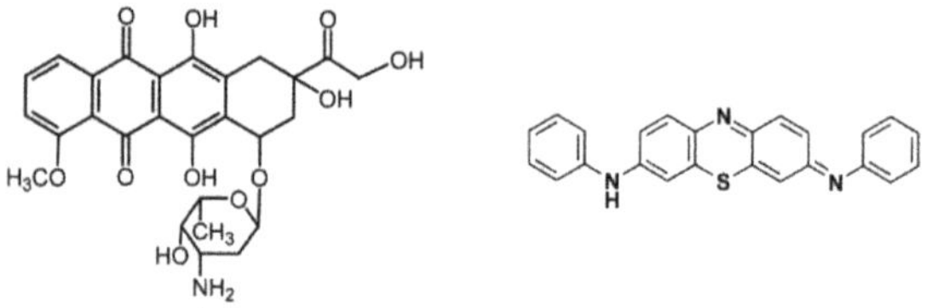

Figure 1. Chemical structures of doxorubicin (a) and N-phenyl-3-(phenylimino)-3H-phenothiazin-7-amine (b).

The PhTz (chemical structure in Figure 1b was synthesized as described elsewhere [57]. For electropolymerization, it was dissolved in acetone and then mixed with 0.4 M H_2SO_4 in 1:1 (v/v) ratio. In electropolymerization, the working concentration of PhTz (72 µM) corresponded to its maximal solubility. DNA was dissolved in deionized water.

2.1. Apparatus

Electrochemical measurements were performed with a µSTAT 400 (Metrohm DropSens, Oviedo, Spain) potentiostat/galvanostat and Autolab PGSTAT302 N equipped with the FRA32M module (Metrohm Autolab b.v., Utrecht, The Netherlands) at ambient temperature in a three-electrode working cell. A glassy carbon electrode (GCE, 2 mm in diameter, OhmLiberScience, Saint-Petersburg, Russia) was used as a working electrode and transducer in biosensor assembling. The Ag/AgCl/3.0 M NaCl reference electrode (ALS Co., Ltd., Tokyo, Japan, Cat. No. 012167) was used in voltametric measurements and Ag/AgCl/3.0 M KCl (Metrohm Autolab b.v. Cat No. 6.0733.100) in impedimetric measurements. Pt wire (ALS Co., Ltd., Cat. No. 002233, or Metrohm Autolab b.v., Cat. No. 6.1248.000) was applied as a counter electrode.

Cyclic voltammetry was chosen for the study of the DNA–polyphenothiazine composite formation and DNA–doxorubicin interaction. This method has a well-developed theory, an intuitively understandable explanation of the peak changes on voltammograms, and it is compatible with portable biosensors' requirements in the framework of point-of-care testing (POCT). Here, 25.5 cycles were used for electropolymerization assuming finalizing the synthesis at the highest anodic potential. All the measurements were performed with five individual sensors. Calculation of the peak currents was performed using the NOVA Software (Metrohm Autolab b.v.) after baseline correction. Geometric electrode area was used in calculation of the current densities.

Electrochemical impedance spectra (EIS) were recorded at the equilibrium potential in the frequency range from 100 kHz to 0.04 Hz, with 50 frequencies in the scan, at an amplitude of 5 mV. Equilibrium potential was determined as a half-sum of the peak potentials recorded in the equimolar mixture of $K_3[Fe(CN)_6]$ and $K_4[Fe(CN)_6]$. The impedance parameters were determined by fitting data with the Randles equivalent circuit $[R_s(Q[R_{ct}W])]$, where R_s represents the resistance of electrode material, of all electric contacts, and the uncompensated ohmic resistance of solution, Q is the constant-phase element (CPE) representing the non-ideal capacitive behavior of the electrical double-layer; R_{ct} is the charge transfer resistance (interfacial electron transfer), and W the Warburg element (a virtual electronic component that models the diffusion of electroactive species). Equivalent circuit fitting was performed with the NOVA software (Metrohm Autolab b.v.).

Scanning electron microscopy (SEM) images of the electrode coatings were obtained with the high-resolution field emission scanning electron microscope Merlin™ (Carl Zeiss, Jena, Germany). ZeissSmartSEM software was used for image processing.

Statistical data treatment was performed using OriginPro 8.1 software (OriginLab Corporation, Northampton, MA, USA).

2.2. Electrode Modification

Before use, the GCE was mechanically cleaned with silicon carbide abrasive paper (P5000) and washed with deionized water and ethanol. After that, it was electrochemically activated in 0.2 M sulfuric acid containing 50 vol.% acetone by cycling the potential until stabilization of the voltammogram. Then, an aliquot of the PhTz dissolved in the acetone was added to its final concentration of 72 µM. The electropolymerization was performed by multiple cycling of the potential in the range from -0.5 to 1.3 V, scan rate of 0.1 V/s. The DNA immobilization was performed by two various approaches, i.e., by addition of the DNA aliquot to the working solution on the electropolymerization step or by drop casting of the DNA solution onto the polyPhTz/GCE surface followed by drying the electrode on air at ambient temperature and its rinsing with deionized water. Both methods differ in the distribution of the DNA molecules in the self-association products on the electrode interface and in accessibility of the DNA molecules for the intercalator (doxorubicin as model analyte).

3. Results and Discussion

3.1. PhTz Electropolymerization

The presence of both phenyl imine and phenothiazine fragments in the PhTz structure is interesting from the point of view of the electropolymerization because they can alter the redox activity and an electron–ion conductivity depending on the centers involved in the formation of the polymeric product. Figure 2a represents cyclic voltammograms recorded in the H_2SO_4–acetone (1:1 v/v) solution containing PhTz.

The irreversible anodic peak at 1.1 V corresponds to the formation of the radical initiating the polymerization. If the maximal potential was chosen below this value, no significant changes were observed on multiple cycling. Similar activation of the polymerization at high anodic potential was reported for the electropolymerization of aniline and phenothiazine dyes [47,56]. Increasing the number of the cycles of the potential between -0.5 and 1.3 V consecutively increased the peaks attributed to the products of electrosynthesis (0.4 V on anodic branch of the voltammogram and 0.2 V at the cathodic branch). These changes were observed until the 25th cycle, after which the currents started decreasing due to blocking the GCE surface with the polymerization products, which complicated access of the monomers to the electrode interface. Such non-regular changes in the peak currents are typical for the electropolymerization of the phenothiazine and phenoxazine derivatives [58,59]. The dependence of the oxidation/reduction peak currents on the number of electropolymerization cycles is presented in Figure 2b.

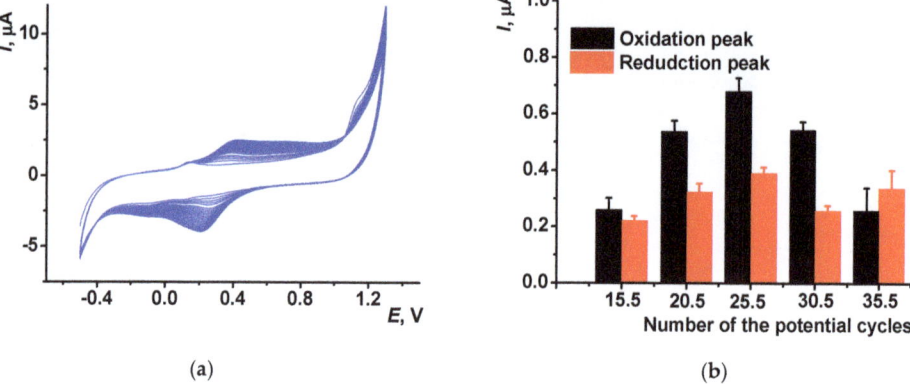

Figure 2. (**a**) Cyclic voltammograms of 72 µM PhTz dissolved in H$_2$SO$_4$_acetone (1:1 v/v) mixture, 100 mV/s; (**b**) The dependence of the peak currents recorded on the modified GCE in 0.1 M HEPES + 0.1 M NaNO$_3$ in the absence of PhTz monomer on the number of the potential cycles. Average ± S.D. for five individual sensors.

In the following experiments, 25.5 cycles of electropolymerization were used. A half-cycle corresponds to finishing the potential scan at the highest anodic potential required for the accumulation of the maximally oxidized product able to electrostatically interact with negatively charged DNA molecules.

The redox activity of the product of electropolymerization studied in HEPEs solution with no monomer corresponded to the reversible reduction/oxidation of the phenothiazine core, as shown in Scheme 1, for a single unit of the polymer present in a neutral (non-protonated) form. The formation of linear polymers for aminated phenothiazine derivatives was recently proved by ATR FTIR based on the signals related to phenylene diamine and quinone fragments [60].

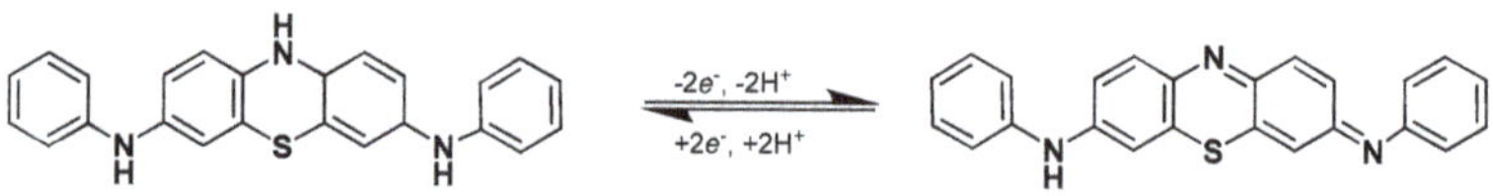

Scheme 1. Redox equilibrium of the monomeric form (PhTz) and the fragment of the polymer (polyPhTz) responsible for the first peak pair on voltamogram.

The dependence of the peak currents related to the PhTz redox conversion on the scan rate was assessed in 0.1 M HEPES. The slopes of plots in bi-logarithmic coordinates were near 1, indicating a surface-confined limiting step of the electron transfer (Figure 3).

Appropriate equations are presented in the following Equations (1) and (2):

$$\log(j_{pa}, \mu A/cm^2) = -(0.66 \pm 0.14) + (0.95 \pm 0.08) \times \log(\nu, mV/S), R^2 = 0.965 \quad (1)$$

$$\log(j_{pc}, \mu A/cm^2) = -(0.85 \pm 0.10) + (0.97 \pm 0.06) \times \log(\nu, mV/s), R^2 = 0.986 \quad (2)$$

where j_{pa}, j_{pc} are the peak current densities, µA/cm^2, of the anodic and cathodic peaks, respectively, and ν is the scan rate, mV/s. When electropolymerized in neutral media, polyPhTz showed mixed adsorption–diffusion control of the electron transfer [53].

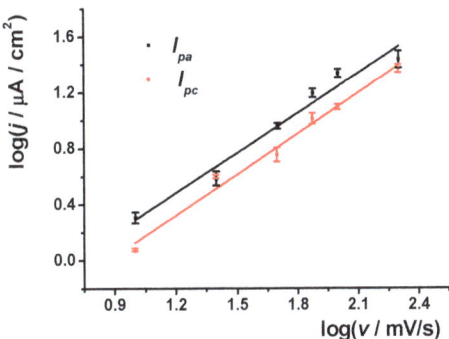

Figure 3. The dependence of the polyPhTz peak potentials on the scan rate. Average ± S.D. for five individual sensors.

The pH dependence of the peaks is shown in Figure 4a. In acidic media, the peak currents are similar to each other, and the peak potentials regularly shifted with the pH to their lower values (Figure 4b). The cathodic peaks overlapped the peaks of oxygen reduction so that maximal pH available was limited to 8.0. Within the range from pH = 2.0 to 8.0, the slope of the dependence (55 mV/pH for cathodic and 41 mV/pH for anodic peaks) corresponded to the transfer of equal number of electrons and hydrogen ions.

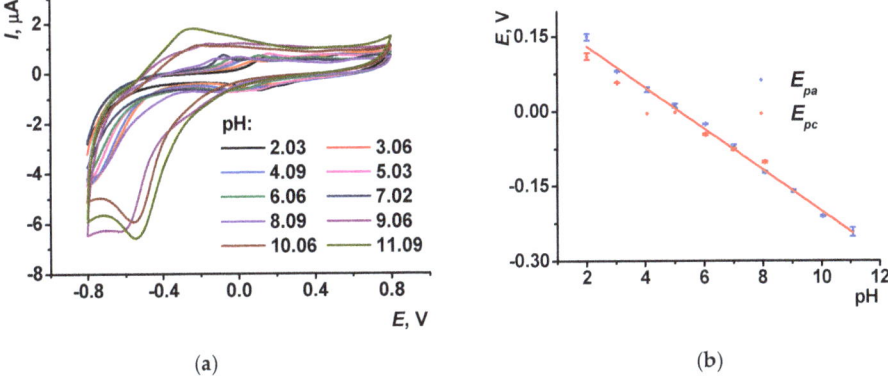

Figure 4. (a) Cyclic voltammograms recorded on the GCE covered with polyPhTz in 0.1 M HEPES + 0.1 M NaNO$_3$, 100 mV/s, at various pH; (b) the dependence of the peak potentials of anodic (E_{pa}) and cathodic (E_{pc}) peak potentials on the pH. Average ± S.D. for five individual sensors.

3.2. DNA Immobilization

3.2.1. Voltammetric Study

If DNA was added to the reaction media in the electropolymerization step, appropriate sensors are denoted as GCE/(polyPhTz + DNA). For them, implementation of DNA is promoted by electrostatic interactions between positively charged PhTz and negatively charged phosphate residues of the DNA helix.

As an example, Figure 5a illustrates electropolymerization performed in the presence of 12.5 µg/mL DNA1 from salmon sperm. The addition of DNA1 increased the currents on cyclic voltammograms against those shown in Figure 2a, where no DNA was added. In the course of electropolymerization, changes in the peak pair attributed to the monomeric PhTz at about 0.2–0.4 V are supported with by the appearance of a comparable in heigh anodic peak at 0.6 V, which can be due to the redox activity of the polyPhTz. The DNA's influence increases with its quantities and is especially obvious in the range from 12.5 to

18.7 µg/mL (Figure 5b). To reach a maximally accurate signal, 12.5 µg/mL DNA addition was used in the following experiments.

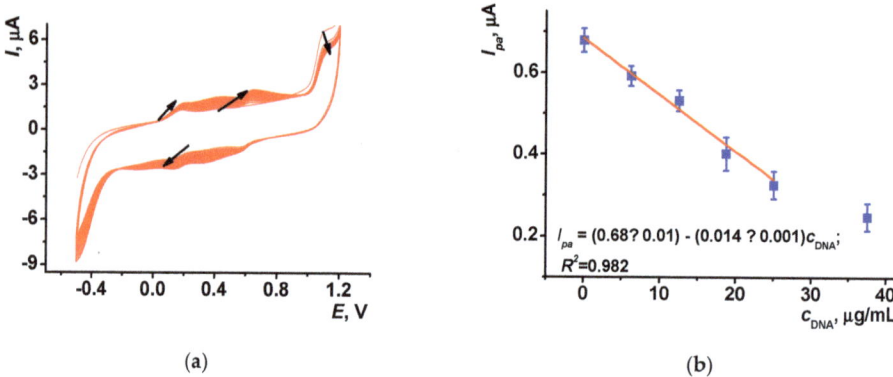

(a) (b)

Figure 5. (**a**) Multiple cyclic voltammograms recorded on GCE in a solution of 72 µM PhTz and 12.5 µM DNA1. Arrows show the direction of the changes with an increasing number of the cycles; (**b**) the dependence of the anodic peak current of the GCE modified with polyPhTz/DNA1 on the concentration of DNA in the solution for electropolymerization. Average ± S.D. for five individual sensors.

Incubation of the GCE modified with polyPhTz in the DNA solution resulted in its electrostatic accumulation on the polymer surface. The appropriate DNA sensor is denoted as GCE/polyPhTz/DNA. In this protocol, 10 µL aliquot of 0.1 µg/mL DNA was drop-casted on the electrode and left for a certain time period for the adsorption of the biopolymer. It was established in preliminary experiments that 20 min incubation was sufficient to reach a stable reproducible response toward DNA. Being simpler than previously described DNA entrapment methods into the growing electrosynthesized film, this protocol can be complicated by leaching adsorbed DNA from the surface layer. It was checked by monitoring the signal by alternating the incubation of the sensor in deionized water and HEPES buffer (Figure S1). The deviation in the peak currents for six repeated cycles measurement washing (measurement-to-measurement repeatability) was about 2.3%, and for six individual sensors (sensor-to-sensor repeatability) it was 3.3%. Similar experiments with the GCE/polyPhTz containing no DNA showed a small but regular decrease in the polyPhTz peak currents, probably due to partial desorption of the soluble products of oligomerization. Thus, DNA adsorption stabilized the surface film obtained. A further increase in the incubation period to 30–60 min, as well as higher DNA loading (up to 10 mg/mL in 10 µL casted per electrode) decreased the peaks on the resulting voltammograms (Figure S2). These changes are hardly attributed to the leakage of the surface components and are rather related to recharging the electrode interface and/or lower reversibility of the PhTz redox conversion caused by increased quantities of the highly charged DNA molecules transferred.

The DNA source affected the polyPhTz currents both in the case of GCE/(polyPhTz + DNA) and GCE/polyPhTz/DNA sensors (Figure 6). We have not found any evidence of DNA damage or denaturation during the electrolysis in acidic media for all the DNA sources. Probably, this was due to a short electrolysis time and electrostatic stabilization of the native DNA structure by depositing polyPhTz molecules.

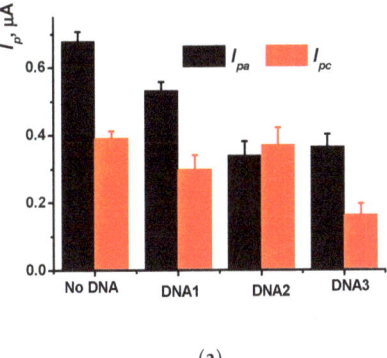 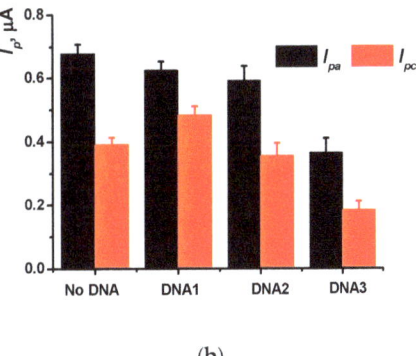

Figure 6. Influence of the DNA source on the PhTz peak currents: (**a**) GCE/(polyPhTz + DNA) sensor; (**b**) GCE/polyPhTz/DNA sensor. Average ± S.D. for five individual sensors.

In both protocols, the maximal changes in the peak currents were obtained with the application of high-molecular DNA3 from chicken erythrocytes. This might result from lower solubility of the poly(PhTz)–DNA3 complexes and from the accumulation of a larger negative charge affecting the stability of the coating. The difference in the appropriate peak currents was higher for the layered coating (GCE/polyPhTz/DNA sensor). DNA1 and DNA3 caused changes by about 50%, while those for the DNA implemented in the growing polymer film (GCE/(polyPhTz + DNA)) were about 30%. This indirectly provides evidence in favor of partial coverage of the electrode with non-conductive DNA molecules. The above parameter is sensitive to the specific square of the electrode occupied with DNA molecules. Regarding DNA1 and DNA2, their implementation into the polymer film mostly affected oxidation peaks, whereas DNA deposition onto the electrode moderately influenced reduction peaks, too. This might result from the molar mass distribution of both DNA samples. DNA2 from fish sperm contains a larger variety of double-stranded and single-stranded DNA fragments [60] that can more effectively block the access to the redox active sites on the electrode surface in comparison with DNA1 of a more regular structure. Difference in the DNA accumulation is mostly obvious on the reverse branch of voltammograms, where the product of primary electron transfer is re-reduced and the coordination of oppositely charged functional groups of the layer interface affects the shape of cyclic voltammogram to a higher extent.

The electrochemical behavior of the PhTz was compared with that of the derivatives containing additional charged groups at the phenothiazine core, i.e., the diaminated and dicarboxylated derivatives studied in [59]. All the compounds mentioned showed the ability of electropolymerization, but the quantities of the products deposited and the pH dependence of redox activity depended on the number and charge of terminal groups. The introduction of carboxylic groups suppressed the accumulation of polymeric products against aminated derivatives. Diaminated derivatives demonstrated more complicated voltammograms with the peaks of monomeric and polymeric forms comparable in height. Structurally relative thionine and azure dyes exerted the activity in electropolymerization sensitive to the steric loading of the amino groups [29,32,33,37,58]. Cyclic voltammograms of Azure A and Azure B demonstrated similar broad peaks of the polymer accumulation but the quantities of appropriate products assessed from the currents were significantly lower. In all cases mentioned, the electropolymerization allowed for DNA deposition. The thinner the electropolymerized film, the higher the sensitivity of the voltametric parameters to the DNA intercalators was. This might be explained by compatibility of the charge distribution in positively charged polymer and DNA molecules expected for thin films and to some extent by lower solubility of the polymerized dyes providing their dense contact with the underlying electrode surface. The possibility of the PhTz electropolymerization on bare

GCE with no additional modifiers is another advantage of the sensor developed versus the assemblies based on pre-concentration of the monomeric precursors on carbon black. The influence of amino groups of the PhTz molecule on the reversibility of the redox reactions estimated by ferricyanide redox probe was also lower than that of the phenothiazines with amino groups directly attached to the phenothiazine core. This partially compensates for exhausting pH sensitivity of the response mentioned for the analogs studied. Similar trends were discussed in the application of phenothiazine dyes for mediation of the oxidation of small molecules (see reviews [31,36,58]).

3.2.2. EIS Study

The DNA accumulation onto the polyPhTZ layer was then studied by the EIS with 0.01 M $[Fe(CN)_6]^{3-/4-}$ as the redox probe. The redox probe concentration was specified to reach well-resolved peak currents on the modified electrodes. The Nyquist diagrams obtained with the GCE modified with DNA from various sources are presented in Figure 7. The EIS parameters obtained by fitting the experimental data with a $[R_s(Q[R_{ct}W])]$ equivalent circuit (see the description of variables in the Section 2.1 Apparatus) are presented in Table 1. The Randles equivalent circuit used reflects the electrochemical phenomena on the electrode interface in terms of charge distribution at the electric double layer and kinetics of electron exchange with the redox probe.

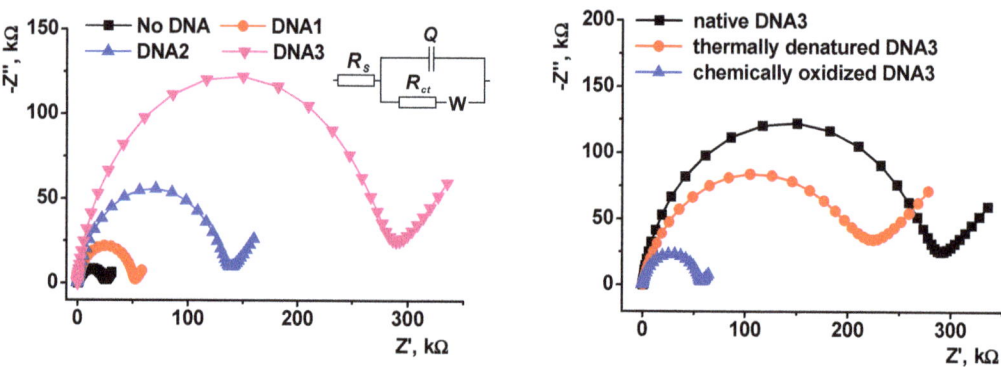

Figure 7. The Nyquist diagrams obtained with GCE covered with polyPhTz and DNA from various sources. Inset: equivalent circuit and semi-circle area of the diagram).

Table 1. EIS data of GCE modified with polyPhTz and drop-casted DNA from various source. Average ± S.D. for five sensors.

Modifier	R_S, Ω	R_{ct}, kΩ	Q, μF	N
PolyPhTz	162 ± 8	24.5 ± 3	0.27 ± 0.04	0.86
PolyPhTz/DNA1	191 ± 8	51.2 ± 4	0.32 ± 0.05	0.91
PolyPhTz/DNA2	171 ± 10	134 ± 6	0.08 ± 0.03	0.88
PolyPhTz/DNA3	172 ± 12	277 ± 6	0.04 ± 0.01	0.90

The semi-circle area on the Nyquist diagram corresponds to the kinetic control of the electron exchange whereas the linear piece in the range of low frequencies corresponds to the diffusion control of the reaction. The depressed form of the semi-circles is caused by non-ideal behavior of the CPE and coincides well with the roughness factor N, which differs from 1 corresponded to capacitance behavior. Such a behavior can be explained by uneven distribution of the modifiers along the electrode, or by high roughness and porosity of the surface layer on the electrode [61]. With the increased average molar mass of the DNA (from DNA1 to DNA3), the charge transfer resistance R_{ct} increased because of the accumulation of a higher negative charge and repulsion of anionic ferro- and ferricyanide ions of the redox probe on the electrode interface. A sharp decrease in the Q value observed for DNA3 with maximal molar mass is due to the same reason. The EIS data agrees with the voltammetry studies of the surface layer assembling and with the conclusion on the importance of negative charge of DNA helix and its alteration after the DNA deposition (see Figure 6b) with minimal peak currents recorded for DNA3 loading.

It is interesting that the redox signals of the DNA sensors were sensitive to thermally denatured and chemically oxidized DNA. Such a differentiation has already been described for similar DNA sensors [29,32,33,62] and was referred to changes in the flexibility of the DNA molecules. High-molecular DNA3 was heated for 30 min to 95 °C and then sharply cooled for 5 min in the ice bath prior to its introduction in the surface layer of the GCE/polyPhTz/DNA sensor. This resulted in partial unwinding of the DNA strands and primary structure distortion (formation of abasic sites etc.) [63]. The EIS experiment showed the decrease in the R_{ct} value to 212 ± 8 kΩ, indicating satisfactory better electroconductivity of the polymer layer [64].

3.2.3. SEM Study

The assembling of the polyPhTz/DNA surface layer was proved by SEM. On bare GCE, electropolymerization resulted in the formation of a dense film with a wavy surface (Figure 8A). The addition of DNA either to the PhTz solution or on the surface of the polymer film resulted in the formation of the structured elements with microcrystalline inclusions and cavities randomly distributed on the surface (Figure 8B,C). The morphology of the polyPhTz/DNA1 layer was similar for both protocols of the DNA introduction but the depth of the cavities and probably the thickness of the layer were bigger for the DNA deposited on the film. One could see the roundish particles within the surface layer, mostly obvious for the side walls of the cavities and after the DNA implementation (Figure 8C). Their size was in between 40–80 nm and tended to increase with the DNA molar mass.

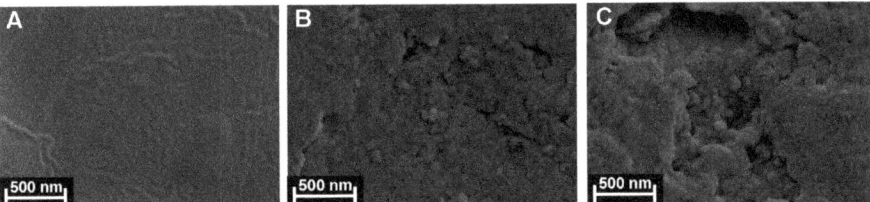

Figure 8. SEM images of GCE covered with electropolymerized PhTz (**A**) (25.5 cycles, 72 μM PhTz) and image obtained with DNA1 added to the solution (37.5 μg/mL) (**B**) or drop-casted on the polymer film (10 μL of 100 ng/mL per electrode) (**C**).

3.3. Measurement Conditions

3.3.1. Cyclic Voltammetry

The dependence of the PhTz peak currents on the DNA loading and electrostatic interactions within the layer as a driving force of the changes made it possible to propose the DNA sensor for the determination of low-molecular species able to specific interaction with DNA. In this work, doxorubicin was used as a kind of model intercalator because of the importance of its sensitive detection in biological fluids and well-elaborated mechanism of interaction with DNA.

Indeed, incubation of the GCE/polyPhTz/DNA sensor in the doxorubicin solution resulted in an increase in the peak currents to the values obtained for the polyPhTz coating with no DNA (Figure 9 for DNA1, the oxidation peak current for the GCE/polyPhTz was equal to 0.62 μA).

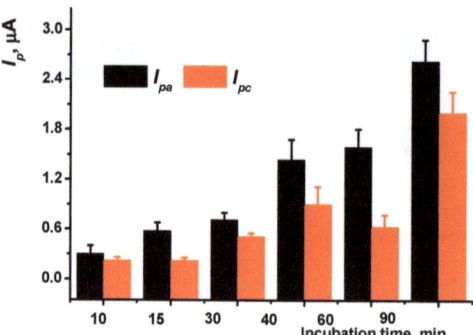

Figure 9. The polyPhTz peak currents recorded after incubation of the GCE/polyPhTz/DNA1 sensor in 0.1 mM doxorubicin solution for 10–90 min. Average ± S.D. for five individual sensors.

Intercalation of the drug in the DNA helix results in an increase in the biopolymer specific volume, partial unwinding, and spatial separation of the negative charges of the phosphate residues of the DNA backbone. These changes promote transfer of the phenothiazine units in oxidized (positively charged) form and hence increase their currents on voltammogram. The influence of the drug on the DNA sensor signals was rather stable within the 10–30 min incubation period in the doxorubicin solution. A further increase in the incubation period resulted in a lower reproducibility of the response, compensated for by higher currents.

Changes in the oxidation peak currents caused by doxorubicin were more reproducible than those of cathodic peaks and they were used for the quantitative analysis of the drug. Figure 10 shows changes in the voltammograms for 15 min incubation of the GCE/polyPhTz/DNA1 sensor in the doxorubicin solution. The results obtained with DNA2 and DNA3 are shown in Figure S3.

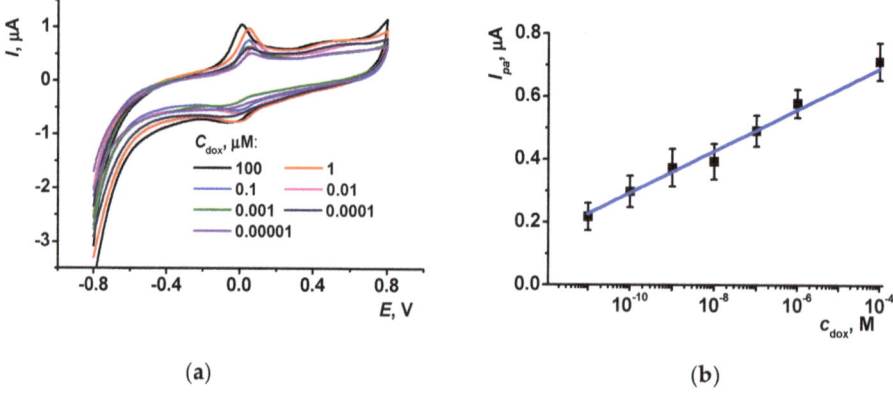

Figure 10. Influence of the doxorubicin on the PhTz peak currents: (**a**) cyclic voltammograms recorded after 15 min incubation with the GCE/polyPhTz/DNA1 sensor; (**b**) calibration curve of doxorubicin. Average ± S.D. for five individual sensors.

The analytical characteristics of the doxorubicin determination are summarized in Table 2. The limit of detection (LOD) of doxorubicin was assessed with S/N = 3 criterion. As one can see, application of DNA from salmon testes showed the broadest range of concentrations determined. The use of DNA2 from fish sperm was found to be less effective. The slope of the linear part of the plot was similar that that obtained with DNA1, but the interval was narrower and the relative deviation of the peak current was higher. This might result from the rather high content of single-stranded DNA fragments. They do not interact with doxorubicin in accordance with the intercalation mechanism. High-molecular DNA3 from chicken erythrocytes was less sensitive to the contact with doxorubicin, probably due to a much larger number of binding sites and the higher stability of less soluble polyelectrolyte complexes on the electrode interface.

Table 2. Analytical characteristics of doxorubicin determination with GCE/polyPhTz/DNA, I_{pa}, µA = a + b × (c_{dox}, M). Average ± S.D. for five sensors, where n is the number of experimental points within the linear range of concentrations determined.

DNA Source	$a \pm \Delta a$	$b \pm \Delta b$	R^2	n	Linear Concentration Range	LOD
DNA1	0.99 ± 0.02	0.070 ±0.003	0.9898	7	10 pM–0.2 mM	5 pM
DNA2	1.12 ± 0.11	0.078 ± 0.012	0.9784	5	20 pM–12 nM	10 pM
DNA3	1.06 ± 0.06	0.089 ±0.009	0.9529	5	15 nM–0.3 mM	50 nM

The results obtained are comparable or better than those reported for other electrochemical sensors (Table S1). It should be noted that doxorubicin itself is electrochemically active and shows its own peaks on voltammograms but at much higher concentrations so that they do not interfere with the redox signals of the polyPhTz layer.

The use of redox active polymers offers sub-nanomolar detection limits sufficient for the drug residues determination in urine and blood serum. The common concentrations of doxorubicin reported for patients with gastrointestinal cancer were on the level of 10 nM within 60–100 min after the drug administration [65]. In plasma of superficial bladder cancer patients, doxorubicin was detected at a maximal level of 0.9–8 nM (mean 2 nM for six weekly treatments of 40 mg doxorubicin in 20 mL physiological saline [66]). The concentration of doxorubicin in urea depends on the dose and excretion period. Within two weeks, about 50% of the drug is released with urine. Its stability depends on the pH and other conditions but can be assessed as 0.1–10 µM [67,68]. Thus, the comparison of the DNA sensor performance with the drug levels makes it possible to conclude that possible interference expected from the matrix components can be easily eliminated by the sample dilution.

The only DNA sensor showing lower detectable concentrations contained layered polyaniline and DNA layers (polyaniline–DNA–polyaniline). Polyaniline offered higher electrostatic attraction of DNA than polyPhTz and is more sensitive to DNA-specific interactions but had limitations in terms of pH working region and stability in neutral media. Electrochemical sensors mostly utilize mediated oxidation of doxorubicin and/or its preliminary accumulation in the surface layer. High working potential prevents their application in complex media containing oxidizable species (drug stabilizers, food additives) though the stability of the response in protein-containing media is rather high. It should be noted that the approach to the layered deposition of polyPhTz utilized in this work offers a very simple and reproducible protocol of sensor assembling and makes it possible to obtain inexpensive but reliable sensors for single use in medical diagnostics, pharmaceutics, chemotherapy monitoring, and other purposes related to oncology and antitumor drug design.

3.3.2. Measurement Precision and Sensor Lifetime

Sensor-to-sensor repeatability was assessed using a set of five electrodes modified with the same reagents. The R.S.D. was found to be 5.6% (0.1 nM doxorubicin, 20 min incubation) and the period of 25% decay of the signal was equal to 15 days (Figure S4a). In dry conditions, polyPhTz retains its electrochemical activity for at least 60 days with no DNA and 90 days when covered with 0.1 mg/mL DNA (Figure S4b). The R.S.D. of the signal toward 0.1 nM doxorubicin increased to 9% at the end of the storage period. In working solution, the reliable detection of doxorubicin (R.S.D. < 10%) was achieved within three days.

3.3.3. Selectivity and Real Sample Assay

The possibility of direct determination of doxorubicin in serum is important due to the high risks of cardiotoxicity and individual variability in the pharmacokinetics of the drug. We have checked the signals of other antitumor drugs, including the same anthracycline core as well as commercial medications with different stabilizers and common components of the fluids to be sure that the complex character of the samples did not interfere with the doxorubicin determination.

Determination of doxorubicin is based on the general mechanism of intercalation, which is common for all the anthracycline preparations. To compare the influence of other representatives of this group of drugs, the concentrations exerting a 15% shift in the signal (IC_{15}) were determined with the GCE/polyPhTz/DNA1 sensor. They were equal to 5 pM for doxorubicin, 20 pM for daunorubicin, 30 pM for idarubicin, and 18 nM for valrubicin for GCE/polyPhTz/DNA1 sensor, at 15 min incubation. A 15% shift was chosen assuming a 5% deviation of the signal typical for electrochemical sensors. Relative changes in the anthracycline detection coincide with the previously reported results obtained with other electrochemical DNA sensors. Thus, the LODs of 0.01 nM doxorubicin, 0.1 nM daunorubicin, and 0.2 nM idarubicin were obtained using the EIS technique with the DNA sensors based on thin polyaniline films [69]. Valrubicin showed lower sensitivity with DNA sensors with carbon nanomaterials and Au nanoparticles (LOD 18 nM [70]). It should be noted that anthracycline drugs are mostly applied separately, and medications differ in the nature of auxiliary components providing target delivery of the substances to the solid tumor (lipids, stabilizers, etc.). Regarding other drugs, IC_{15} = 0.01 mM was obtained for sulfamethoxazole. Cyclophosphamide, prednisone, and dacarbazine, commonly applied together with doxorubicin in chemotherapy regimens, showed an insignificant irregular influence on the GCE/polyPhTz/DNA1 signal at concentrations exceeding 1.0 mM.

Some matrix components capable of redox conversion on the electrode can also interfere with the doxorubicin signal. In Figure 11, the relative changes in the signal toward 1.0 nM doxorubicin are compared with those obtained in the presence of 1.0 mM interferences (glucose, uric acid, catechol, ascorbate). Furthermore, urea and KCl as macro components of urine were tested in the same conditions. Here, 100% corresponded to no interference, and the higher is the difference from 100%, the bigger the undesirable contribution of the additives. One can see that the effect is below the standard deviation of the signal for all the model compounds tested.

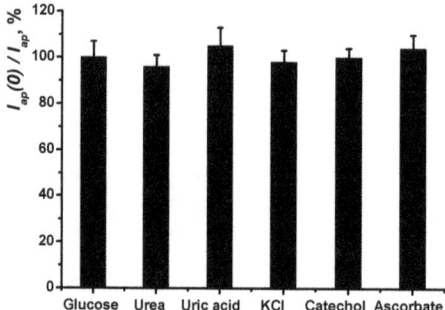

Figure 11. Relative changes in the polyPhTz anodic peak current recorded with the GCE/polyPhTz/DNA1 sensor after 15 min incubation in 1.0 nM doxorubicin solution ($I_{pa}(0)$) and in the presence of 1.0 mM interferences (I_{pa}). Measurements in 0.1 M HEPES + 0.1 M $NaNO_3$, 100 mV/s. Average ± S.D. for five individual sensors.

The influence of serum proteins was modeled by bovine serum albumin added to the HEPES buffer. Moderate level of albumin typical for the adults' blood serum was selected for model experiments (41.4 mg/mL).

Furthermore, Ringer–Locke's solution (0.45 g NaCl, 0.021 g KCl, 0.016 g $CaCl_2 \cdot 2H_2O$, 0.005 g $NaHCO_3$, 0.015 g of $MgSO_4$, and 0.025 g of $NaH_2PO_4 \cdot 2H_2O$ per 50 mL of water [71]) was used to estimate the influence of plasma electrolytes. The recoveries of $105 \pm 10\%$ and $105 \pm 8\%$ were found for 0.1 nM doxorubicin in both cases.

Doxorubicin-TEVA® and Doxorubicin-LANS® were dissolved in 0.1 M HEPES buffer and applied for the incubation as described above for a standard doxorubicin solution. The recovery was calculated from the calibration plot obtained in HEPES buffer, pH = 7.0. Three nominal concentrations of doxorubicin (10 and 100 pM, and 1.0 nM) were tested, and average recovery was equal to $90 \pm 10\%$ (Doxorubicin-TEVA) and $95 \pm 10\%$ (Doxorubicin-LANS). Thus, stabilizers present in the medications (lactose and mannitol) did not interfere with the doxorubicin determination.

4. Concluding Remarks

Fast and reliable determination of anthracycline drugs in biological fluids is demanded in two areas related to the anticancer drugs' administration and to the selection of new drugs with reduced toxic effects. The rather high cardiotoxicity of doxorubicin and its analogs, together with hepatotoxicity, limit their application and increase health risks in the population. Although they are effective, especially in the solid tumor family, anthracycline drugs need monitoring for an effective dose depending on metabolism specificity and other influencing factors. DNA-based biosensors represent an approach to the quantification of DNA–drug interactions because they utilize biological targets, i.e., native DNA molecules, as both recognition elements and signal forming components. Common approaches based on direct analysis of accumulated drugs or detection of the products of DNA damage are rather sensitive but require sophisticated equipment and skilled staff. Electrochemical DNA sensors can offer attractive opportunities, e.g., very simple assembling and signal measurement protocol, and intuitively understandable interpretation of the results.

Electropolymerization is one of modern trends broadly applied for assembling biorecognition layers of biosensors. Regarding DNA–drug interaction, this protocol allows self-assembling of the surface layer consisting of oppositely charged counterparts (polymers and DNA molecules), the electrochemical properties of which depend on both the drug nature and conditions of their interaction. The electropolymerization protocol is easily modified for particular tasks and can control the growth and assembling of the surface film. Furthermore, the redox properties of the nanoaggregates formed are sensitive to the DNA structure and intercalation of the anthracycline drugs. One-pot synthesis and the possibility of entrapping DNA in the polymer film via weak multipoint interactions are the advantages of the approach over the use of conventional modifiers, e.g., carbon nanoparticles or quantum dots. The low cost of consumables and rather high reproducibility of main biosensor parameters should also be mentioned.

However, the advantages of electropolymerization can be applied only in the case of careful control of the assembling conditions and specifying the factors that can influence the sensitivity of drug detection. Variation in the monomer structure and implementation of charged terminal groups near the redox centers makes it possible to specify the main requirements to the assembling of the biorecognition layer for particular analytes and real sample assays.

In this work, a new derivative of phenothiazine has been investigated for the first time in the conditions of polymerization in acidic conditions. The choice of the phenothiazine monomer was governed by the existence of two alternative ways of polymerization that would result in the polymeric films with various distribution of charge and ability to interact with DNA. In basic media, phenothiazine-like polymerization takes place, as was shown for similar structures of phenothiazine dyes [48,54]. In acidic media, electropolymerization via aniline fragment can take place, resulting in the formation of an electroconductive coating, such as polyaniline. We have found that multiple cycling the potential in the monomer solution performed in sulfuric acid mixed with an equal volume of acetone resulted in the deposition of a stable film easily adsorbing DNA both on the surface and by entrapment from the reaction media. The polymers obtained from acidic media exert higher sensitivity toward the DNA intercalator and lower deviation in the signal in repeating records of cyclic voltammograms than similar DNA-sensors assembled in a neutral solution.

The implementation of DNA was proved by EIS data and the dependency of the redox activity of the coating on the DNA source and biopolymer deposition protocol. Indeed, the influence of DNA is based on two simultaneous reactions, i.e., the limitation of electron transfer by partial blocking of the electrode surface and the promotion of the formation of oxidized (positively charged) polymers due to electrostatic interactions with the phosphate residues of the DNA backbone. The balance between these two processes makes the recorded redox peaks on voltammograms sensitive to the DNA source and interaction with model intercalator (doxorubicin). SEM images confirmed the nanostructuring of the surface film because of the uneven distribution of the polyelectrolyte molecules and different distances between the charge locations. The roughness of the coating promotes the DNA implementation and increases the active surface area. The latter one positively affects the amperometric signal and redox conversion of the polymer fragments responsible for drug detection.

Three different DNA preparations were used. High-molecular DNA from chicken erythrocytes formed the most stable complexes with polyPhTz. However, the response of the appropriate biosensor toward doxorubicin was lower than that for the biosensors implementing DNA from other sources. DNA from fish sperm with maximal variety of the molar mass and rather high content of single-stranded DNA strands was found to be less applicable in the DNA sensor assembly, both from the point of view of the redox characteristics of the surface layer and its sensitivity toward the intercalator.

Regarding the DNA implementation, both protocols suggested (addition of DNA to the reaction media and its adsorption onto the polymer layer after electropolymerization) showed similar effects on the electrochemical characteristics of the composites formed. This confirms the idea of the predominant influence of charge separation on the redox equilibrium of polyPhTz. Meanwhile, the influence of doxorubicin was higher when DNA was placed onto the polymer film. This can result from better accessibility of the DNA active sites for the drug molecules and/or from the difference in the roughness of the layer indirectly shown by SEM. On the other hand, DNA adsorption complicates the protocol of the biosensor assembling and slightly decreases the accuracy of the signal measurement.

Intercalation of doxorubicin resulted in partial unwinding of double-stranded DNA and changes in the negative charge separation and was found to be selective against sulfanilamide and some drugs applied together with anthracyclines. Relative changes in the sensitivity toward anthracycline preparations tested were similar to those already described in the literature for similar DNA sensors [29,69,70]. Common components of biological fluids, stabilizers of commercial medications, and drugs commonly applied together with doxorubicin in chemotherapy regimens (cyclophosphamide, prednisone, and dacarbazine) did not regularly affect the doxorubicin determination.

Polyphenothiazine derivatives show some advantages over other electropolymerized materials used in biosensors. Their redox activity is less pH sensitive and can be monitored both in acidic (like polyaniline) and neutral media, including physiological pH range. Then, the high density of the positive charge of oxidized forms promoted self-assembling with polyanilines including DNA molecules. Lower sensitivity against three-layered polyaniline–DNA–polyaniline biosensors can be explained by the application of diffusionally free mediator, methylene blue, that was accumulated in the DNA and substituted with doxorubicin molecules during the incubation step. For this reason, such a biosensor cannot be named as reagent free and needs more steps for its manufacture and signal measurement. Other DNA sensors described for doxorubicin determination exploit nanomaterials that are synthesized and implemented in the sensing layer by several steps complicating the biosensor operation. Single-walled carbon nanotubes and Ag and Pt nanoparticles improve the conditions of electron exchange and increase the specific electrode surface accessible for electropolymerized monomers and DNA molecules. The function of TiO_2 and MgO is less understandable. They do not exert redox activity and are non-conductive. Probably, they adsorb species involved in the electrode reactions and increase their local concentration on the electrode. Whatever the case, the more complicated the layer, the more steps there are in its assembling and the more sources of measurement deviations.

High sensitivity of the doxorubicin determination observed for the DNA sensor developed can be attributed to a combination of various factors, i.e., accessibility of DNA in the nanocomposite with a polymeric form of dye for low-molecular intercalators, a well-developed surface of the film accumulating both the DNA and drug molecules, and high quantities of DNA implemented in the surface layer.

Nano- and picomolar detectable concentrations made it possible to avoid the interference of serum proteins and plasma electrolytes and to monitor typical levels of the anthracycline drugs currently used in cancer therapy.

The progress in the development of such DNA sensors assumes further efforts in establishing optimal structure of phenothiazine derivatives as the redox polymer precursors and improvement of metrological characteristics of the signals, especially during the storage period. The electrochemical DNA sensor developed can be applied for monitoring drug release in chemotherapy and screening new antitumor drugs able to intercalate DNA. Furthermore, the electropolymerization protocol elaborated can be useful for preparing organic films suitable for optical of sensing organic species using nanoplasmonic approaches and metasurfaces, artificial 2D materials with peculiar electromagnetic properties. Electropolymerized films can serve as photosensitizers in the catalytic reaction of the oxidation of organic matter and as components of electrochromic devices. Although such an application is beyond the topic of this investigation, the electrochemical characterization of the polymeric films formed can stimulate their application in the above areas.

Supplementary Materials: The following supporting information can be downloaded at: https://www.mdpi.com/article/10.3390/nano13162369/s1, Figure S1: The dependence of the polyPhTz peak currents on the incubation of the GCE/polyPhTz sensors in water and HEPES; cyclic voltammograms recorded with intermediate stirring the solution; Figure S2: The dependence of the polyPhTz peak currents on the incubation time; Figure S3: Cyclic voltammograms recorded after incubation of the GCE/polyPhTz/DNA sensor in doxorubicin solution; Figure S4: Anodic peak currents on cyclic voltammograms recorded after incubation of the GCE/polyPhTz/DNA sensor in doxorubicin solution within the storage period in the buffer and in dry conditions; Table S1: Analytical characteristics of the determination of doxorubicin with electrochemical sensors and DNA sensors. References [72–87] are cited in the supplementary materials.

Author Contributions: Conceptualization, G.E.; methodology, I.S.; validation, Y.K.; investigation, A.M., K.S. and A.K.; writing—original draft preparation, G.E.; project administration, P.P. All authors have read and agreed to the published version of the manuscript.

Funding: This work was financially supported by Russian Science Foundation, Russian Federation (grant № 19-73-10134, https://rscf.ru/en/project/19-73-10134/ (accessed on 13 August 2023)).

Institutional Review Board Statement: Not applicable.

Informed Consent Statement: Not applicable.

Data Availability Statement: The data presented in this study are available in supplementary materials.

Conflicts of Interest: The authors declare no conflict of interest.

References

1. *World Health Organization Model List of Essential Medicines: 21st List 2019*; World Health Organization: Geneva, Switzerland, 2019.
2. Minotti, G.; Menna, P.; Salvatorelli, E.; Cairo, G.; Gianni, L. Anthracyclines: Molecular advances and pharmacologic developments in antitumor activity and cardiotoxicity. *Pharmacol. Rev.* **2004**, *56*, 185–229. [CrossRef] [PubMed]

3. Berthiaume, J.M.; Wallace, K.B. Adriamycin-induced oxidative mitochondrial cardiotoxicity. *Cell Biol. Toxicol.* **2007**, *23*, 15–25. [CrossRef] [PubMed]
4. Chatterjee, K.; Zhang, J.; Honbo, N.; Karliner, J.S. Doxorubicin cardiomyopathy. *Cardiology* **2010**, *115*, 155–162. [CrossRef]
5. Swain, S.M.; Whaley, F.S.; Ewer, M.S. Congestive heart failure in patients treated with doxorubicin: A retrospective analysis of three trials. *Cancer* **2003**, *97*, 2869–2879. [CrossRef] [PubMed]
6. Kondori, T.; Tajik, S.; Akbarzadeh, T.N.; Beitollahi, H.; Graiff, C. Screen-printed electrode modified with co-NPs, as an electrochemical sensor for simultaneous determination of doxorubicin and dasatinib. *J. Iran. Chem. Soc.* **2022**, *19*, 4423–4434. [CrossRef]
7. Zhang, W.; Ma, R.; Gu, S.; Zhang, L.; Li, N.; Qiao, J. Nitrogen and phosphorus co-doped carbon dots as an effective fluorescence probe for the detection of doxorubicin and cell imaging. *Opt. Mater.* **2022**, *128*, 112323. [CrossRef]
8. Del Bonis-O'Donnell, J.T.; Pinals, R.L.; Jeong, S.; Thakrar, A.; Wolfinger, R.D.; Landry, M.P. Chemometric approaches for developing infrared nanosensors to image anthracyclines. *Biochemistry* **2019**, *58*, 54–64. [CrossRef]
9. Fleury-Souverain, S.; Maurin, J.; Guillarme, D.; Rudaz, S.; Bonnabry, P. Development and application of a liquid chromatography coupled to mass spectrometry method for the simultaneous determination of 23 antineoplastic drugs at trace levels. *J. Pharm. Biomed. Anal.* **2022**, *221*, 115034. [CrossRef]
10. Haq, N.; Alanazi, F.K.; Salem-Bekhit, M.M.; Rabea, S.; Alam, P.; Alsarra, I.A.; Shakeel, F. Greenness estimation of chromatographic assay for the determination of anthracycline-based antitumor drug in bacterial ghost matrix of salmonella typhimurium. *Sustain. Chem. Pharm.* **2022**, *26*, 100642. [CrossRef]
11. Estakhri, N.M.; Edwards, B.; Engheta, N. Inverse-designed metastructures that solve equations. *Science* **2019**, *363*, 1333–1338. [CrossRef]
12. Lalegani, Z.; Ebrahimi, S.A.S.; Hamawandi, B.; La Spada, L.; Batili, H.; Toprak, M.S. Targeted dielectric coating of silver nanoparticles with silica to manipulate optical properties for metasurface applications. *Mater. Chem. Phys.* **2022**, *287*, 126250. [CrossRef]
13. Lalegani, Z.; Ebrahimi, S.A.S.; Hamawandi, B.; La Spada, L.; Toprak, M.S. Modeling, design, and synthesis of gram-scale monodispersed silver nanoparticles using microwave-assisted polyol process for metamaterial applications. *Opt. Mater.* **2020**, *108*, 110381. [CrossRef]
14. Pacheco-Peña, V.; Beruete, M.; Rodríguez-Ulibarri, P.; Engheta, N. On the performance of an ENZ-based sensor using transmission line theory and effective medium approach. *New J. Phys.* **2019**, *21*, 043056. [CrossRef]
15. Akbari, M.; Shahbazzadeh, M.J.; La Spada, L.; Khajehzadeh, A. The graphene field effect transistor modeling based on an optimized ambipolar virtual source model for DNA detection. *Appl. Sci.* **2021**, *11*, 8114. [CrossRef]
16. Zhao, H.; Shi, K.; Zhang, C.; Ren, J.; Cui, M.; Li, N.; Ji, X.; Wang, R. Spherical COFs decorated with gold nanoparticles and multiwalled carbon nanotubes as signal amplifier for sensitive electrochemical detection of doxorubicin. *Microchem. J.* **2022**, *182*, 107865. [CrossRef]
17. Mehmandoust, M.; Khoshnavaz, Y.; Karimi, F.; Çakar, S.; Özacar, M.; Erk, N. A novel 2-dimensional nanocomposite as a mediator for the determination of doxorubicin in biological samples. *Environ. Res.* **2022**, *213*, 113590. [CrossRef]
18. Cunha, C.E.P.; Rodrigues, E.S.B.; De Oliveira Neto, J.R.; Somerset, V.; Taveira, S.; Sgobbi, L.; De Souza Gil, E. Voltammetric glassy carbon sensor approach for the extended stability studies of doxorubicin in lyophilized dosage form. *Eclet. Quim.* **2022**, *47*, 32–38. [CrossRef]
19. Berneth, H.; Bayer, A.G. *Ullmann's Encyclopedia of Industrial Chemistry*; Wiley-VCH Press: Baden-Wurttemberg, Germany, 2003; 585p. [CrossRef]
20. Brown, J.S. Treatment of cancer with antipsychotic medications: Pushing the boundaries of schizophrenia and cancer. *Neurosci. Biobehav. Rev.* **2022**, *141*, 104809. [CrossRef]
21. Mitchell, S.C. Phenothiazine: The parent molecule. *Curr. Drug Targets* **2006**, *7*, 1181–1189. [CrossRef]
22. Stanković, D.; Dimitrijević, T.; Kuzmanović, D.; Krstić, M.P.; Petković, B.B. Voltammetric determination of an antipsychotic agent trifluoperazine at a boron-doped diamond electrode in human urine. *RSC Adv.* **2015**, *5*, 107058–107063. [CrossRef]
23. Vinothkumar, V.; Sakthivel, R.; Chen, S.-M. Rare earth dysprosium nickelate nanospheres for the selective electrochemical detection of antipsychotic drug perphenazine in biological samples. *Mater. Today Chem.* **2022**, *24*, 100883. [CrossRef]
24. Pulikkutty, S.; Manjula, N.; Chen, T.-W.; Chen, S.-M.; Lou, B.-S.; Siddiqui, M.R.; Wabaidur, S.M.; Ali, M.A. Fabrication of gadolinium zinc oxide anchored with functionalized-SWCNT planted on glassy carbon electrode: Potential detection of psychotropic drug (phenothiazine) in biotic sample. *J. Electroanal. Chem.* **2022**, *918*, 116521. [CrossRef]
25. Abad-Gil, L.; Brett, C.M.A. Poly(methylene blue)-ternary deep eutectic solvent/Au nanoparticle modified electrodes as novel electrochemical sensors: Optimization, characterization and application. *Electrochim. Acta* **2022**, *434*, 1141295. [CrossRef]
26. Manusha, P.; Yadav, J.; Satija, J.; Senthilkumar, S. Designing electrochemical NADH sensor using silver nanoparticles/phenothiazine nanohybrid and investigation on the shape dependent sensing behavior. *Sens. Actuators B* **2021**, *347*, 130649. [CrossRef]
27. Sree, V.G.; Sohn, J.I.; Im, H. Pre-anodized graphite pencil electrode coated with a poly(thionine) film for simultaneous sensing of 3-nitrophenol and 4-nitrophenol in environmental water samples. *Sensors* **2022**, *22*, 1151. [CrossRef]

28. Li, Y.; Liu, D.; Meng, S.; Dong, N.; Liu, C.; Wei, Y.; You, T. Signal-enhanced strategy for ratiometric aptasensing of aflatoxin B1: Plasmon-modulated competition between photoelectrochemistry-driven and electrochemistry-driven redox of methylene blue. *Biosens. Bioelectron.* **2022**, *218*, 114759. [CrossRef]
29. Goida, A.; Kuzin, Y.; Evtugyn, V.; Porfireva, A.; Evtugyn, G.; Hianik, T. Electrochemical sensing of idarubicin—DNA interaction using electropolymerized azure B and methylene blue mediation. *Chemosensors* **2022**, *10*, 33. [CrossRef]
30. Ivanov, A.; Stoikov, D.; Shafigullina, I.; Shurpik, D.; Stoikov, I.; Evtugyn, G. Flow-through acetylcholinesterase sensor with replaceable enzyme reactor. *Biosensors* **2022**, *12*, 676. [CrossRef]
31. Barsan, M.M.; Ghica, M.E.; Brett, C.M.A. Electrochemical sensors and biosensors based on redox polymer/carbon nanotube modified electrodes: A review. *Anal. Chim. Acta* **2015**, *881*, 1–23. [CrossRef]
32. Porfireva, A.; Plastinina, K.; Evtugyn, V.; Kuzin, Y.; Evtugyn, G. Electrochemical DNA sensor based on poly(Azure A) obtained from the buffer saturated with chloroform. *Sensors* **2021**, *21*, 2949. [CrossRef]
33. Stoikov, D.I.; Porfir'eva, A.V.; Shurpik, D.N.; Stoikov, I.I.; Evtyugin, G.A. Electrochemical DNA sensors on the basis of electropolymerized thionine and Azure B with addition of pillar[5]arene as an electron transfer mediator. *Russ. Chem. Bull.* **2019**, *68*, 431–437. [CrossRef]
34. Diaz, A.F.; Logan, J.A. Electroactive polyaniline films. *J. Electroanal. Chem.* **1980**, *111*, 111–114. [CrossRef]
35. Diaz, A.F.; Kanazawa, K.K.; Gardini, G.P. Electrochemical polymerization of pyrrole. *J. Chem. Soc. Chem. Commun.* **1979**, 635–636. [CrossRef]
36. Barsan, M.M.; Pinto, E.M.; Brett, C.M.A. Electrosynthesis and electrochemical characterisation of phenazine polymers for application in biosensors. *Electrochim. Acta* **2008**, *53*, 3973–3982. [CrossRef]
37. Topçu, E.; Alanyalıoğlu, M. Electrochemical formation of poly(thionine) thin films: The effect of amine group on the polymeric film formation of phenothiazine dyes. *J. Appl. Polym. Sci.* **2014**, *131*, 39686. [CrossRef]
38. Gligor, D.; Dilgin, Y.; Popescu, I.C.; Gorton, L. Poly-phenothiazine derivative-modified glassy carbon electrode for NADH electrocatalytic oxidation. *Electrochim. Acta* **2009**, *54*, 3124–3128. [CrossRef]
39. Puskás, Z.; Inzelt, G. Formation and redox transformations of polyphenazine. *Electrochim. Acta* **2005**, *50*, 1481–1490. [CrossRef]
40. Karyakin, A.A. Chapter 5. Electropolymerized azines: A new group of electroactive polymers. In *Electropolymerization: Concepts, Materials and Applications*; Cosnier, S., Karyakin, A., Eds.; Willey VCH: Weinheim, Germany, 2010. [CrossRef]
41. Seifi, A.; Afkhami, A.; Madrakian, T. Highly sensitive and simultaneous electrochemical determination of lead and cadmium ions by poly(thionine)/MWCNTs-modified glassy carbon electrode in the presence of bismuth ions. *J. Appl. Electrochem.* **2022**, *52*, 1513–1523. [CrossRef]
42. Prado, N.S.; Silva, L.A.J.; Takeuchi, R.M.; Richter, E.M.; Santos, A.L.D.; Falcão, E.H.L. Graphite sheets modified with poly(methylene blue) films: A cost-effective approach for the electrochemical sensing of the antibiotic nitrofurantoin. *Microchem. J.* **2022**, *177*, 107289. [CrossRef]
43. Soranzo, T.; Ben Tahar, A.; Chmayssem, A.; Zelsmann, M.; Vadgama, P.; Lenormand, J.-L.; Cinquin, P.; Martin, D.K.; Zebda, A. Electrochemical biosensing of glucose based on the enzymatic reduction of glucose. *Sensors* **2022**, *22*, 7105. [CrossRef]
44. Jiménez-Fiérrez, F.; González-Sánchez, M.I.; Jiménez-Pérez, R.; Iniesta, J.; Valero, E. Glucose biosensor based on disposable activated carbon electrodes modified with platinum nanoparticles electrodeposited on poly(Azure A). *Sensors* **2020**, *20*, 4489. [CrossRef] [PubMed]
45. Fang, J.; Qi, B.; Yang, L.; Guo, L. Ordered mesoporous carbon functionalized with poly-azure B for electrocatalytic application. *J. Electroanal. Chem.* **2010**, *643*, 52–57. [CrossRef]
46. Lin, K.-C.; Yin, C.-Y.; Chen, S.-M. An electrochemical biosensor for determination of hydrogen peroxide using nanocomposite of poly(methylene blue) and FAD hybrid film. *Sens. Actuators B* **2011**, *157*, 202–210. [CrossRef]
47. Ding, M.; Hou, T.; Niu, H.; Zhang, N.; Guan, P.; Hu, X. Electrocatalytic oxidation of NADH at graphene-modified electrodes based on electropolymerized poly(thionine-methylene blue) films from nature deep eutectic solvents. *J. Electroanal. Chem.* **2022**, *920*, 116602. [CrossRef]
48. Tiravia, M.; Sabuzi, F.; Cirulli, M.; Pezzola, S.; Di Carmine, G.; Cicero, D.O.; Floris, B.; Conte, V.; Galloni, P. 3,7-Bis(N-methyl-N-phenylamino)phenothiazinium salt: Improved synthesis and aggregation behavior in solution. *Eur. J. Org. Chem.* **2019**, *2019*, 3208–3216. [CrossRef]
49. Peterson, B.M.; Shen, L.; Lopez, G.J.; Gannett, C.N.; Ren, D.; Abruña, H.D.; Fors, B.P. Elucidation of the electrochemical behavior of phenothiazine-based polyaromatic amines. *Tetrahedron* **2019**, *75*, 4244–4249. [CrossRef]
50. Padnya, P.L.; Khadieva, A.I.; Stoikov, I.I. Current achievements and perspectives in synthesis and applications of 3,7-disubstituted phenothiazines as methylene blue analogues. *Dyes Pigment.* **2022**, *208*, 110806. [CrossRef]
51. Khadieva, A.; Rayanov, M.; Shibaeva, K.; Piskunov, A.; Padnya, P.; Stoikov, I. Towards asymmetrical methylene blue analogues: Synthesis and reactivity of 3-N′-arylaminophenothiazines. *Molecules* **2022**, *27*, 3024. [CrossRef]
52. Wainwright, M.; McLean, A. Rational design of phenothiazinium derivatives and photoantimicrobial drug discovery. *Dyes Pigment.* **2017**, *136*, 590–600. [CrossRef]
53. Kuzin, Y.I.; Padnya, P.L.; Stoikov, I.I.; Gorbatchuk, V.V.; Stoikov, D.I.; Khadieva, A.I.; Evtugyn, G.A. Electrochemical behavior of the monomeric and polymeric forms of N-phenyl-3-(phenylimino)-3H-phenothiazin-7-amine. *Electrochim. Acta* **2020**, *345*, 136195. [CrossRef]

54. Salehan, P.; Ensafi, A.A.; Mousaabadi, K.Z.; Ghasemi, J.B.; Aghaee, E.; Rezaei, B. A theoretical and experimental study of polyaniline/GCE and DNA G-quadruplex conformation as an impedimetric biosensor for the determination of potassium ions. *Chemosphere* **2022**, *292*, 133460. [CrossRef] [PubMed]
55. Ramanavicius, S.; Deshmukh, M.A.; Apetrei, R.-M.; Ramanaviciene, A.; Plikusiene, I.; Morkvenaite-Vilkonciene, I.; Thorat, H.N.; Shirsat, M.D.; Ramanavicius, A. Chapter 15—Conducting polymers—Versatile tools in analytical systems for the determination of biomarkers and biologically active compounds. The detection of biomarkers: Past, present, and the future prospects. In *The Detection of Biomarkers. Past, Present and the Future Prospects*; Academic Press: Cambridge, MA, USA, 2022; pp. 407–434. [CrossRef]
56. Tran, L.T.; Tran, H.V.; Cao, H.H.; Tran, T.H.; Huynh, C.D. Electrochemically effective surface area of a polyaniline nanowire-based platinum microelectrode and development of an electrochemical DNA sensor. *J. Nanotechnol.* **2022**, *2022*, 8947080. [CrossRef]
57. Khadieva, A.; Gorbachuk, V.; Shurpik, D.; Stoikov, I. Synthesis of tris-pillar[5]arene and its association with phenothiazine dye: Colorimetric recognition of anions. *Molecules* **2019**, *24*, 1807. [CrossRef] [PubMed]
58. Evtugyn, G.; Porfireva, A.; Hianik, T. Electropolymerized materials for biosensors. In *Advanced Bioelectronics Materials*; Tiwari, A., Patra, H.K., Turner, A.P.F., Eds.; Wiley: New York, NY, USA, 2015; pp. 89–185. [CrossRef]
59. Kuzin, Y.I.; Khadieva, A.I.; Padnya, P.L.; Khannanov, A.A.; Kutyreva, M.P.; Stoikov, I.I.; Evtugyn, G.A. Electrochemistry of new derivatives of phenothiazine: Electrode kinetics and electropolymerization conditions. *Electrochim. Acta* **2021**, *375*, 137985. [CrossRef]
60. Suprun, E.V.; Kutdusova, G.R.; Khmeleva, S.A.; Radko, S.P. Towards deeper understanding of DNA electrochemical oxidation on carbon electrodes. *Electrochem. Commun.* **2021**, *124*, 106947. [CrossRef]
61. Brett, C.M.A. Electrochemical impedance spectroscopy in the characterisation and application of modified electrodes for electrochemical sensors and biosensors. *Molecules* **2022**, *27*, 1497. [CrossRef]
62. Kuzin, Y.; Kappo, D.; Porfireva, A.; Shurpik, D.; Stoikov, I.; Evtugyn, G.; Hianik, T. Electrochemical DNA sensor based on carbon black—Poly(neutral red) composite for detection of oxidative DNA damage. *Sensors* **2018**, *18*, 3489. [CrossRef]
63. Chatterjee, N.; Walker, G.C. Mechanisms of DNA damage, repair, and mutagenesis. *Environ. Mol. Mutagen.* **2017**, *58*, 235–263. [CrossRef]
64. Malanina, A.N.; Kuzin, Y.I.; Ivanov, A.N.; Ziyatdinova, G.K.; Shurpik, D.N.; Stoikov, I.I.; Evtugyn, G.A. Polyelectrolyte polyethylenimine-DNA complexes in the composition of voltammetric sensors for detecting DNA damage. *J. Anal. Chem.* **2022**, *77*, 185–194. [CrossRef]
65. Gunvén, G.; Theve, N.O.; Peterson, C. Serum and tissue concentrations of doxorubicin after IV administration of doxorubicin or doxorubicin-DNA complex to patients with gastrointestinal cancer. *Cancer Chemother. Pharmacol.* **1986**, *17*, 153–156. [CrossRef]
66. Chai, M.; Wientjes, M.G.; Badalament, R.A.; Burgers, J.K.; Au, J.L.-S. Pharmacokinetics of intra vesical doxorubicin in superficial bladder cancer patients. *J. Urol.* **1994**, *152*, 374–378. [CrossRef] [PubMed]
67. Krarup-Hansen, A.; Wassermann, K.; Rasmussen, S.N.; Dalmark, M. Pharmacokinetics of doxorubicin in man with induced acid or alkaline urine. *Acta Oncol.* **1988**, *27*, 25–30. [CrossRef] [PubMed]
68. Maudens, K.E.; Stove, C.P.; Lambert, W.E. Quantitative liquid chromatographic analysis of anthracyclines in biological fluids. *J. Chromatogr. B* **2011**, *879*, 2471–2486. [CrossRef] [PubMed]
69. Shamagsumova, R.; Porfireva, A.; Stepanova, V.; Osin, Y.; Evtugyn, G.; Hianik, T. Polyaniline–DNA based sensor for the detection of anthracycline drugs. *Sens. Actuators B* **2015**, *220*, 573–582. [CrossRef]
70. Hajian, R.; Mehrayin, Z.; Mohagheghian, M.; Zafari, M.; Hosseini, P.; Shams, N. Fabrication of an electrochemical sensor based on carbon nanotubes modified with gold nanoparticles for determination of valrubicin as a chemotherapy drug: Valrubicin-DNA interaction. *Mater. Sci. Eng. C* **2015**, *49*, 769–775. [CrossRef]
71. Hongpaisan, J.; Roomans, G.M. Retaining ionic concentrations during in vitro storage of tissue for microanalytical studies. *J. Microsc.* **1999**, *193*, 257–267. [CrossRef]
72. Alavi-Tabari, S.A.R.; Khalilzadeh, M.A.; Karimi-Maleh, H. Simultaneous determination of doxorubicin and dasatinib as two breast anticancer drugs uses an amplified sensor with ionic liquid and ZnO nanoparticle. *J. Electroanal. Chem.* **2018**, *811*, 84–88. [CrossRef]
73. Liu, J.; Bo, X.; Zhou, M.; Guo, L. A nanocomposite prepared from metal-free mesoporous carbon nanospheres and graphene oxide for voltammetric determination of doxorubicin. *Microchim. Acta* **2019**, *186*, 639. [CrossRef]
74. Vacek, J.; Havran, L.; Fojta, M. Ex situ voltammetry and chronopotentiometry of doxorubicin at a pyrolytic graphite elec-trode: Redox and catalytic properties and analytical applications. *Electroanalysis* **2009**, *21*, 21399–22144. [CrossRef]
75. Skalová, Š.; Langmaier, J.; Barek, J.; Vyskočil, V.; Navrátils, T. Doxorubicin determination using two novel voltammetric approaches: A comparative study. *Electrochim. Acta* **2020**, *330*, 135180. [CrossRef]
76. Ali, A.-M.B.H.; Rageh, A.H.; Abdel-aal, F.A.M.; Mohamed, A.-M.I. Anatase titanium oxide nanoparticles and multi-walled carbon nanotubes-modified carbon paste electrode for simultaneous determination of avanafil and doxorubicin in plasma samples. *Microchem. J.* **2023**, *185*, 108361. [CrossRef]
77. Singh, T.A.; Sharma, V.; Thakur, N.; Tejwan, N.; Sharma, A.; Das, J. Selective and sensitive electrochemical detection of doxorubicin via a novel magnesium oxide/carbon dot nanocomposite based sensor. *Inorg. Chem. Commun.* **2023**, *150*, 110527. [CrossRef]
78. Abbasi, M.; Ezazi, M.; Jouyban, A.; Lulek, E.; Asadpour-Zeynali, K.; Ertas, Y.N.; Houshyar, J.; Mokhtarzadeh, A.; Soleymani, J. An ultrasensitive and preprocessing-free electrochemical platform for the detection of doxorubicin based on tryptophan/polyethylene glycol-cobalt ferrite nanoparticles modified electrodes. *Microchem. J.* **2022**, *183*, 108055. [CrossRef]

79. Porfireva, A.; Vorobev, V.; Babkina, S.; Evtugyn, G. Electrochemical sensor based on poly(Azure B)-DNA composite for doxorubicin determination. *Sensors* **2019**, *19*, 2085. [CrossRef] [PubMed]
80. Kulikova, T.; Porfireva, A.; Evtugyn, G.; Hianik, T. Electrochemical DNA sensors with layered polyaniline-DNA coating for detection of specific DNA interactions. *Sensors* **2019**, *19*, 469. [CrossRef]
81. Peng, A.; Xu, H.; Luo, C.; Ding, H. Application of a disposable doxorubicin sensor for direct determination of clinical drug concentration in patient blood. *Int. J. Electrochem. Sci.* **2016**, *11*, 6266–6278. [CrossRef]
82. Evtugyn, G.; Porfireva, A.; Stepanova, V.; Budnikov, H. Electrochemical biosensors based on native DNA and nanosized mediator for the detection of anthracycline preparations. *Electroanalysis* **2015**, *27*, 629–637. [CrossRef]
83. Kulikova, T.; Porfireva, A.; Rogov, A.; Evtugyn, G. Electrochemical DNA sensor based on acridine yellow adsorbed on glassy carbon electrode. *Sensors* **2021**, *21*, 7763. [CrossRef]
84. Kappo, D.; Shurpik, D.; Padnya, P.; Stoikov, I.; Rogov, A.; Evtugyn, G. Electrochemical DNA sensor based on carbon black-poly(methylene blue)-poly(neutral red) composite. *Biosensors* **2022**, *12*, 329. [CrossRef]
85. Karadurmus, L.; Dogan-Topal, B.; Kurbanoglu, S.; Shah, A.; Ozkan, S.A. The interaction between DNA and three intercalating anthracyclines using electrochemical DNA nanobiosensor based on metal nanoparticles modified screen-printed electrode. *Micromachines* **2021**, *12*, 1337. [CrossRef]
86. Moghadam, F.H.; Taher, M.A.; Karimi-Maleh, H. Doxorubicin anticancer drug monitoring by ds-DNA-based electrochemical biosensor in clinical samples. *Micromachines* **2021**, *12*, 808. [CrossRef] [PubMed]
87. Asai, K.; Yamamoto, T.; Nagashima, S.; Ogata, G.; Hibino, H.; Einaga, Y. An electrochemical aptamer-based sensor prepared by utilizing the strong interaction between a DNA aptamer and diamond. *Analyst* **2020**, *145*, 544–549. [CrossRef] [PubMed]

Disclaimer/Publisher's Note: The statements, opinions and data contained in all publications are solely those of the individual author(s) and contributor(s) and not of MDPI and/or the editor(s). MDPI and/or the editor(s) disclaim responsibility for any injury to people or property resulting from any ideas, methods, instructions or products referred to in the content.

Review

Supramolecular Nanostructures Based on Perylene Diimide Bioconjugates: From Self-Assembly to Applications

Nadjib Kihal [1,2,3], Ali Nazemi [1,3,*] and Steve Bourgault [1,2,*]

1 Department of Chemistry, Université du Québec, Montreal, QC H2X 2J6, Canada; kihal.nadjib@courrier.uqam.ca
2 Quebec Network for Research on Protein Function, Engineering and Applications, PROTEO, Quebec City, QC G1V 0A6, Canada
3 Centre Québécois sur les Matériaux Fonctionnels/Québec Centre for Advanced Materials, CQMF/QCAM, Montreal, QC H3A 2A7, Canada
* Correspondence: nazemi.ali@uqam.ca (A.N.); bourgault.steve@uqam.ca (S.B.)

Abstract: Self-assembling π-conjugated systems constitute efficient building blocks for the construction of supramolecular structures with tailored functional properties. In this context, perylene diimide (PDI) has attracted attention owing to its chemical robustness, thermal and photo-stability, and outstanding optical and electronic properties. Recently, the conjugation of PDI derivatives to biological molecules, including oligonucleotides and peptides, has opened new avenues for the design of nanoassemblies with unique structures and functionalities. In the present review, we offer a comprehensive summary of supramolecular bio-assemblies based on PDI. After briefly presenting the physicochemical, structural, and optical properties of PDI derivatives, we discuss the synthesis, self-assembly, and applications of PDI bioconjugates.

Keywords: perylene diimide; self-assembly; nanostructures; peptides; oligonucleotides

Citation: Kihal, N.; Nazemi, A.; Bourgault, S. Supramolecular Nanostructures Based on Perylene Diimide Bioconjugates: From Self-Assembly to Applications. *Nanomaterials* 2022, 12, 1223. https://doi.org/10.3390/nano12071223

Academic Editor: Pavel Padnya

Received: 6 March 2022
Accepted: 2 April 2022
Published: 5 April 2022

Publisher's Note: MDPI stays neutral with regard to jurisdictional claims in published maps and institutional affiliations.

Copyright: © 2022 by the authors. Licensee MDPI, Basel, Switzerland. This article is an open access article distributed under the terms and conditions of the Creative Commons Attribution (CC BY) license (https://creativecommons.org/licenses/by/4.0/).

1. Introduction

Mother Nature has always been a unique source of inspiration for the design of nanostructures and materials for application in various fields spanning from biomedicine to electronics. Living organisms are hierarchically built on the organized self-assembly of biomacromolecules, such as lipids, proteins, deoxyribonucleic acids (DNA), ribonucleic acids (RNA), and polysaccharides, controlled by a delicate balance of non-covalent interactions [1,2]. For example, the cellular plasma membrane results from the spontaneous organization of a diversity of lipids into a fluid and complex bilayer, while the exceptional mechanical properties of spider silk arise from the coordinated self-recognition of proteins. Over the last decades, biological macromolecules have been continuously harnessed as building blocks for the construction of (nano)structures with atomic-scale precision. Owing to their biocompatibility, functionality, and ability to undergo self-assembly, the conjugation of biological macromolecules to organic self-assembling molecules with interesting (photo)chemical properties has supported the construction of hybrid supramolecular architectures [3,4]. Of such small organic molecules, the family of perylene diimide (PDI; perylene-3,4:9,10-tetracarboxylic acid diimide) dyes has attracted tremendous attention. Particularly, PDI-based chromophores are known for their electron mobility, high fluorescence quantum yields, photo- and thermal stability, and semi-conductivity [5,6]. These properties benefit them in applications such as pigments, dye lasers, sensors, bioprobes, and photovoltaics [5,7,8]. Interestingly, the physicochemical, optical, and structural properties of PDI derivatives can be tuned by the modification of their substituents. Particularly, conjugating PDIs to biological macromolecules can allow the conception of supramolecular structures with unique architectures and properties. Albeit these chimeric PDI-biomolecules have been exploited for two decades, their self-assembly as well as the resulting morphologies remain difficult to predict from the monomeric building blocks, still precluding their full potential. This review aims at providing a comprehensive overview of PDI-bioconjugates by highlighting the most recent studies and by emphasizing the relationships existing between the physicochemical properties of the bioconjugates, the conditions of self-assembly, and the resulting supramolecular structures. While excellent reviews regarding PDI-based assemblies have been published [9–11], including the comprehensive reviews of Guo [9], and of Würthner [5], the present review focusing exclusively on PDIs conjugated to peptides and oligonucleotides will hopefully offer a novel perspective on this flourishing field of research.

2. Synthesis and Properties of PDI Derivatives

Owing to their stability and unique optical properties, organic π-conjugated systems constitute an important class of molecules for different applications in various fields, from biomedicine to (nano)material sciences. In particular, PDI and its derivatives are π-conjugated systems that have been widely exploited as building blocks for the preparation of stable and functional nanostructures with tailored optical and physicochemical properties [5,10,11]. To this end, several groups have developed innovative and reliable methods for the efficient synthesis of PDI derivatives and their subsequent self-assembly into tailored structures. In this first section, we will offer a brief overview of the main synthetic approaches commonly used to prepare PDI and its derivatives and we will present the optical, physicochemical, and self-assembly properties of PDI-based π-conjugated systems.

2.1. Physicochemical, Optical, and Self-Assembling Properties of PDIs

PDI derivatives have been recognized as an ideal π-conjugated system for chemical, colorimetric and fluorescent sensors [12,13], organic semiconductors and optoelectronic devices [14–16], phototheranostics, [17,18] as well as bioimaging and gene/drug delivering agents [19]. Structurally, the PDI molecule is composed of one rigid, planar, and stable perylene core with two *imide* groups at both ends of the polycyclic aromatic scaffold. The perylene core has 12 characteristic positions, known as the *bay* (1, 6, 7 and 12 positions), *ortho* (2, 5, 8 and 11 positions), and *peri* (3, 4, 9 and 10 positions). The attractiveness of PDIs as chromophores is associated with their chemical robustness, excellent thermal and photostabilities as well as their unique optical and electronic properties [9,10]. Soluble monomeric PDIs without substituents on the *ortho* and *bay* positions of the perylene core show prototypical UV-Vis absorption spectra characterized with three vibronic peaks and a typical mirror image emission spectrum with high fluorescence quantum yields in most organic solvents [5,20]. Interestingly, substitutions at the two *imide* positions have a negligible effect on the optical properties of PDIs, while they significantly modulate the self-assembling processes as well as the architecture of the resulting supramolecular structures [11,21]. In contrast, substituents at the *bay* positions have a considerable effect on the morphology as well as the optical, chiroptical [22], and electronic properties of PDI-based assemblies, such as exciton diffusion, charge transfer, and high fluorescence quantum yield with red-shifted emission bands, which are associated with applications in organic solar cells [23], organic semiconductor devices [24], organic light-emitting diodes [25], vapor sensors [26–28], and theranostic agents [29,30]. These *bay*-substitutions can lead to a distortion of the planar PDI that affects solubility and weakens the π–π staking interactions [11]. Consequently, the twisting of the PDI cores can be precisely tuned through *bay*-substitutions, leading to some control over the self-assembly and the desired morphology of the nano- and mesoscopic structures [30] and liquid-crystalline materials [31,32].

Generally, PDIs form weakly emissive H-aggregates in aqueous solutions via non-covalent π–π stacking interactions, leading to the linear orientation of aromatic systems, i.e., one top of each other (Figure 1). In contrast, the modulation of the planar structure by introducing bulky substituents into the *bay* positions of the perylene core can generate strongly emissive core-twisted self-assembled supramolecular structures [5], in which neighboring chromophores are oriented in a head-to-tail fashion, known as J-aggregates (Figure 1) [33,34].

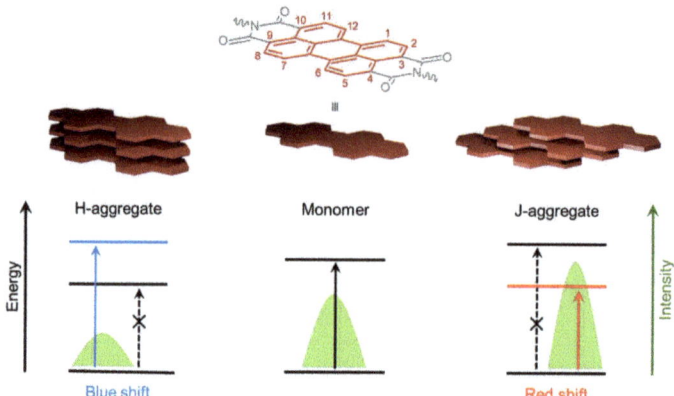

Figure 1. Molecular structure of PDI with the 12 positions of the perylene core. Arrangements of H- (left) and J- (right) aggregates resulting from the self-assembly of PDI monomers (center) with the corresponding effects on fluorescence emission spectra and electronic π–π * transitions. Full arrows indicate the permitted transitions and dashed arrows the forbidden ones.

2.2. Synthetic Strategies to Access Symmetrical PDIs

Perylene-3,4,9,10-tetracarboxylic dianhydride (PTCDA) is considered the parent compound of PDIs and is easily accessible through several well-defined multistep processes, including the one represented in Scheme 1 [20,21,35]. In this example, the first step includes the synthesis of naphthalene-1,8-dicarboxanhydride via the oxidation of acenaphthene, followed by ammonia treatment to convert the anhydride to an *imide* functionality. *Imide* dimerization facilitated by molten potassium hydroxide provides perylene-3,4,9,10-tetracarboxylic diimide (PTCDI) that can be hydrolyzed using concentrated sulfuric acid at high temperature to obtain PTCDA. Imidization of PTCDA with primary amines or anilines, in imidazole and in the presence of anhydrous zinc acetate as catalyst, leads to the formation of appropriate symmetrically N,N'-substituted PDIs in high yields [20,21,35,36].

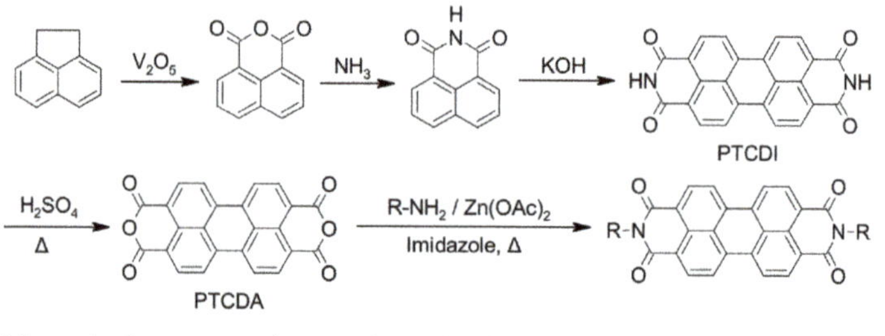

Scheme 1. Common synthetic pathway for the preparation of PTCDA and symmetrically N,N'-substituted PDIs.

2.3. Synthetic Strategies to Access Asymmetrical PDIs

Asymmetrical PDIs can be obtained by the partial hydrolysis of PDIs to perylene monoimide monoanhydride followed by condensation of this mixed *imide–anhydride* compound with primary amines to access the desired asymmetrical PDIs, as illustrated in Scheme 2 (Approach A) [21,37]. Moreover, the partial hydrolysis of PTCDA, which provides a mixed anhydride dicarboxylate salt, is another common approach to access asymmetrical PDIs via successive imidization reactions (Scheme 2, Approach B) [21,35,37,38].

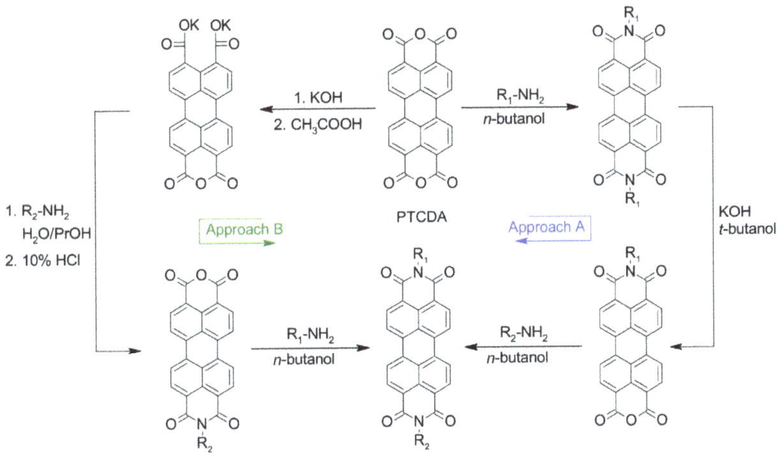

Scheme 2. Synthetic routes for the preparation of asymmetrical PDIs with different substituents on *imide* positions.

3. Peptide-PDI Bioconjugates

Over the last two decades, amino acids and peptides have been harnessed to control the solubility, optical properties and/or self-assembly of PDI derivatives by taking advantage of highly directional hydrogen-bonding interactions as well as other non-covalent interactions such as ionic and π–π interactions [39]. In addition, polypeptides are known for their biocompatibility, biodegradability and molecular specificity [40], ultimately supporting the use of peptide-based PDI assemblies for different biomedical applications. Chemically, peptides can be considered as linear polymers/oligomers assembled from the condensation of amino acid building blocks [41]. The 20 natural amino acids, as well as countless unnatural amino acids, allow virtually infinite combinations of sequences and offer an unlimited diversity of physicochemical and structural properties of the resulting polypeptide chains [42]. Herein, we present relevant examples of the use of amino acids and short peptide sequences to modulate the self-assembly of PDI derivatives into tailored nanostructures. The self-assembly of PDI–peptide conjugates and the resulting supramolecular morphologies are modulated by a fine balance of complex intermolecular interactions, such as PDI's π–π stacking interactions and numerous non-covalent interactions involving side chains and the polyamide backbone, as well as by the conditions of the microenvironment, including solvent polarity, solution ionic strength, pH, and temperature [9,11,39].

3.1. Synthesis of Peptide-Conjugated PDIs

Amino acid– and peptide–PDI conjugates are commonly prepared by respectively introducing amino acids and short peptide sequences at one or both *imide* positions of PDI. Peptides are usually synthesized on solid support, which involves the attachment of the C-terminal amino acid to the polymeric resin by a covalent bond followed by the sequential addition of individual preactivated amino acids, and the final cleavage of the polypeptide from the solid support [43,44]. In contrast to solution synthesis, solid-phase peptide synthesis (SPPS) simplifies dramatically the purification steps between each reaction, thus reducing synthesis time and increasing yield. Consequently, not only does SPPS offer an efficient route towards automation, but also allows the routine access to high-molecular weight polypeptides of up to 50-residues [45]. Due to the insolubility of PTCDA as PDI precursor in common solvents used in SPPS, Kim et al. developed an alternative method to overcome this problem by initially conjugating the PTCDA to amino acids, accessing soluble PDI–amino acid derivatives [46]. Afterwards, the soluble PDI–amino acid derivative can be conjugated to the target peptide sequence that was elongated on the polymeric support by the common SPPS method (Scheme 3) [46].

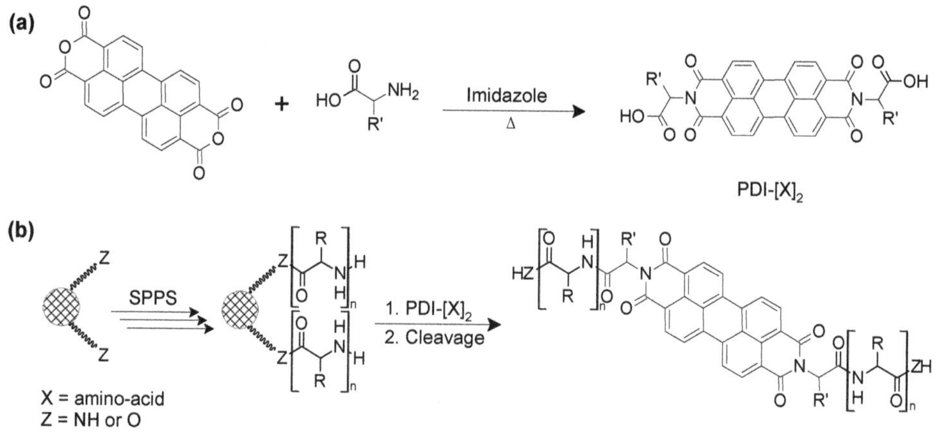

Scheme 3. (a) Synthetic route to amino acid-conjugated PDIs and (b) preparation of peptide-conjugated PDI through the SPPS method followed by their cleavage from the solid support.

3.2. Sequence-Dependent Self-Assembly

Specific variations within peptide sequences have been used to control the self-assembly process of PDI conjugates and to modulate the morphology of the resulting nanostructures through a delicate balance of non-covalent interactions (hydrogen bonding, hydrophobicity, ionic bonding) involving specific residue side chains. Short dipeptides GX (where X = D or Y) were used to enhance the solubility and to modulate the self-assembly properties of PDI in polar organic solvents and aqueous solution [47]. It was observed that, by varying the X residue of the PDI–[GX]$_2$ *bola*-amphiphile conjugates from hydrophobic Tyr to hydrophilic Asp, the balance between hydrogen bonding and π–π stacking interactions were altered, ultimately affecting the morphology of the assemblies and their optical properties. For instance, in aqueous sodium bicarbonate buffer (pH 10.8), dimethyl sulfoxide (DMSO), dimethylformamide (DMF), tetrahydrofuran (THF), or acetone, the symmetric PDI–[GY]$_2$ formed chiral nanofibers, whereas PDI–[GD]$_2$ assembled into achiral spherical aggregates in buffer and DMSO. In addition, PDI–[GY]$_2$ formed a gel in DMF, while organogels were observed for the PDI–[GD]$_2$ derivative in this polar aprotic solvent. An exhaustive study of structure-assembly relationships revealed how peptide's physicochemical properties and length, asymmetric substitution at the *imide* positions, and stereocenter inversion can affect the thermodynamics of the self-assembly of peptide–PDI hybrid molecules [39]. A set of peptide–PDI conjugates were synthetized, all encompassing three units: (i) a glycine residue at the *N*-terminal position used as a low-steric hindrance linker; (ii) a central variable region composed of three L or D amino acids to evaluate the impact of increasing the peptide hydrophobicity and the role of stereocenters; and (iii) a C-terminal charged region composed of one to three Glu residues to enhance the hydrosolubility and examine the effect of the charge density on the assembly process (Figure 2) [39]. Moreover, to induce a strong amphiphilic character, one of the peptide sequences was replaced by a hydrophobic hexyl chain. It was observed that peptide hydrophobicity and an asymmetrical hexyl substitution induce significant changes on the aggregation thermodynamics of the bioconjugates (Figure 2a,b). In contrast, varying the peptide length, the C-terminal charged region length or the stereocenter inversion induced a significantly lower impact on the aggregation thermodynamics, while having an effect on the peptide-driven self-assembly of PDI nanofibers [39]. Overall, these studies revealed that the physicochemical properties of the residue side chains, the configuration of stereocenters, and the symmetric/asymmetric conjugation can be exploited to dictate the non-covalent interactions that drive the self-assembly of peptide-conjugated PDIs and the final morphology of the assemblies.

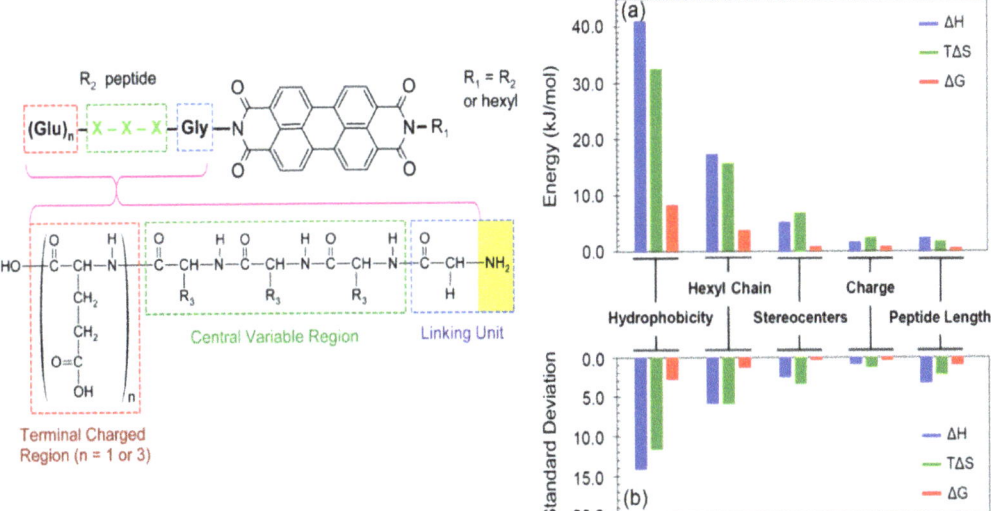

Figure 2. Structure of the peptide–PDI conjugates and impact of structural modulation on the aggregation thermodynamics. The histogram shows (**a**) the deviation between the lowest and the highest values for each parameter, and (**b**) the standard deviation as a more collective measure of variability [39]. Copyright © 2014 American Chemical Society.

Inspired by the β-continuous interface of the bovine peroxiredoxin-3 protein, a short heptapeptide (IKHLSVN) was conjugated to PDI in order to control self-assembly into organic semiconductor nanostructures [48]. The designed self-assembling peptide encompassed three different regions: (i) a glycine or ethylamino linker attached to an *imide* position of the PDI to reduce steric hindrance between the PDI core and the peptide sequence, (ii) a β-sheet-forming peptide, and (iii) a terminal unit composed of glutamic acid residues to assist solubility and to trigger the assembly of peptide–PDI conjugates by pH jump. Two groups of peptide–PDI derivatives were prepared. The first group results from the symmetrical substitution of PDI with peptide sequences, whereas for the second group the PDI core was replaced with perylene *imide* bis-ester to generate asymmetrical derivatives. Furthermore, for one derivative of each group, the peptide core was attached to the PDI via the amino terminus using a glycine linker, while for the other derivative the peptide core was attached via the carboxyl terminus using an ethylamino linker. The symmetrically substituted PDI showed spectral profiles characteristic of monomers in DMSO. However, UV-visible spectral profiles of bis-ester-functionalized PDI displayed some aggregation, with predominantly monomeric species. In aqueous media, these peptide–PDI conjugates self-assembled into H-aggregate suprastructures. Particularly, all peptide–PDI derivatives self-assembled into extensive fibril networks in aqueous solution, except for the bis-ester-functionalized PDI for which the peptide was attached via the C-terminal position. This compound formed amorphous, plate-like accretions. Furthermore, reversing the peptide sequence, i.e., *N*- to *C*-, for the symmetric derivatives led to the formation of short fibrils and thread-like assemblies instead of ribbon-like structures. This observation highlights the importance of the attachment mode of peptide to the PDI core towards self-assembly and final morphology of the assemblies.

Similarly, the N-(tetra (L-alanine) glycine)-N'-(1-undecyldodecyl) functionalized perylene-3,4,9,10-tetracarboxyl diimide was designed as an asymmetric amphiphilic derivative in order to elucidate how molecular-scale interactions govern the overall self-assembly process [49]. The oligopeptide block on one of the *imide* nodes of the PDI core provided aggregation directionality through hydrogen bonding and π–π stacking interactions. In chloroform, which was chosen to strengthen inter-peptide hydrogen bonds, this asymmetric amphiphilic PDI derivative adopted a right-handed helical arrangement due to the delicate balance between π–π stacking involving PDI cores and the network of hydrogen bonds between β-sheet-forming peptides. Interestingly, the addition of trifluoroacetic acid (TFA) to the self-assembling media, which was used as a hydrogen-bonding breaking agent, induced the transition of the nanofibers into small aggregates. These aggregates could be brought back into nanofibers by the addition of triethylamine (TEA), which was used to neutralize TFA, favoring the formation of H-bonding between the peptide blocks [49].

Computational simulation and experimental studies have been combined to understand the relationships between the sequence and the self-assembly process of π-conjugated peptides to ultimately predict the resulting supramolecular organisations and photophysical properties from the peptide sequence. For instance, it was shown that increasing the hydrophobicity of the closest residue attached to the PDI core can modulate the photophysical responses in aqueous solution via the conversion of J aggregates, or liquid-crystalline-type materials, to H-type aggregates [50]. In addition, the relationships between the resulting morphologies and the molecular structure of a small library of peptide–PDI derivatives bearing a variable number of L-alanine units as well as methylene, ethylene, and propylene spacers were investigated [51]. It was revealed that the number of L-alanine units in the β-strand peptide segments and the length of the spacer affected the morphology of the resulting suprastructures. In addition, it was shown, through molecular dynamic simulations, that there is a complex interplay between the translation of molecular chirality into supramolecular helicity and the inherent propensity for well-defined one-dimensional aggregation into β-sheet-like superstructures in the presence of a central chromophore [51]. Finally, a symmetric PDI–tripeptide conjugate, which was obtained by introducing a KPA tripeptide block at the 1 and 7 *bay* positions of PDI via a 2-(2-aminoethoxy)-ethoxy linker, self-assembled into β-sheet nanohelices directed by hydrogen-bonding [52]. These resulting supramolecular structures were particularly sensitive to thermal and ultrasound stimuli. For instance, upon heating/cooling and sonication of the peptide–PDI sample, an interconversion of the supramolecular chirality between left- and right-handed nanostructures was observed [52].

Except for glycine, which is achiral, α-amino acids have an S configuration and are designated as *L* using the Fischer configurational system [42]. It has been reported that the presence of chiral proximal residues in close proximity to the achiral PDI, for symmetrical peptide–PDI derivatives, influences π–π stacking interactions and induces helical chirality to the PDI core. However, when chiral residues were located distant to the PDI core, or when an isolated stereocenter was introduced in proximal distance of PDI, no effect of chirality was observed during self-assembly [53]. The designed peptide sequences with stereogenic positions and stereochemical configurations included three blocks: (i) an achiral glycine used as a spacer between PDI and the peptide, (ii) three variable residues forming the central block, and (iii) a terminal block of three ionizable glutamic acid residues to assist solubility and pH-triggered aggregation. It was observed that the self-assembly process is modulated by the β-sheet-forming potential of the peptide moieties and the π–π stacking interactions of PDI units. Interestingly, an inversion of the stereocenter within the proximal residues revealed chiral influence. In contrast, an asymmetrical peptide–PDI derivative obtained by the introduction of an alkyl chain at one of the amide nodes, generated an amphiphilic PDI conjugate and disrupted the chiral-mediated self-assembly [53].

Besides, it was observed that the symmetrical conjugation of FF dipeptide to the PDI core leads to a helical assembly due to the chirality of amino acids as well as the co-facial π–π stacking of PDI units [54]. Furthermore, it was shown that there is a close relationship between the translation of molecular chirality into supramolecular helicity and the one-dimensional assembly into well-defined β-sheet-like suprastructures [51]. Overall, these structure–assembly relationship studies have indicated that the morphology of the resulting peptide-conjugated PDI nanostructures can be, to some extent, controlled by modulating the peptide sequence. Moreover, the stereocenters embedded in the peptide backbone can be exploited, under specific conditions, to induce a chiral morphology to the resulting assemblies.

3.3. Solvent-Dependent Self-Assembly

As described above, several substituted water-soluble PDIs have been obtained by functionalizing the PDI core at their *imide* positions with amino acids and short peptides. Not only do the physicochemical properties of the conjugated moieties dictate the self-assembly behaviour of the PDI core as well as the morphology of the final aggregates, but the solvent also strongly influences the thermodynamics and kinetics of self-assembly [54–56]. It is generally assumed that the aggregation constant of PDI derivatives decreases with increasing solvent polarity [57]. The photophysical and aggregation propensities of PDI–[X]$_2$ symmetrical derivatives, where X is a residue with an aromatic group (Y, W or F), were evaluated in various organic solvents [58]. This study revealed that all derivatives self-assemble into amorphous aggregates, excepted for PDI–[Y]$_2$ that forms J-type aggregates in methanol. Although it is still unclear why this effect manifested only in methanol, the authors suggested that the formation of J-aggregates may be possible due to the network of hydrogen bonds involving the hydroxyl groups of tyrosine residues and the carbonyl groups of the PDI core. In pyridine and acetone, PDI–[F]$_2$ and PDI–[Y]$_2$ showed a higher propensity to aggregate than PDI–[W]$_2$, however, the origin of this tendency was unclear. Furthermore, the NMR data indicated a large degree of aggregation in DMSO for all three PDI–peptide derivatives, although the absorption and the fluorescence spectra were both characteristics of soluble and monomeric PDIs [58]. In another study, it was reported that the relatively polar nature of chloroform, in contrast to THF and DMF, facilitates the formation of intermolecular H-bonding of PDI–[F]$_2$, which ultimately leads to J-aggregates [59]. In chloroform, PDI–[F]$_2$ assembled into vesicular suprastructures through the formation of right-handed helix, involving intermolecular H-bonding in lateral and π–π stacking in longitudinal growth directions. As shown in Figure 3, the increase in solvent polarity correlated well with a decrease in the diameter of the vesicles assembled from the PDI–[F]$_2$.

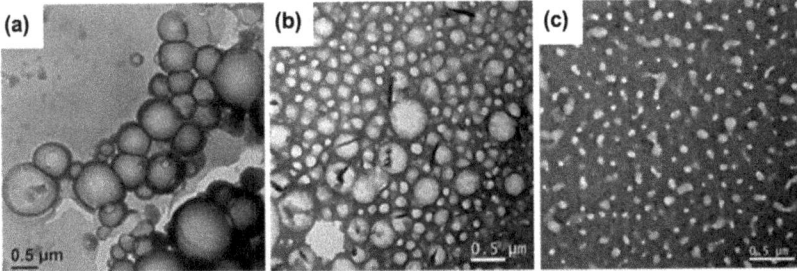

Figure 3. Transmission electron microscopy images of PDI-[F]$_2$ assembled in (**a**) CHCl$_3$, (**b**) THF and (**c**) DMF [59]. Copyright © 2018 Wiley-VCH Verlag GmbH & Co. KGaA, Weinheim.

Similarly, the self-assembly of the symmetrical PDI–[FF]$_2$ conjugate and the morphology of the resulting suprastructures were dramatically modulated by the polarity of the solvent [55]. For instance, fibrillar nanostructures were obtained in relatively non-polar solvents, such as THF and CHCl$_3$, while in more polar solvents (HFIP, MeOH, ACN, acetone), spherical morphologies were observed, for which the diameter correlates inversely with the polarity of the solvent (Figure 4) [55]. By density-functional theory (DFT) calculations combined with experimental studies, the kinetic and thermodynamic parameters driving the self-assembly of PDI–[FF]$_2$ were investigated in the THF/water mixed solvent system [54]. In THF, PDI–[FF]$_2$ formed right-handed helical nanofibrils under kinetic control, while in 10% THF the helicity switched to left-handed orientation governed by thermodynamics, leading to the formation of nanorings [54].

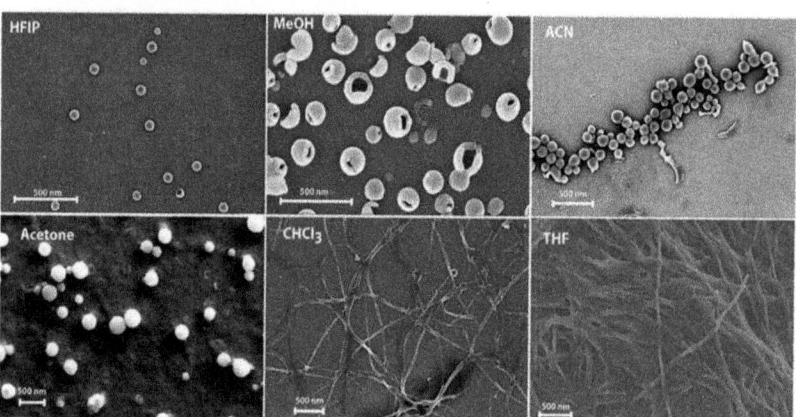

Figure 4. Field emission scanning electron microscopy (FESEM) images obtained after 72 h incubation of PDI-[FF]$_2$ in different solvents (10 mM) [55]. Copyright © 2018 American Chemical Society.

By harnessing N,N'-α-carbobenzyloxy-L-lysine-functionalized 1,6,7,12-tetrachloroperylene diimide, it was further demonstrated that the polarity of the solvent strongly modulates the morphology of the nanostructures assembled from PDI–amino acid chimeric derivatives [56]. These PDI derivatives were assembled via a phase transfer strategy for which the compounds were first dissolved in acetone followed by the gradual addition of water to reach the desired ratios. When the -COOH group of the lysine was placed five carbon–carbon bonds away from the *imide* node position of PDI (i.e., PDI-A), increasing the H$_2$O content up to 55% did not affect the morphology of the suprastructure, but caused an important decrease in nanobelt length with an increase in their width. (Figure 5a). Interestingly, conversion into spherical aggregates was observed when the water percentage increased up to 75% (Figure 5c). The effect of the solvent nature on PDI–peptide self-assembly was attributed to a delicate balance between reassembly at low H$_2$O fractions to minimize the system's free energy (thermodynamic control) and fast nucleation (kinetics control) in presence of high water content to adjust toward the optimal conformation. Moreover, by moving the carboxylate of the lysine directly on the next carbon attached to the PDI's *imide* node (i.e., PDI-B), it was observed that the position of the COOH group is a critical factor governing the nanostructure morphology. The PDI-B derivative assembled into condensed, layered structures characterized by a mixture of nanosheets and spheres in mixed acetone/water solution. The presence of small spheres became progressively more prevalent as the water content increased (Figure 5d–f). It was suggested that the formation of these ill-ordered structures was triggered by the steric hindrance of the carboxylate next to the *imide* node, which obstructs intermolecular H-bond formation between neighboring -COOH groups and deteriorates their long-range ordering [56].

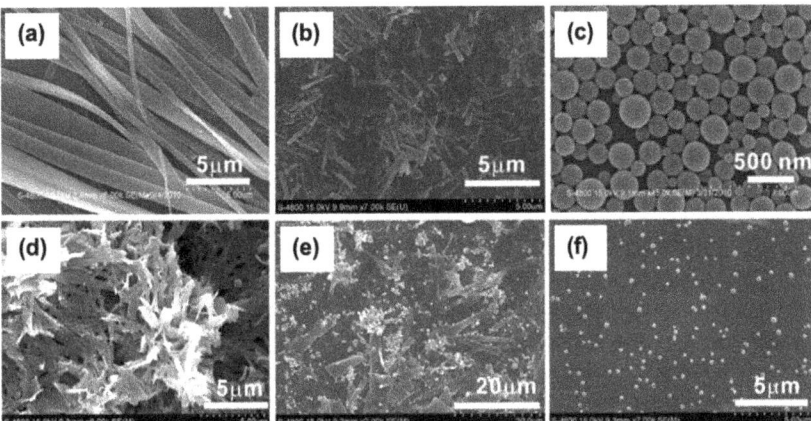

Figure 5. SEM images of (**a–c**) nanostructures assembled from PDI-A (300 μM), for which -COOH group of the lysine was positioned five carbons–carbons away from the *imide* node position of PDI, in acetone/H$_2$O mixture with (**a**) 45%, (**b**) 55%, and (**c**) 75% H$_2$O respectively. SEM images of (**d–f**) nanostructures assembled from PDI-B (300 μM), for which the carboxylate of the lysine was located on the carbon attached to the PDI *imide* node, in acetone/H$_2$O mixture with 50%, 55%, and 80% H$_2$O respectively [56]. Copyright © 2011, American Chemical Society.

3.4. PH-Dependent Self-Assembly

In addition to the peptide sequence and solvent polarity, pH plays a critical role in dictating the hierarchical organization of amino acid–PDI and peptide–PDI derivatives in aqueous solution. Using A-, H-, F-, and V-functionalized PDIs, it was reported that symmetrical PDI-[A]$_2$, PDI-[H]$_2$ and PDI-[V]$_2$ conjugates formed worm-like micelles at high pH (~10), whereas transparent gels were obtained at low pH [60]. The PDI-[F]$_2$ conjugate formed a turbid gel at low pH and a disordered structure upon drying of the solutions in air. Films obtained by drying solutions and gels of the other residue–PDI derivatives showed similar structures, with thin entangled fibers that are less aligned in xerogel state [60]. Other studies have also explored the role of pH in controlling the self-assembly of the symmetrical PDI-[X]$_2$ derivatives. While the carboxylate groups at both ends provide good solubility in basic aqueous solution, aggregation can be triggered by the protonation of the carboxylate groups at pH values below their pKa [61,62]. Moreover, addition of carboxylic acid groups at the two terminal positions was also used to increase the hydrosolubility of a bola-amphiphilic system composed of a PDI core conjugated on their *imide* positions with phenylalanine via an oligomethylene glycol spacer [63]. Designed to self-assemble through π–π staking, H-bonds and van der Waals interactions, this peptide–PDI conjugate self-assembled into hydrogels in phosphate buffer within a pH range of 7–9 and showed interesting photoswitching properties in xerogel state, characterized by a nanofibrous network [63]. A bola-amphiphilic system composed of a PDI core conjugated at its *imide* positions with histidine (PDI–[H]$_2$) demonstrated a pH-responsive protonation-deprotonation that controlled non-covalent interactions between the imidazole ring of adjacent monomers and modulated their self-assembly behavior in water [64]. At pH 10, PDI–[H]$_2$ self-assembled into long, thin fibers with a 12 ± 1 nm width. At pH 7, thick fiber bundles with a 22 ± 1 nm width were observed, which evolved into interconnected belt structures with a 120 ± 5 nm width at pH 2 [64]. In summary, according to the peptide sequence, the pH of the medium can be exploited to control the net charge of the system and the non-covalent interactions between side chains in order to drive self-assembly of the peptide-conjugated PDIs and the resulting nanostructures.

3.5. Concentration-Dependent Self-Assembly

Numerous studies have shown that the concentration of peptide-PDI conjugates can also influence their self-assembly as well as the final supramolecular architecture by modulating enthalpically and/or entropically driven self-recognition processes. Interestingly, it was shown that the symmetrical PDI-[F]$_2$ derivative exhibited distinct concentration-dependent photophysical properties in different organic solvents, such as THF, DMF, and CHCl$_3$ [59]. Increasing PDI–[F]$_2$ concentration from 0.45 mM to 14.6 mM in CHCl$_3$ caused a 43 nm red shift of the fluorescence emission peak, indicative of formation of J-aggregates. As expected, the self-assembly of PDI-[F]$_2$ was stimulated with increasing its concentration. Additionally, the self-organization propensity of PDI-[F]$_2$ derivative correlated closely with the polarity of the solvent (CHCl$_3$ > THF > DMF). Another study on concentration-dependent aggregation property of symmetrical water soluble PDI-[X]$_2$ systems (where X = T or D) has shown an inversion of the maximum absorption intensities and gradual photoluminescence quenching with increasing concentrations, from 10 µM to 250 µM [61]. The N,N'-α-carbobenzyloxy-L-lysine-functionalized 1,6,7,12-tetrachloroperylene diimide compound formed short nanofibrils at low concentrations ranging from 100 to 300 µM, whereas it self-assembled into 16 µm long sheet-like materials at concentrations over 800 µM. Further concentration increases up to 1 mM induced the formation of a mixture of nanotubes and large nanosheets, highlighting how the concentration of PDI-bioconjugates can dramatically modulate the resulting architectures [56].

4. Oligonucleotide–PDI Bioconjugates

Oligonucleotides are well known for their capacity to self-associate in a programed manner into organized supramolecular structures of various sizes and shapes, which offer exceptional opportunities in nanomedicine and nanotechnology [65–67]. DNA-based technology exploits non-canonical base pairing to easily access a large diversity of tailored three-dimensional DNA complexes, also known as DNA origami [66–68]. As previously described for peptides, oligonucleotide sequences have been exploited to modulate the solubility, optical properties, pathways of self-assembly, and/or supramolecular structures of PDI derivatives by means of specific base-pairing as well as hydrophobic and π–π interactions between the PDI core moieties and the oligonucleotide chains. In this section, we will briefly present the main synthesis routes to access oligonucleotide–PDI conjugates before introducing the key studies regarding their self-assembly into supramolecular structures.

4.1. Synthesis of Oligonucleotide–PDI Bioconjugates

Many synthetic approaches used to incorporate PDI as DNA base surrogate have been described to date. The modified oligonucleotides can be synthetized via automated phosphoramidite chemistry and be used as DNA building blocks that allow incorporation of the PDI moiety as an artificial nucleoside surrogate, either at the 5'-terminus or at internal positions of duplex DNA [69]. This approach starts with the preparation of the mixed bisimide from the protected enantiomerically pure (S)-aminopropane-2,3-diol, followed by the coupling of the phospiteamide group to the free hydroxy functionality of the mixed diimide intermediate, as illustrated in Scheme 4.

A convenient click chemistry approach has also been used for the preparation of PDI–oligonucleotide conjugates. The reaction was carried out in a mixture of water and DMSO by means of a copper-catalyzed Huisgen [3 + 2] cycloaddition reaction involving alkyne and azide, as shown in Scheme 5 [70].

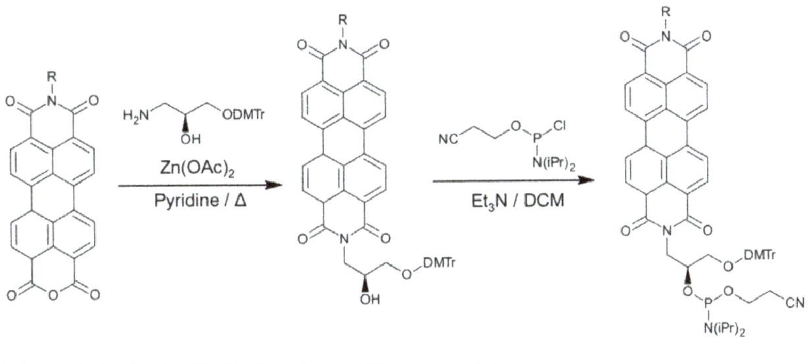

Scheme 4. Synthesis of PDI–phosphoramidites conjugates.

Scheme 5. Synthesis of PDI- oligonucleotide conjugates using [3 + 2] cycloaddition.

4.2. Structure-Dependent Self-Assembly of DNA–PDI Bioconjugates

The oligonucleotide sequence, as well as its length and the positioning of the incorporated PDI(s), are known to significantly modulate the self-assembly process as well as the properties and the morphologies of oligonucleotide–PDI archi. For instance, it was observed that when the PDI chromophore is internally inserted as a synthetic nucleoside surrogate, it favors strong π-stacking interactions with the adjacent DNA base pair. In contrast, dimerization of two whole DNA duplexes is induced when the PDI is attached to the 5′-end [69]. Interestingly, it was observed that the incorporation of multiple PDI moieties into the DNA backbone generates folded nanostructures that exhibit unique hyperthermophilic properties (Figure 6) [71]. It has been proposed that hydrophobic attractions between PDI moieties play an important role in stabilizing the folded structure at high temperatures, and that the DNA loops do not contribute to the inverse-temperature folding behavior [71].

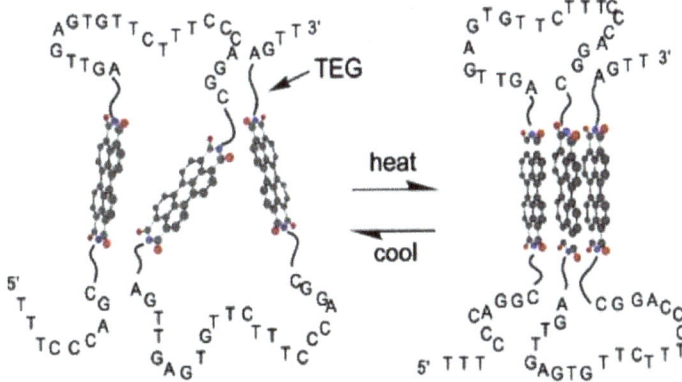

Figure 6. Thermophilic property of chromophoric DNA–perylene trimer, where TEG is a tetraethylene glycol [71]. Copyright © 2003, American Chemical Society.

Furthermore, it was reported that the parent PDI and pyrrolidine-substituted PDI pair are comparable to the natural base pairs and can lead to the formation of stable and stacked heterodimers within DNA duplexes [72]. These modified DNA duplexes were obtained by using enzymatically generated abasic sites that create reactive sites for PDI. The influence of the hydrophobic π–π stacking of PDI groups and base-pairing of oligo tails on the self-assembly behavior of oligonucleotide-conjugated PDI have been explored by varying the length and composition of the oligonucleaotides [73]. Helical as well as nonhelical fibers were obtained from π-stacked PDI groups conjugated with pendent short oligo tails in aqueous solution. The oligonucleotide tails provided solubility in water, whereas the hydrophobic π–π interactions involving PDIs governed self-recognition. In contrast, long oligonucleotide tails led to a diversity of large and ordered assemblies governed by means of base-pairing as well as hydrophobic π–π stacking induced by PDI head groups [73].

Molecular dynamics simulation was employed to understand the self-assembly dynamics and kinetics of the oligonucleotide backbone covalently linked to PDI moieties [74]. It was shown that the mechanism of formation of PDI trimers requires two steps. According to this model, two PDI molecules initially stack into a dimer and the third PDI moiety is then aligned on top of this dimer to finally form a trimer. This study also revealed that each PDI pair interrelates through attractive van der Waals interactions and repulsive electrostatic interactions to drive their self-assembly into ordered structures [74]. Fascinatingly, up to six PDI dyes were covalently incorporated into an oligonucleotide sequence, yielding a DNA duplex characterized by excimer-type fluorescence [75]. The presence of thymines in abasic opposite site of the counter strand impacted PDI-based hydrophobic interactions and the authors proposed a zipper-like recognition motif of the PDI hexamers formed inside the DNA duplex (Figure 7) [75].

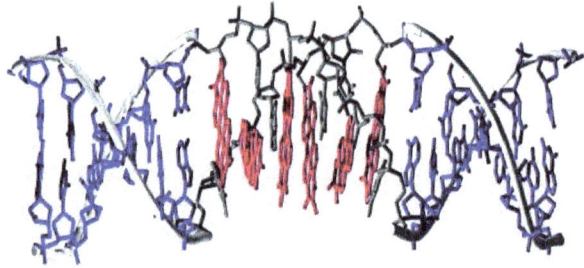

Figure 7. Minimized geometry of DNA duplex with the PDI hexamers drawn in red [75]. Copyright © 2008, WILEY-VCH Verlag GmbH & Co. KGaA, Weinheim.

Moreover, the 1,7-dibromo-PDI derivative was 5′-conjugated to a short coding sequence derived from the human telomeric and it was observed that the PDI moieties drive dimerization of these DNA strands [76]. The topology and stoichiometry of the structures assembled from this PDI–oligonucleotide were modulated by the ionic strength of the solution as well as by the concentration of the bioconjugates. Oligonucleotide strands quickly self-assembled into G-quadruplexes in desalted solutions, whereas Q-assemblies were obtained in the presence of Na^+ or K^+. In this system, the PDI moiety formed dimers instead of extended aggregates, allowing the formation of antiparallel Q species that ultimately converted into parallel species [76].

In another study, PDI was used as a hairpin linker to design short synthetic DNA hairpins in which the PDI core and a guanine–cytosine (G–C) base pair, which serves as a hole trap, are systematically separated by adenine–thymine (A–T) base pairs of various lengths (Figure 8) [77].In this elegant study, it was observed that the presence of stacked PDIs leads to the dimerization of the PDI–oligonucleotide hairpins in buffer solutions and their photoexcitation resulted in subpicosecond formation of a lower exciton state, followed by the formation of an excimer-like state.

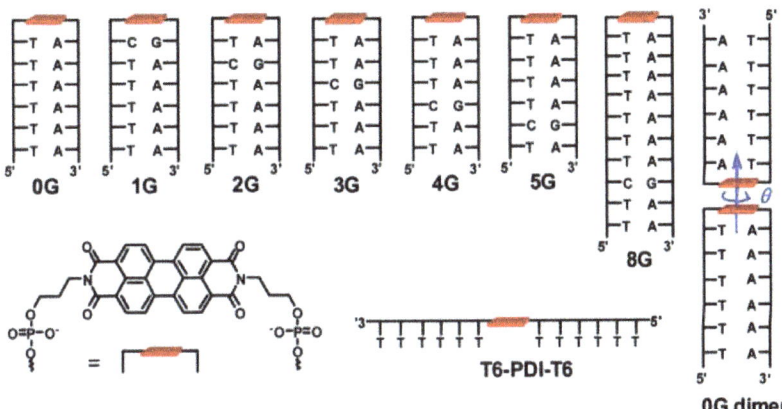

Figure 8. Schematic representation of PDI–DNA conjugates and structures of the PDI chromophore modified with phosphate at *imide* positions [77]. Copyright © 2009, American Chemical Society.

4.3. pH-Dependent Self-Assembly of DNA–PDI Bioconjugates

pH modulation offers an interesting approach to control the self-assembly behavior of PDI molecules conjugated to oligonucleotides as well as the resulting morphology of the chimeric nanostructures. In order to elucidate the roles of PDI aggregation on the formation of G-quadruplex DNA, two symmetrical PDI derivatives respectively bis-substituted in *imide* position by piperidino–ethyl or morpholino–propyl groups were synthesized and their pH-dependent self-assembly was investigated [78]. Under acidic conditions, the PDI–DNA derivatives were not aggregating and were bound avidly to both duplex and G-quadruplex DNA. In sharp contrast, under basic conditions, the ligands extensively aggregated with a high degree of selectivity to bind the G-quadruplex DNA over the DNA duplex. The pH-dependent self-assembly of these PDI derivatives differs according to the nature of the amide substituents. For instance, the more basic piperidino side chain requires higher pH (> 8–9) compared to the derivative bearing the less basic morpholino group, which extensively aggregated at pH > 7–8 [78]. Thus, the pH-dependent self-assembly of DNA-PDI bioconjugates can be exploited to control non-covalent interactions and the resulting morphologies of the nanomaterials.

4.4. Temperature-Dependent Self-Assembly of DNA–PDIs

As for double-stranded DNA, temperature plays a critical role in the self-recognition of PDI–oligonucleotide conjugates. The temperature-dependent self-assembly of synthetic DNA dumbbells having 6 to 16 A–T base pairs connected at their extremities by two PDI linkers at the *imide* nodes was investigated [79]. This PDI-linked bis(oligonucleotide) conjugate was present predominantly as a monomer at room temperature in low-ionic-strength aqueous buffer. Upon heating, a stack of PDIs and a fusion of the base pairs was observed as noticed with the UV and fluorescence spectra (Figure 9a). Transition from DNA capped hairpin to collapsed dumbbell structures with intramolecular π–π stacking of PDIs involved an intermediate state for which all base pairs must be dissociated (Figure 9b). Molecular dynamic simulations highlighted electronic interactions between PDI moieties and the adjacent base pair in the DNA for the capped hairpin while π–π interactions between parallel PDIs were inferred for the dumbbell form [79].

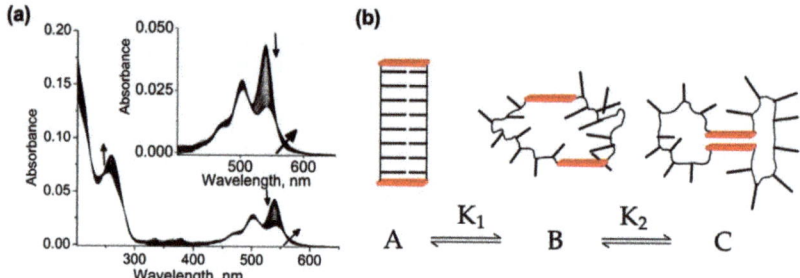

Figure 9. (a) Temperature dependent absorption spectra of the conjugate encompassing 13 A–T base pairs in 10 mM phosphate at pH 7.2. Arrows indicate the changes occurring upon heating from 5 to 95 °C. (b) Model of the three-state generated from base pair dissociation and PDI–PDI association of dumbbell possessing 8 A–T base pairs. PDI is represented in red [79]. Copyright © 2010, American Chemical Society.

The PDI scaffold was also used as a hairpin linker to produce a hairpin-forming bis(oligonucleotide) conjugate that remains predominantly in its monomeric form at room temperature in low-ionic-strength aqueous solution, whereas it dimerizes to head-to-head hairpins under high salt concentrations (>50 mM NaCl) [80]. This effect of salt on the assembly state was attributed to the increase in cation condensation in the hairpin dimer vs. monomer. In contrast, upon heating and in the presence of a low salt concentration, the hydrophobic association between the two PDI units was disrupted, leading to the formation of a monomeric hairpin followed by dissociation of base pairs to reach a random coil structure (Figure 10) [80].

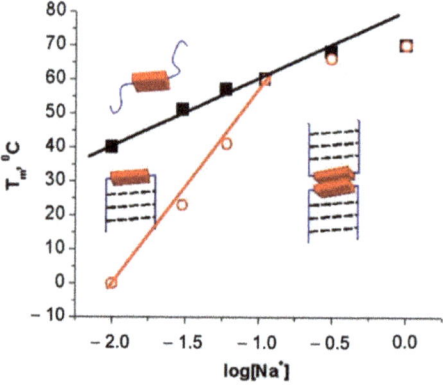

Figure 10. Effect of sodium concentration on temperature-dependent dissociation of DNA base pairs (black) and PDI–PDI dimer (red) [80]. Copyright © 2009, American Chemical Society.

5. Potential Applications of PDI Bioconjugates

5.1. Applications of Peptide–PDI Bioconjugates

As for other PDI-based derivatives, peptide-functionalized PDI conjugates were developed for a variety of applications spanning from microelectronics to biomedicine. A number of studies have reported interesting electronic and optoelectronic properties of PDI–peptide assemblies and showed that these hybrid nanomaterials can be used to fabricate organic light-emitting diode devices and bioelectronic materials [48,59,81]. Interestingly, a sensor probe for Pd^{2+} and CN^- was designed by incorporating two pyridine groups in the PDI core through an Asp residue linker [82]. The asymmetrical PDI-based probe showed high selectivity toward Pd^{2+} ions over other metal cations including Mg^{2+}, Sr^{2+}, Al^{3+}, Cr^{3+},

Pb^{2+}, Mn^{2+}, Fe^{3+}, Co^{2+}, Ni^{2+}, Zn^{2+}, Cd^{2+}, Hg^{2+} and Cu^{2+}, with the formation of Pd^{2+}-ligand complexes triggering the aggregation process and quenching PDI's emission. The presence of CN^- ions stimulated disaggregation of the Pd^{2+}-induced assemblies (Figure 11). Particularly, this system was found to be extremely selective for CN^- over other anions, including OAc^-, ClO_4^-, HSO_4^-, $H_2PO_4^-$, SCN^-, BF_4^-, PF_6^-, NO_3^-, OH^-, F^-, Cl^-, Br^-, I^-, SO_4^{2-}, and PO_4^{3-}. This sensor probe that efficiently measures the concentration of Pd^{2+} and CN^- with detection limits at 0.55 ppb and 0.226 ppb, respectively, constitutes a relevant example of potential applications for PDI-peptide conjugates.

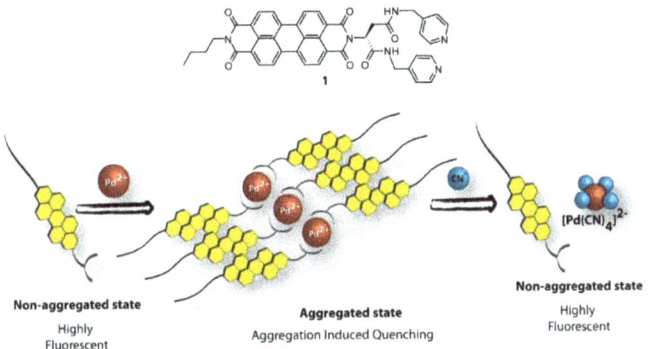

Figure 11. Chemical structure of the PDI sensor probe with a schematic representation of the sensing mechanism of Pd^{2+} and CN^- ions [82]. Copyright © 2017, Wiley-VCH Verlag GmbH & Co. KGaA, Weinheim.

In addition, PDI–amino acid derivatives were used as fluorescent probes to selectively detect anions [6]. It was demonstrated that PDIs respectively functionalized with L-alanine, L-glutamic acid, L-phenylalanine or L-tyrosine exhibit selectivity and high sensitivity for the anions F^- and OH^-, which involve a synergetic effect of H-bonding and anion–π interaction. It was revealed that films obtained by drying solutions and gels of the L-alanine, L-histidine, L-phenylalanine and L-valine functionalized PDIs are photoconductive, and this photoconductivity correlates with the formation of the radical anion [60]. Interestingly, the photoreduction of PDI gelator was associated with the formation of a stable radical anion that induces a change in the packing of the PDI assemblies. This ultimately led to structural modification of the fibrous network, and changes in the rheological properties of the gels [81].

Pyrophosphate (PPi) plays crucial roles in numerous biochemical reactions such as the inhibition or activation of some enzymes, cellular metabolism, protein and nucleic acid synthesis, cell proliferation and cellular iron transport, but abnormal levels can lead to physiopathological conditions [83]. A PDI–peptide derivative was developed as a selective and sensitive fluorescent probe for detecting PPi by exploiting the on/off fluorescence. Complexed to cupric acid ion, the PDI–[GD]$_2$ self-assembled into PDI–[GD]$_2$/Cu^{2+} aggregates, which led to the fluorescence quenching of the PDI units [84]. The displacement of Cu^{2+} by pyrophosphate, for which the PDI–[GD]$_2$ had higher affinity, induced the disassembly of the aggregates and fluorescence recovery in which the intensity correlates directly to the concentration of pyrophosphate in solution [84]. Moreover, symmetrical cysteine-modified PDI (PDI–[C]$_2$) was developed as a simple, low-cost and selective sensing probe for mercury ions, a toxic heavy metal pollutant, in aqueous media [85]. Besides, it was revealed that the thin film formed by the assembly of symmetrical PDI conjugated to histidine (PDI–[H]$_2$) can be used as a reliable, cost-effective, and selective sensor device for the detection of NH_3 vapors under ambient conditions [86]. A simple strategy was developed for the detection and rapid clearance of bacterial lipopolysaccharides (LPS) by combining a specific targeting ligand based on a PDI-conjugated LPS-recognition peptide, with magnetic $Fe_3O_4@SiO_2$ core–shell structures [87].

Interestingly, PDI-peptide conjugates have been evaluated as supramolecular structures for photocatalytic hydrogen production. It was observed that the self-assembly of PDI–[H]$_2$ is needed for hydrogen evolution, i.e., the production of hydrogen through water electrolysis, which occurs at pH 4.5 [62]. Furthermore, another study has demonstrated the stability, phototoxicity and biocompatibility of PDI–[ARGD]$_2$ symmetrical derivatives, suggesting potential usage in the field of photodynamic therapy [46]. Finally, a biocompatible PDI[H]$_2$ probe was described as a unique system towards the modulation of the amyloid fibril formation process [88]. This probe co-assembled with amyloid-β peptide (Aβ), which is associated with Alzheimer's disease, via H-bonding that led to the enhancement of the π–π interactions between Aβ and PDI–[H]$_2$. This interaction accelerated the aggregation process of Aβ into large micron-sized co-assembled structures. Since the oligomeric forms of amyloidogenic peptides are reported to have higher toxicity as compared to the fibrillar aggregates [89,90], these PDI conjugates could lead to potential design and development of drugs targeted toward Alzheimer's disease treatment [88].

5.2. Applications of Oligonucleotide-Conjugated PDIs

Numerous studies have shown that oligonucleotide-conjugated PDIs can be used in biological applications and imaging tools owing to their self-assembly and programmability. For instance, high quenching efficiency of the parent PDI/pyrrolidine-substituted PDI heterodimers, which were arranged in a DNA stack, was exploited to design fluorescent probes toward specific nucleic acid sequences [71]. In this fluorescent reporter, the closed form of the molecular beacon (MB) induces π–π interactions of the pair in the stem region of the molecular beacon sequence, which leads to fluorescence quenching (Figure 12). This closed form is converted to the open form in the presence of a specific target oligonucleotide strand that is complementary to the loop region of the molecular beacon, leading to fluorescence recovery of parent PDI (PH)/pyrrolidine-substituted PDI (PN) heterodimers [72].

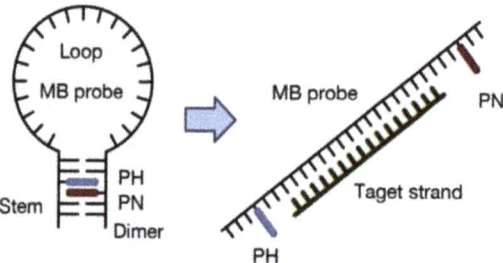

Figure 12. Structure of molecular beacon probe (MB) with a heterodimer based on a parent perylene diimide (PH) and pyrrolidine-substituted PDI in the stem region [72]. © 2018 Wiley-VCH Verlag GmbH & Co. KGaA, Weinheim.

The binding selectivity of cationic PDI to thymine–thymine (T-T) mismatches and the formation of PDI dimer at the mismatch site were exploited to design biosensors [91]. In fact, it was observed that the presence of Hg^{2+} ions produced switching of cationic PDI emission from excimer to monomer due to the disruption of the dimeric form and the conversion of T-T mismatch into T–Hg–T base pair in DNA. Thus, the binding and optical properties of cationic PDIs can be exploited to develop fluorescent probes for T-containing DNA mismatches as well as for Hg^{2+} ions [91]. Moreover, a light-up fluorescent sensor for nucleic and ribonucleic acids with nanomolar sensitivity was developed by conjugating a PDI derivative, harboring an alkoxy group at the *bay* position, to an artificial pocket made by replacing nucleosides with deoxyribospacers inside the DNA structure [92]. In contrast to the assembled form, the monomeric form of PDI linked to the DNA pocket exhibited distinct photochemical properties.

Furthermore, it was proposed that PDI–triplex conjugates can be used to specifically target single-stranded nucleic acid sequences and to modulate genetic expression by the formation of complexes with segments involved in transcriptional processes [93]. In addition, self-assembled DNA monolayers conjugated with PDI base surrogate were used to develop a redox-based probe [94]. It was observed that the rate of electron transfer, mediated by stacks of DNA base pairs, is dependent on the stability of the DNA bridge and the distance between the PDI moiety and the electrode surface [94]. Finally, a cyclic PDI ligand was designed and evaluated as an anticancer drug owing to its high selectivity and efficiency in inhibiting telomerase activity [95].

6. Conclusions and Perspectives

PDIs are some of the most relevant organic dyes that have been largely studied and extensively used, owing to their chemical robustness, thermal and photo-stability as well as their physicochemical and optoelectronic properties. In addition, once rendered amphiphilic, PDI-based molecules undergo self-organization to form nanostructures of varying morphologies. However, the precise control of their molecular self-assembly into ordered supramolecular nanomaterials remains a highly active research area. Conjugation of PDIs to oligonucleotides and peptides has opened new avenues for the design of nanomaterials with unique structures, properties, and functionalities. Such an achievement is largely due to the advances made in the synthetic approaches to (a)symmetric PDI derivatives, and the development of cutting-edge technologies for characterization. The self-assembly process of these PDI-bioconjugates as well as the morphology of the resulting superstructures are dictated by internal non-covalent interactions and numerous external microenvironment conditions. This balance of internal and external factors complexifies the prediction of the morphology of the resulting supramolecular architectures. Nonetheless, as highlighted in this review, recent advances regarding the controlled assembly of peptide–PDI and DNA–PDI bioconjugates will most likely lead to a plethora of innovative functional biomaterials that can be exploited for various biomedical and nanotechnological applications.

Author Contributions: N.K. wrote the initial draft. N.K., A.N. and S.B. revised and corrected the manuscript. All authors have contributed substantially to this review. All authors have read and agreed to the published version of the manuscript.

Funding: This research was funded by the Natural Sciences and Engineering Research Council of Canada (NSERC) grants RGPIN-2018-06209 to S.B. and RGPIN-2018-05799 to A.N. N.K. acknowledges a scholarship from the Center of Excellence in Research on Orphan Diseases—Fondation Courtois (CERMO-FC).

Acknowledgments: The authors apologize to all research groups whose contributions could not be cited because of space limitations.

Conflicts of Interest: The authors declare no conflict of interest.

References

1. Yadav, S.; Sharma, A.K.; Kumar, P. Nanoscale Self-Assembly for Therapeutic Delivery. *Front. Bioeng. Biotechnol.* **2020**, *8*, 127. [CrossRef] [PubMed]
2. Ekiz, M.S.; Cinar, G.; Khalily, M.A.; Guler, M.O. Self-assembled peptide nanostructures for functional materials. *Nanotechnology* **2016**, *27*, 402002. [CrossRef] [PubMed]
3. Farkaš, P.; Bystrický, S. Chemical conjugation of biomacromolecules: A mini-review. *Chem. Pap.* **2010**, *64*, 683–695. [CrossRef]
4. Cigánek, M.; Richtár, J.; Weiter, M.; Krajčovič, J. Organic π-Conjugated Molecules: From Nature to Artificial Applications. Where are the Boundaries? *Isr. J. Chem.* **2021**, *61*, 1–15. [CrossRef]
5. Würthner, F.; Saha-Möller, C.R.; Fimmel, B.; Ogi, S.; Leowanawat, P.; Schmidt, D. Perylene Bisimide Dye Assemblies as Archetype Functional Supramolecular Materials. *Chem. Rev.* **2016**, *116*, 962–1052. [CrossRef] [PubMed]
6. Chen, C.-Y.; Wang, K.; Gu, L.-L.; Li, H. The study of perylene diimide–amino acid derivatives for the fluorescence detection of anions. *RSC Adv.* **2017**, *7*, 42685–42689. [CrossRef]

7. Würthner, F.; Stepanenko, V.; Chen, Z.; Saha-Möller, C.R.; Kocher, N.; Stalke, D. Preparation and Characterization of Regioisomerically Pure 1,7-Disubstituted Perylene Bisimide Dyes. *J. Org. Chem.* **2004**, *69*, 7933–7939. [CrossRef]
8. Abd-Ellah, M.; Cann, J.; Dayneko, S.V.; Laventure, A.; Cieplechowicz, E.; Welch, G.C. Interfacial ZnO Modification Using a Carboxylic Acid Functionalized N-Annulated Perylene Diimide for Inverted Type Organic Photovoltaics. *ACS Appl. Electron. Mater.* **2019**, *1*, 1590–1596. [CrossRef]
9. Guo, Z.; Zhang, X.; Wang, Y.; Li, Z. Supramolecular Self-Assembly of Perylene Bisimide Derivatives Assisted by Various Groups. *Langmuir* **2019**, *35*, 342–358. [CrossRef] [PubMed]
10. Nowak-Król, A.; Würthner, F. Progress in the synthesis of perylene bisimide dyes. *Org. Chem. Front.* **2019**, *6*, 1272–1318. [CrossRef]
11. Chen, S.; Slattum, P.; Wang, C.; Zang, L. Self-Assembly of Perylene Imide Molecules into 1D Nanostructures: Methods, Morphologies, and Applications. *Chem. Rev.* **2015**, *115*, 11967–11998. [CrossRef]
12. Singh, P.; Sharma, P.; Kaur, N.; Mittal, L.S.; Kumar, K. Perylene diimides: Will they flourish as reaction-based probes? *Anal. Methods* **2020**, *12*, 3560–3574. [CrossRef] [PubMed]
13. Chen, S.; Xue, Z.; Gao, N.; Yang, X.; Zang, L. Perylene Diimide-Based Fluorescent and Colorimetric Sensors for Environmental Detection. *Sensors* **2020**, *20*, 917. [CrossRef] [PubMed]
14. Steinbrück, N.; Kickelbick, G. Perylene polyphenylmethylsiloxanes for optoelectronic applications. *J. Polym. Sci. Part. B Polym. Phys.* **2019**, *57*, 1062–1073. [CrossRef]
15. Yuen, J.D.; Pozdin, V.A.; Young, A.T.; Turner, B.L.; Giles, I.D.; Naciri, J.; Trammell, S.A.; Charles, P.T.; Stenger, D.A.; Daniele, M.A. Perylene-diimide-based n-type semiconductors with enhanced air and temperature stable photoconductor and transistor properties. *Dyes Pigments* **2020**, *174*, 108014. [CrossRef]
16. Song, C.; Liu, X.; Li, X.; Wang, Y.-C.; Wan, L.; Sun, X.; Zhang, W.; Fang, J. Perylene Diimide-Based Zwitterion as the Cathode Interlayer for High-Performance Nonfullerene Polymer Solar Cells. *ACS Appl. Mater. Interfaces* **2018**, *10*, 14986–14992. [CrossRef] [PubMed]
17. Yang, Z.; Chen, X. Semiconducting Perylene Diimide Nanostructure: Multifunctional Phototheranostic Nanoplatform. *Accounts Chem. Res.* **2019**, *52*, 1245–1254. [CrossRef] [PubMed]
18. Wang, H.; Xue, K.-F.; Yang, Y.; Hu, H.; Xu, J.-F.; Zhang, X. In Situ Hypoxia-Induced Supramolecular Perylene Diimide Radical Anions in Tumors for Photothermal Therapy with Improved Specificity. *J. Am. Chem. Soc.* **2022**, *144*, 2360–2367. [CrossRef] [PubMed]
19. Sun, M.; Müllen, K.; Yin, M. Water-soluble perylenediimides: Design concepts and biological applications. *Chem. Soc. Rev.* **2016**, *45*, 1513–1528. [CrossRef] [PubMed]
20. Langhals, H. Cyclic Carboxylic Imide Structures as Structure Elements of High Stability. Novel Developments in Perylene Dye Chemistry. *Heterocycles* **1995**, *40*, 477. [CrossRef]
21. Huang, C.; Barlow, S.; Marder, S.R. Perylene-3,4,9,10-tetracarboxylic Acid Diimides: Synthesis, Physical Properties, and Use in Organic Electronics. *J. Org. Chem.* **2011**, *76*, 2386–2407. [CrossRef] [PubMed]
22. Liu, B.; Böckmann, M.; Jiang, W.; Doltsinis, N.L.; Wang, Z. Perylene Diimide-Embedded Double [8]Helicenes. *J. Am. Chem. Soc.* **2020**, *142*, 7092–7099. [CrossRef] [PubMed]
23. Tang, F.; Wu, K.; Zhou, Z.; Wang, G.; Zhao, B.; Tan, S. Alkynyl-Functionalized Pyrene-Cored Perylene Diimide Electron Acceptors for Efficient Nonfullerene Organic Solar Cells. *ACS Appl. Energy Mater.* **2019**, *2*, 3918–3926. [CrossRef]
24. Zhang, L.; Song, I.; Ahn, J.; Han, M.; Linares, M.; Surin, M.; Zhang, H.-J.; Oh, J.H.; Lin, J. π-Extended perylene diimide double-heterohelicenes as ambipolar organic semiconductors for broadband circularly polarized light detection. *Nat. Commun.* **2021**, *12*, 142. [CrossRef] [PubMed]
25. Dayneko, S.V.; Rahmati, M.; Pahlevani, M.; Welch, G.C. Solution processed red organic light-emitting-diodes using an N-annulated perylene diimide fluorophore. *J. Mater. Chem. C* **2020**, *8*, 2314–2319. [CrossRef]
26. Wang, J.; He, E.; Liu, X.; Yu, L.; Wang, H.; Zhang, R.; Zhang, H. High performance hydrazine vapor sensor based on redox mechanism of twisted perylene diimide derivative with lower reduction potential. *Sens. Actuators B Chem.* **2017**, *239*, 898–905. [CrossRef]
27. Seo, J.; Khazi, M.I.; Kim, J.-M. Highly responsive triethylamine vapor sensor based on a perylene diimide-polydiacetylene system via heat-induced tuning of the molecular packing approach. *Sens. Actuators B Chem.* **2021**, *334*, 129660. [CrossRef]
28. Yang, Z.; Song, J.; Tang, W.; Fan, W.; Dai, Y.; Shen, Z.; Lin, L.; Cheng, S.; Liu, Y.; Niu, G.; et al. Stimuli-Responsive Nanotheranostics for Real-Time Monitoring Drug Release by Photoacoustic Imaging. *Theranostics* **2019**, *9*, 526–536. [CrossRef]
29. Gong, Q.; Xing, J.; Huang, Y.; Wu, A.; Yu, J.; Zhang, Q. Perylene Diimide Oligomer Nanoparticles with Ultrahigh Photothermal Conversion Efficiency for Cancer Theranostics. *ACS Appl. Bio Mater.* **2020**, *3*, 1607–1615. [CrossRef]
30. Würthner, F. Perylene bisimide dyes as versatile building blocks for functional supramolecular architectures. *Chem. Commun.* **2004**, *35*, 1564–1579. [CrossRef]
31. Arantes, J.T.; Lima, M.P.; Fazzio, A.; Xiang, H.; Wei, S.-H.; Dalpian, G.M. Effects of Side-Chain and Electron Exchange Correlation on the Band Structure of Perylene Diimide Liquid Crystals: A Density Functional Study. *J. Phys. Chem. B* **2009**, *113*, 5376–5380. [CrossRef]
32. Polkehn, M.; Tamura, H.; Eisenbrandt, P.; Haacke, S.; Méry, S.; Burghardt, I. Molecular Packing Determines Charge Separation in a Liquid Crystalline Bisthiophene–Perylene Diimide Donor–Acceptor Material. *J. Phys. Chem. Lett.* **2016**, *7*, 1327–1334. [CrossRef] [PubMed]

33. Spano, F.C.; Silva, C. H- and J-Aggregate Behavior in Polymeric Semiconductors. *Annu. Rev. Phys. Chem.* **2014**, *65*, 477–500. [CrossRef] [PubMed]
34. Ghosh, S.; Li, X.-Q.; Stepanenko, V.; Würthner, F. Control of H- and J-Type π Stacking by Peripheral Alkyl Chains and Self-Sorting Phenomena in Perylene Bisimide Homo- and Heteroaggregates. *Chem. A Eur. J.* **2008**, *14*, 11343–11357. [CrossRef]
35. Pasaogullari, N.; Icil, H.; Demuth, M. Symmetrical and unsymmetrical perylene diimides: Their synthesis, photophysical and electrochemical properties. *Dyes Pigments* **2006**, *69*, 118–127. [CrossRef]
36. Nagao, Y.; Misono, T. Synthesis and Reactions of Perylenecarboxylic Acid Derivatives. VII. Hydrolysis ofN,N′-Dialkyl-3,4: 9,10-Perylenebis(dicarboximide) with Sulfuric Acid. *Bull. Chem. Soc. Jpn.* **1981**, *54*, 1269–1270. [CrossRef]
37. Nagao, Y. Synthesis and properties of perylene pigments. *Prog. Org. Coat.* **1997**, *31*, 43–49. [CrossRef]
38. Tröster, H. Untersuchungen zur Protonierung von Perylen-3,4,9,10-tetracarbonsäurealkalisalzen. *Dyes Pigments* **1983**, *4*, 171–177. [CrossRef]
39. Eakins, G.L.; Gallaher, J.K.; Keyzers, R.A.; Falber, A.; Webb, J.E.A.; Laos, A.; Tidhar, Y.; Weissman, H.; Rybtchinski, B.; Thordarson, P.; et al. Thermodynamic Factors Impacting the Peptide-Driven Self-Assembly of Perylene Diimide Nanofibers. *J. Phys. Chem. B* **2014**, *118*, 8642–8651. [CrossRef]
40. Zottig, X.; Côté-Cyr, M.; Arpin, D.; Archambault, D.; Bourgault, S. Protein Supramolecular Structures: From Self-Assembly to Nanovaccine Design. *Nanomaterials* **2020**, *10*, 1008. [CrossRef] [PubMed]
41. Makishima, A. (Ed.) Fundamental Knowledges and Techniques in Biochemistry. In *Biochemistry for Materials*; Elsevier Science: Amsterdam, The Netherlands, 2019; pp. 35–51. [CrossRef]
42. Ouellette, R.J.; Rawn, J.D. Amino Acids, Peptides, and Proteins. In *Organic Chemistry Study Guide*; Elsevier: Amsterdam, The Netherlands, 2015; p. 569. [CrossRef]
43. Vadehra, G.S.; Wall, B.D.; Diegelmann, S.R.; Tovar, J.D. On-resin dimerization incorporates a diverse array of π-conjugated functionality within aqueous self-assembling peptide backbones. *Chem. Commun.* **2010**, *46*, 3947–3949. [CrossRef]
44. Godin, E.; Nguyen, P.T.; Zottig, X.; Bourgault, S. Identification of a hinge residue controlling islet amyloid polypeptide self-assembly and cytotoxicity. *J. Biol. Chem.* **2019**, *294*, 8452–8463. [CrossRef]
45. Merrifield, R.B. Solid Phase Peptide Synthesis. I. The Synthesis of a Tetrapeptide. *J. Am. Chem. Soc.* **1963**, *85*, 2149–2154. [CrossRef]
46. Kim, Y.-O.; Park, S.-J.; Jung, B.Y.; Jang, H.-S.; Choi, S.K.; Kim, J.; Kim, S.; Jung, Y.C.; Shin, D.-S.; Lee, Y.-S. Solid-Phase Synthesis of Peptide-Conjugated Perylene Diimide Bolaamphiphile and Its Application in Photodynamic Therapy. *ACS Omega* **2018**, *3*, 5896–5902. [CrossRef] [PubMed]
47. Bai, S.; Debnath, S.; Javid, N.; Frederix, P.W.J.M.; Fleming, S.; Pappas, C.; Ulijn, R.V. Differential Self-Assembly and Tunable Emission of Aromatic Peptide Bola-Amphiphiles Containing Perylene Bisimide in Polar Solvents Including Water. *Langmuir* **2014**, *30*, 7576–7584. [CrossRef] [PubMed]
48. Eakins, G.L.; Pandey, R.; Wojciechowski, J.P.; Zheng, H.Y.; Webb, J.E.A.; Valéry, C.; Thordarson, P.; Plank, N.O.V.; Gerrard, J.A.; Hodgkiss, J.M. Functional Organic Semiconductors Assembled via Natural Aggregating Peptides. *Adv. Funct. Mater.* **2015**, *25*, 5640–5649. [CrossRef]
49. Wei, D.; Ge, L.; Wang, Z.; Wang, Y.; Guo, R. Self-Assembled Dual Helical Nanofibers of Amphiphilic Perylene Diimides with Oligopeptide Substitution. *Langmuir* **2019**, *35*, 11745–11754. [CrossRef] [PubMed]
50. Panda, S.S.; Shmilovich, K.; Herringer, N.S.M.; Marin, N.; Ferguson, A.L.; Tovar, J.D. Computationally Guided Tuning of Peptide-Conjugated Perylene Diimide Self-Assembly. *Langmuir* **2021**, *37*, 8594–8606. [CrossRef] [PubMed]
51. Marty, R.; Nigon, R.; Leite, D.; Frauenrath, H. Two-Fold Odd–Even Effect in Self-Assembled Nanowires from Oligopeptide-Polymer-Substituted Perylene Bisimides. *J. Am. Chem. Soc.* **2014**, *136*, 3919–3927. [CrossRef]
52. Ke, D.; Tang, A.; Zhan, C.; Yao, J. Conformation-variable PDI@β-sheet nanohelices show stimulus-responsive supramolecular chirality. *Chem. Commun.* **2013**, *49*, 4914–4916. [CrossRef]
53. Eakins, G.L.; Wojciechowski, J.P.; Martin, A.D.; Webb, J.E.A.; Thordarson, P.; Hodgkiss, J.M. Chiral effects in peptide-substituted perylene imide assemblies. *Supramol. Chem.* **2015**, *27*, 746–756. [CrossRef]
54. Ahmed, S.; Pramanik, B.; Sankar, K.N.A.; Srivastava, A.; Singha, N.; Dowari, P.; Srivastava, A.; Mohanta, K.; Debnath, A.; Das, D. Solvent Assisted Tuning of Morphology of a Peptide-Perylenediimide Conjugate: Helical Fibers to Nano-Rings and Their Differential Semiconductivity. *Sci. Rep.* **2017**, *7*, 9485. [CrossRef]
55. Ahmed, S.; Sankar, K.N.A.; Pramanik, B.; Mohanta, K.; Das, D. Solvent Directed Morphogenesis and Electrical Properties of a Peptide–Perylenediimide Conjugate. *Langmuir* **2018**, *34*, 8355–8364. [CrossRef]
56. Sun, Y.; He, C.; Sun, K.; Li, Y.; Dong, H.; Wang, Z.; Li, Z. Fine-Tuned Nanostructures Assembled from l-Lysine-Functionalized Perylene Bisimides. *Langmuir* **2011**, *27*, 11364–11371. [CrossRef]
57. Chen, Z.; Lohr, A.; Saha-Möller, C.R.; Würthner, F. Self-assembled π-stacks of functional dyes in solution: Structural and thermodynamic features. *Chem. Soc. Rev.* **2009**, *38*, 564–584. [CrossRef]
58. Farooqi, M.J.; Penick, M.A.; Burch, J.; Negrete, G.R.; Brancaleon, L. Characterization of novel perylene diimides containing aromatic amino acid side chains. *Spectrochim. Acta Part A Mol. Biomol. Spectrosc.* **2015**, *153*, 124–131. [CrossRef]
59. Chal, P.; Shit, A.; Nandi, A.K. Optoelectronic Properties of Supramolecular of Aggregates Phenylalanine Conjugated Perylene Bisimide. *ChemistrySelect* **2018**, *3*, 3993–4003. [CrossRef]
60. Draper, E.R.; Walsh, J.J.; McDonald, T.O.; Zwijnenburg, M.A.; Cameron, P.J.; Cowan, A.J.; Adams, D.J. Air-stable photoconductive films formed from perylene bisimide gelators. *J. Mater. Chem. C* **2014**, *2*, 5570–5575. [CrossRef]

61. Kozma, E.; Grisci, G.; Mróz, W.; Catellani, M.; Eckstein-Andicsovà, A.; Pagano, K.; Galeotti, F. Water-soluble aminoacid functionalized perylene diimides: The effect of aggregation on the optical properties in organic and aqueous media. *Dyes Pigments* **2016**, *125*, 201–209. [CrossRef]
62. Nolan, M.C.; Walsh, J.J.; Mears, L.L.E.; Draper, E.R.; Wallace, M.; Barrow, M.; Dietrich, B.; King, S.M.; Cowan, A.J.; Adams, D.J. pH dependent photocatalytic hydrogen evolution by self-assembled perylene bisimides. *J. Mater. Chem. A* **2017**, *5*, 7555–7563. [CrossRef]
63. Roy, S.; Basu, K.; Gayen, K.; Panigrahi, S.; Mondal, S.; Basak, D.; Banerjee, A. TiO$_2$ Nanoparticles Incorporated Peptide Appended Perylene Bisimide-Based Nanohybrid System: Enhancement of Photo-Switching Behavior. *J. Phys. Chem. C* **2017**, *121*, 5428–5435. [CrossRef]
64. Pandeeswar, M.; Govindaraju, T. Engineering molecular self-assembly of perylene diimide through pH-responsive chiroptical switching. *Mol. Syst. Des. Eng.* **2016**, *1*, 202–207. [CrossRef]
65. Li, J.; Pei, H.; Zhu, B.; Liang, L.; Wei, M.; He, Y.; Chen, N.; Li, D.; Huang, Q.; Fan, C. Self-Assembled Multivalent DNA Nanostructures for Noninvasive Intracellular Delivery of Immunostimulatory CpG Oligonucleotides. *ACS Nano* **2011**, *5*, 8783–8789. [CrossRef]
66. Oliviero, G.; D'Errico, S.; Pinto, B.; Nici, F.; Dardano, P.; Rea, I.; De Stefano, L.; Mayol, L.; Piccialli, G.; Borbone, N. Self-Assembly of G-Rich Oligonucleotides Incorporating a 3′-3′ Inversion of Polarity Site: A New Route Towards G-Wire DNA Nanostructures. *ChemistryOpen* **2017**, *6*, 599–605. [CrossRef]
67. Yang, D.; Campolongo, M.J.; Nhi Tran, T.N.; Ruiz, R.C.H.; Kahn, J.S.; Luo, D. Novel DNA materials and their applications. *WIREs Nanomed. Nanobiotechnol.* **2010**, *2*, 648–669. [CrossRef]
68. Marras, A.E.; Zhou, L.; Su, H.-J.; Castro, C.E. Programmable motion of DNA origami mechanisms. *Proc. Natl. Acad. Sci. USA* **2015**, *112*, 713–718. [CrossRef]
69. Wagner, C.; Wagenknecht, H.-A. Perylene-3,4:9,10-tetracarboxylic Acid Bisimide Dye as an Artificial DNA Base Surrogate. *Org. Lett.* **2006**, *8*, 4191–4194. [CrossRef]
70. Ustinov, A.V.; Dubnyakova, V.V.; Korshun, V.A. A convenient 'click chemistry' approach to perylene diimide–oligonucleotide conjugates. *Tetrahedron* **2008**, *64*, 1467–1473. [CrossRef]
71. Wang, W.; Zhou, H.-H.; Niu, A.S.; Li, A.D.Q. Alternating DNA and π-Conjugated Sequences. Thermophilic Foldable Polymers. *J. Am. Chem. Soc.* **2003**, *125*, 5248–5249. [CrossRef]
72. Takada, T.; Ishino, S.; Takata, A.; Nakamura, M.; Fujitsuka, M.; Majima, T.; Yamana, K. Rapid Electron Transfer of Stacked Heterodimers of Perylene Diimide Derivatives in a DNA Duplex. *Chem. A Eur. J.* **2018**, *24*, 8228–8232. [CrossRef]
73. Mishra, A.K.; Weissman, H.; Krieg, E.; Votaw, K.A.; Mccullagh, M.; Rybtchinski, B.; Lewis, F.D. Self-Assembly of Perylenediimide–Single-Strand-DNA Conjugates: Employing Hydrophobic Interactions and DNA Base-Pairing To Create a Diverse Structural Space. *Chem. A Eur. J.* **2017**, *23*, 10328–10337. [CrossRef] [PubMed]
74. Markegard, C.B.; Mazaheripour, A.; Jocson, J.-M.; Burke, A.M.; Dickson, M.N.; Gorodetsky, A.A.; Nguyen, H.D. Molecular Dynamics Simulations of Perylenediimide DNA Base Surrogates. *J. Phys. Chem. B* **2015**, *119*, 11459–11465. [CrossRef] [PubMed]
75. Baumstark, D.; Wagenknecht, H.-A. Fluorescent Hydrophobic Zippers inside Duplex DNA: Interstrand Stacking of Perylene-3,4:9,10-tetracarboxylic Acid Bisimides as Artificial DNA Base Dyes. *Chem. A Eur. J.* **2008**, *14*, 6640–6645. [CrossRef]
76. Franceschin, M.; Borbone, N.; Oliviero, G.; Casagrande, V.; Scuotto, M.; Coppola, T.; Borioni, S.; Mayol, L.; Ortaggi, G.; Bianco, A.; et al. Synthesis of a Dibromoperylene Phosphoramidite Building Block and Its Incorporation at the 5′ End of a G-Quadruplex Forming Oligonucleotide: Spectroscopic Properties and Structural Studies of the Resulting Dibromoperylene Conjugate. *Bioconj. Chem.* **2011**, *22*, 1309–1319. [CrossRef] [PubMed]
77. Carmieli, R.; Zeidan, T.A.; Kelley, R.F.; Mi, Q.; Lewis, F.D.; Wasielewski, M.R. Excited State, Charge Transfer, and Spin Dynamics in DNA Hairpin Conjugates with Perylenediimide Hairpin Linkers. *J. Phys. Chem. A* **2009**, *113*, 4691–4700. [CrossRef] [PubMed]
78. Kerwin, S.M.; Chen, G.; Kern, J.T.; Thomas, P.W. Perylene Diimide G-Quadruplex DNA Binding Selectivity is Mediated by Ligand Aggregation. *Bioorg. Med. Chem. Lett.* **2001**, *12*, 447–450. [CrossRef]
79. Hariharan, M.; Siegmund, K.; Zheng, Y.; Long, H.; Schatz, G.C.; Lewis, F.D. Perylenediimide-Linked DNA Dumbbells: Long-Distance Electronic Interactions and Hydrophobic Assistance of Base-Pair Melting. *J. Phys. Chem. C* **2010**, *114*, 20466–20471. [CrossRef]
80. Hariharan, M.; Zheng, Y.; Long, H.; Zeidan, T.A.; Schatz, G.C.; Vura-Weis, J.; Wasielewski, M.R.; Zuo, X.; Tiede, D.M.; Lewis, F.D. Hydrophobic Dimerization and Thermal Dissociation of Perylenediimide-Linked DNA Hairpins. *J. Am. Chem. Soc.* **2009**, *131*, 5920–5929. [CrossRef]
81. Draper, E.R.; Schweins, R.; Akhtar, R.; Groves, P.; Chechik, V.; Zwijnenburg, M.A.; Adams, D.J. Reversible Photoreduction as a Trigger for Photoresponsive Gels. *Chem. Mater.* **2016**, *28*, 6336–6341. [CrossRef]
82. Pramanik, B.; Ahmed, S.; Singha, N.; Das, D. Self-Assembly Assisted Tandem Sensing of Pd2+ and CN− by a Perylenediimide-Peptide Conjugate. *ChemistrySelect* **2017**, *2*, 10061–10066. [CrossRef]
83. Heinonen, J.K. Regulatory Roles of PPi. In *Biological Role of Inorganic Pyrophosphate*; Springer: Boston, MA, USA, 2001; p. 123. [CrossRef]
84. Feng, X.; An, Y.; Yao, Z.; Li, C.; Shi, G. A Turn-on Fluorescent Sensor for Pyrophosphate Based on the Disassembly of Cu2+-Mediated Perylene Diimide Aggregates. *ACS Appl. Mater. Interfaces* **2012**, *4*, 614–618. [CrossRef]
85. Grisci, G.; Mróz, W.; Catellani, M.; Kozma, E.; Galeotti, F. Off-On Fluorescence Response of a Cysteine-based Perylene Diimide for Mercury Detection in Water. *ChemistrySelect* **2016**, *1*, 3033–3037. [CrossRef]

86. Kalita, A.; Hussain, S.; Malik, A.H.; Subbarao, N.V.V.; Iyer, P.K. Vapor phase sensing of ammonia at the sub-ppm level using a perylene diimide thin film device. *J. Mater. Chem. C* **2015**, *3*, 10767–10774. [CrossRef]
87. Liu, F.; Mu, J.; Wu, X.; Bhattacharjya, S.; Yeow, E.K.L.; Xing, B. Peptide–perylene diimide functionalized magnetic nano-platforms for fluorescence turn-on detection and clearance of bacterial lipopolysaccharides. *Chem. Commun.* **2014**, *50*, 6200–6203. [CrossRef] [PubMed]
88. Muthuraj, B.; Chowdhury, S.R.; Iyer, P.K. Modulation of Amyloid-β Fibrils into Mature Microrod-Shaped Structure by Histidine Functionalized Water-Soluble Perylene Diimide. *ACS Appl. Mater. Interfaces* **2015**, *7*, 21226–21234. [CrossRef]
89. Nguyen, P.T.; Zottig, X.; Sebastiao, M.; Arnold, A.A.; Marcotte, I.; Bourgault, S. Identification of transmissible proteotoxic oligomer-like fibrils that expand conformational diversity of amyloid assemblies. *Commun. Biol.* **2021**, *4*, 939. [CrossRef] [PubMed]
90. Benilova, I.; Karran, E.; De Strooper, B. The toxic Aβ oligomer and Alzheimer's disease: An emperor in need of clothes. *Nat. Neurosci.* **2012**, *15*, 349–357. [CrossRef] [PubMed]
91. Takada, T.; Ashida, A.; Nakamura, M.; Yamana, K. Cationic perylenediimide as a specific fluorescent binder to mismatch containing DNA. *Bioorg. Med. Chem.* **2013**, *21*, 6011–6014. [CrossRef]
92. Takada, T.; Yamaguchi, K.; Tsukamoto, S.; Nakamura, M.; Yamana, K. Light-up fluorescent probes utilizing binding behavior of perylenediimide derivatives to a hydrophobic pocket within DNA. *Analyst* **2014**, *139*, 4016–4021. [CrossRef] [PubMed]
93. Bevers, S.; Schutte, S.; McLaughlin, L.W. Naphthalene- and Perylene-Based Linkers for the Stabilization of Hairpin Triplexes. *J. Am. Chem. Soc.* **2000**, *122*, 5905–5915. [CrossRef]
94. Wohlgamuth, C.H.; McWilliams, M.A.; Mazaheripour, A.; Burke, A.M.; Lin, K.-Y.; Doan, L.; Slinker, J.D.; Gorodetsky, A.A. Electrochemistry of DNA Monolayers Modified With a Perylenediimide Base Surrogate. *J. Phys. Chem. C* **2014**, *118*, 29084–29090. [CrossRef]
95. Vasimalla, S.; Sato, S.; Takenaka, F.; Kurose, Y.; Takenaka, S. Cyclic perylene diimide: Selective ligand for tetraplex DNA binding over double stranded DNA. *Bioorg. Med. Chem.* **2017**, *25*, 6404–6411. [CrossRef] [PubMed]

Article

Self-Healing Thiolated Pillar[5]arene Films Containing Moxifloxacin Suppress the Development of Bacterial Biofilms

Dmitriy N. Shurpik [1,*], Yulia I. Aleksandrova [1], Olga A. Mostovaya [1], Viktoriya A. Nazmutdinova [1], Regina E. Tazieva [1], Fadis F. Murzakhanov [2], Marat R. Gafurov [2], Pavel V. Zelenikhin [3], Evgenia V. Subakaeva [3], Evgenia A. Sokolova [3], Alexander V. Gerasimov [1], Vadim V. Gorodov [4], Daut R. Islamov [5], Peter J. Cragg [6] and Ivan I. Stoikov [1,*]

1. A.M.Butlerov Chemical Institute, Kazan Federal University, Kremlevskaya, 18, 420008 Kazan, Russia; a.julia.1996@mail.ru (Y.I.A.); olga.mostovaya@mail.ru (O.A.M.); n-vika-art@mail.ru (V.A.N.); reginatzv@gmail.com (R.E.T.); alexander.gerasimov@kpfu.ru (A.V.G.)
2. Institute of Physics, Kazan Federal University, Kremlevskaya, 18, 420008 Kazan, Russia; murzakhanov.fadis@yandex.ru (F.F.M.); marat.gafurov@kpfu.ru (M.R.G.)
3. Institute of Fundamental Medicine and Biology, Kazan Federal University, Kremlevskaya, 18, 420008 Kazan, Russia; pasha_mic@mail.ru (P.V.Z.); zs_zs97@mail.ru (E.V.S.); zhenya_mic@mail.ru (E.A.S.)
4. Enikolopov Institute of Synthetic Polymeric Materials, Russian Academy of Sciences (ISPM RAS), Profsoyuznaya, 70, 117593 Moscow, Russia; gorodovvv@ispm.ru
5. Laboratory for Structural Analysis of Biomacromolecules, Kazan Scientific Center of Russian Academy of Sciences, Lobachevskogo, 2/31, 420111 Kazan, Russia; daut1989@mail.ru
6. School of Applied Sciences, University of Brighton, Huxley Building, Brighton BN2 4GJ, UK; p.j.cragg@brighton.ac.uk
* Correspondence: dnshurpik@mail.ru (D.N.S.); ivan.stoikov@mail.ru (I.I.S.)

Citation: Shurpik, D.N.; Aleksandrova, Y.I.; Mostovaya, O.A.; Nazmutdinova, V.A.; Tazieva, R.E.; Murzakhanov, F.F.; Gafurov, M.R.; Zelenikhin, P.V.; Sokolova, E.A.; et al. Self-Healing Thiolated Pillar[5]arene Films Containing Moxifloxacin Suppress the Development of Bacterial Biofilms. *Nanomaterials* **2022**, *12*, 1604. https://doi.org/10.3390/nano12091604

Academic Editor: Vincenzo Amendola

Received: 13 April 2022
Accepted: 6 May 2022
Published: 9 May 2022

Publisher's Note: MDPI stays neutral with regard to jurisdictional claims in published maps and institutional affiliations.

Copyright: © 2022 by the authors. Licensee MDPI, Basel, Switzerland. This article is an open access article distributed under the terms and conditions of the Creative Commons Attribution (CC BY) license (https:// creativecommons.org/licenses/by/ 4.0/).

Abstract: Polymer self-healing films containing fragments of pillar[5]arene were obtained for the first time using thiol/disulfide redox cross-linking. These films were characterized by thermogravimetric analysis and differential scanning calorimetry, FTIR spectroscopy, and electron microscopy. The films demonstrated the ability to self-heal through the action of atmospheric oxygen. Using UV–vis, 2D 1H-1H NOESY, and DOSY NMR spectroscopy, the pillar[5]arene was shown to form complexes with the antimicrobial drug moxifloxacin in a 2:1 composition (logK11 = 2.14 and logK12 = 6.20). Films containing moxifloxacin effectively reduced *Staphylococcus aureus* and *Klebsiella pneumoniae* biofilms formation on adhesive surfaces.

Keywords: pillar[5]arene; self-healing; moxifloxacin hydrochloride; electron spin resonance; polymer films; polythiols; antibacterial activity

1. Introduction

In recent decades, methods of disrupting pathogenic biofilms have been actively developed [1]. These methods are based on finding ways to inhibit and control the bacterial biofilms formation [2]. It is believed that up to 80% of all human bacterial infections (cystic fibrosis pneumonia, otitis media, pathology of teeth and periodontal tissues, osteomyelitis, urinary tract infections, etc.) are associated with the establishment of biofilms by pathogenic and opportunistic microorganisms [3,4]. Along with this, the formation of microbial biofilms speeds up the corrosion of metals, makes medical equipment unusable, and leads to a deterioration in sanitation and hygiene in medical institutions [5].

To date, three main strategies have been proposed in the struggle against pathogenic biofilms: prevention of bacterial adhesion to the surface [6,7]; disruption of biofilm development and/or impact on its structure with an antimicrobial drug; and impact on the creation of the biofilm, followed by its degradation [8–10]. However, all these methods have common limitations, namely a low efficiency on account of the rapid growth and formation of the extracellular bacterial matrix and a short duration of action due to the presence in biofilms of metabolically inactive cells that are insensitive to factors, primarily antimicrobial drugs [11]. The use of polymeric compositions capable of long-term inhibition of biofilm formation will make it possible to eliminate these limitations [12]. Over the past few years, special attention has been given to supramolecular polymer systems formed by macrocyclic structures [13,14]. Such polymer systems can have a number of the required functions, including self-regeneration, controlled adhesion of microbial cells, and the formation of host–guest systems (macrocycle/antibiotic or antiseptic), to inhibit the activity of cells inside the biofilm.

Derivatives of cyclodextrins [15], cucurbit[n]urils [16], and calix[n]arenes [17,18] are frequently used to form supramolecular polymer ensembles with desired properties. Although these macrocycles have a relatively low toxicity, they are quite difficult to modify into polyfunctional polymer structures.

The use of paracyclophane derivatives—pillar[n]arenes—as a macrocyclic system solves this problem [19]. Pillar[n]arenes can be easily functionalized thorough free hydroxyl groups [20,21], and the presence of a hydrophobic cavity promotes the formation of host–guest systems [22–25]. Recent studies [26–30] demonstrate the effectiveness of using pillar[5]arenes as platforms to create self-assembling drug delivery systems, stimulus-responsive polymeric systems, and antibacterial coatings.

In this work, we developed an original strategy for creating disulfide-containing self-healing materials based on the copolymerization of pillar[5]arene-tetrakis(3-mercaptoproprionate) capable of forming host–guest complex with therapeutic drugs that inhibit the development of Gram-positive and Gram-negative bacteria, *Staphylococcus aureus*, and *Klebsiellapneumoniae*.

2. Materials and Methods

2.1. Characterization

^{1}H NMR, ^{13}C NMR, and ^{1}H-^{1}H NOESY spectra were obtained on a Bruker Avance-400 spectrometer (^{13}C{^{1}H}—100 MHz and ^{1}H—400 MHz). Chemical shifts were determined against the signals of residual protons of deuterated solvent (CDCl$_3$).

Attenuated total internal reflectance IR spectra were recorded with Spectrum 400 (Perkin Elmer) Fourier spectrometer. The IR spectra from 4000 to 400 cm^{-1} were considered in this analysis. The spectra were measured with 1 cm^{-1} resolution and 64 scans co-addition.

Elemental analysis was performed with Perkin Elmer 2400 Series II instrument.

Mass spectra (MALDI-TOF) were recorded on an Ultraflex III mass spectrometer in a 4- nitroaniline matrix. Melting points were determined using a Boetius Block apparatus.

Additional control of the purity of compounds and monitoring of the reaction were carried out by thin-layer chromatography using Silica G, 200 μm plates, UV 254.

Stationary electron paramagnetic resonance (EPR) spectra were obtained at a frequency of 9.6 GHz (X-band) on a Bruker Elexsys E580 spectrometer at room temperature (modulation amplitude M = 0.1 Gs, microwave power P = 2 μW). Low-temperature experiments were carried out on a Bruker ESP-300 spectrometer using a flow-through helium cryostat.

The pulsed EPR spectra were recorded by the method of detecting the EPR spectrum from the integrated intensity of the electron spin echo (ESE) as a function of the external magnetic field B$_0$. Khan's sequence was used:

$$\pi/2\text{-}\tau\text{-}\pi\text{-}\tau\text{-ESE} \tag{1}$$

where the duration of a $\pi/2$ pulse was t_p = 16 ns, π pulse t_p = 32 ns, and delay between pulses τ = 200 ns.

A two-pulse Hahn sequence was used to determine the transverse relaxation time T$_2$. The time interval between pulses τ was increased with a step of 4 ns to the required value, and each time, the integrated ESE intensity was recorded at B = B$_0$. Next, we plotted the dependence of the integrated ESE intensity as a function of time 2τ and approximated it with the function:

$$I(2\tau) = I_0 \exp(2\tau/T_2) \tag{2}$$

The transverse relaxation time T_M (phase coherence time) was determined.

Irradiation took place for 1 h on an X-ray unit URS-55 (tungsten anti-cathode W, voltage U = 50 kV). The estimated absorption dose is approximately equal to 10 kGy.

^{1}H diffusion-ordered spectroscopy (DOSY) spectra were recorded on a Bruker Avance 400 spectrometer at 9.4 tesla at a resonating frequency of 400.17 MHz for ^{1}H using a BBO Bruker 5 mm gradient probe. The temperature was regulated at 298 K, and no spinning was applied to the NMR tube. DOSY experiments were performed using the STE bipolar gradient pulse pair (stebpgp1s) pulse sequence with 16 scans of the 16 data points collected. The maximum gradient strength produced in the z direction was 5.35 Gmm^{-1}. The duration of the magnetic field pulse gradients (δ) was optimized for each diffusion time (Δ) in order to obtain a 2% residual signal with the maximum gradient strength. The values of δ and Δ were 1.800 µs and 100 ms, respectively. The pulse gradients were incremented from 2 to 95% of the maximum gradient strength in a linear ramp.

2.2. Fluorescence Spectroscopy

Fluorescence spectra were recorded on a Fluorolog 3 luminescent spectrometer (Horiba Jobin Yvon). The excitation wavelength was selected as 335 nm. The emission scan range was 350–550 nm. Excitation and emission slits were 2 nm for solutions and 2 nm for supramolecular films. Quartz cuvettes with optical path length of 10 mm were used. Fluorescence spectra were automatically corrected by the Fluoressence program. The spectra were recorded in the solvent system (THF:CH$_3$OH = 100:1) with concentration of Moxifloxacin Hydrocloride (moxi) 5 µM. The obtained molar ratio of polymers to moxi **3/3S** or **3/4S** was 1:10. The experiment was carried out at 293 K.

2.3. UV–Visible Spectroscopy

UV–vis spectra were recorded using the Shimadzu UV-3600 spectrometer; the cell thickness was 1 cm, slit width 1 nm. Recording of the absorption spectra of the mixtures of moxifloxacin hydrocloride (moxi) and benzalkonium chloride (BCl) with pillar[5]arenes **3** at 1×10^{-4} M were carried out after mixing the solutions at 298 K. The 1×10^{-4} M solution of pillar[5]arene **3** (100, 120, 150, 200, 400, 600, 800 µL) in THF was added to 10 µL of the solution of guest (moxi) (1.2×10^{-2} M) in methanol and diluted to final volume of 3 mL with THF. The UV spectra of the solutions were then recorded. The stability constant of complexes were calculated as described below. Three independent experiments were carried out for each series. Student's *t*-test was applied in statistical data processing. Experiment was carried out according to the literature method [22].

2.4. Dynamic Light Scattering (DLS)

The particle size and zeta potential was determined by the Zetasizer Nano ZS instrument at 20 °C. The instrument contains 4 mW He-Ne laser operating at a wave length of 633 nm and incorporated noninvasive backscatter optics (NIBS). The measurements were performed at the detection angle of 173°, and the software automatically determined the measurement position within the quartz cuvette. The 1×10^{-4}–1×10^{-6} M THF solutions of **3**, the 1×10^{-3}–1×10^{-5} M solutions of **3/3S** and **3/4S**, the 1×10^{-4} M solutions of antibiotics (moxi, BCl) (dissolved in methanol 1×10^{-2} M), and the complexes of macrocycle **3** or polymers **3/3S** or **3/4S** with antibiotics (moxi, BCl) were prepared. The concentration ratio of macrocycle **3** or polymers **3/3S** or **3/4S** and antibiotics in complexes was 1:10. The experiments were carried out for each solution in triplicate.

2.5. Transmission Electron Microscopy (TEM)

TEM analysis of samples was carried out using the JEOL JEM 100CX II transmission electron microscope. For sample preparation, 10 µL of the suspension 10^{-5} M were placed on the Formvar/carbon-coated 3 mm cuprum grid, which was then dried at room temperature. After complete drying, the grid was placed into the transmission electron microscope using special holder for microanalysis. Analysis was held at the accelerating voltage of 80 kV in

SEM mode by Carl Zeiss Merlin microscope. Additionally, studies of the morphology of the samples were carried out using the Atomic force microscope Dimension FastScan (Bruker).

2.6. Gel Permeation Chromatography (GPC)

GPC studies were carried out on a GTsP chromatograph (Prague, Czech Republic) equipped with a refractometric detector and a 7.8 × 300 mm column. THF was used as the eluent. Phenogel 5 μm, pore size 10 Å (Phenomenex, CA, USA), was utilized as a sorbent. Polystyrene standards were used for calibration.

2.7. Simultaneous Thermogravimetry and Differential Scanning Calorimetry (TG–DSC)

TG–DSC was performed on a Netzsch Jupiter STA 449 C Jupiter analyzer in 40 μL platinum crucibles with a cap having a 0.5 mm hole at constant heating rates (10 and 4 deg/min; heating range 311–500 K) in dynamic argon atmosphere, flow rate 20 mL/min, atmospheric pressure; sample weight 10–20 mg. The results were processed using the NETZSCH Proteus software.

2.8. X-ray Diffraction Analysis

The dataset for single crystal **3** was collected on a Rigaku XtaLab Synergy S instrument with a HyPix detector and a PhotonJet microfocus X-ray tube using Cu Kα (1.54184 Å) radiation at low temperature. Images were indexed and integrated using the CrysAlisPro data reduction package. Data were corrected for systematic errors and absorption using the ABSPACK module: numerical absorption correction based on Gaussian integration over a multifaceted crystal model and empirical absorption correction based on spherical harmonics according to the point group symmetry using equivalent reflections. The GRAL module was used for analysis of systematic absences and space group determination. The structures were solved by direct methods using SHELXT [31] and refined by the full-matrix least-squares on F^2 using SHELXL [32]. Non-hydrogen atoms were refined anisotropically. The hydrogen atoms were inserted at the calculated positions and refined as riding atoms. The positions of the hydrogen atoms of methyl groups were found using rotating group refinement with idealized tetrahedral angles. Disordered parts of the molecule are refined using constraints and restraints. The contribution of the disordered solvent was removed using the SQUEEZE option from PLATON operated the Olex2 interface. The figures were generated using Mercury 4.1 [33] program. Crystals were obtained by slow evaporation method.

Crystal data for $C_{55}H_{70}O_{10}S_{10}$ (M = 1211.71 g/mol): monoclinic, space group $P2_1/n$ (no. 14), a = 13.1087(6) Å, b = 12.0684(3) Å, c = 41.0482(14) Å, β = 92.459(4)°, V = 6487.9(4) Å^3, Z = 4, T = 100.00(10) K, μ(Cu Kα) = 3.559 mm^{-1}, D_{calc} = 1.241 g/cm^3, 46,894 reflections measured (4.31° ≤ 2θ ≤ 154.11°), 13,152 unique (R_{int} = 0.0519, R_{sigma} = 0.0522), which were used in all calculations. The final R_1 was 0.1881 (I > 2σ(I)), and wR_2 was 0.5275 (all data). CCDC refcode: 2163077.

2.9. Computational Method

Models of moxifloxacin and the thiolated pillar[5]arene were generated using Spartan '20 [34]. The lowest energy structures were determined by molecular mechanics and geometry optimized by molecular mechanics using the Merck Molecular Force Field (MMFF). The 2:1 pillar[5]arene:moxifloxacin complex was created using these structures and geometry optimized using the MMFF followed by DFT (B3LYP/6-31G*) to give the final structure.

2.10. Biological Experiments

Films of free and moxi-loaded polymers **3/3S** and **3/4S** were formed in the wells of 8-well glass plates and dried for 72 h until a dry film formed on the adhesive surface of the glass.

Cultures of *Staphylococcus aureus* and *Klebsiella pneumoniae* were grown in L-broth to a density of 1.5×10^{11} cells/mL. Then, 400 μL were added to the wells of culture plates and chambers for microscopy. Cultivated at 37 °C for 48 h until the formation of stable biofilms.

Biofilms formed after 48 h of cultivation were washed with sodium phosphate buffer (pH = 7.2) from planktonic cells, dried under sterile conditions, and stained with 0.1% gentian violet solution for 20 min. Stained biofilms were washed three times with sodium phosphate buffer (pH = 7.2) and dried. Biofilm thickness was determined by washing with 96% ethanol the gentian violet dye from the biofilm matrix.

Light absorption measurements in eluate samples were measured at λ = 570 nm on a BIO-Rad xMark Microplate spectrophotometer.

The data were given in relative units. In the calculations, the light absorption of the eluate was taken as a unit in the variants without surface modification by films.

Most chemicals were purchased from Aldrich and used as received without additional purification. Organic solvents were purified in accordance with standard procedures.

2.11. Synthesis

Pillar[5]aren **1** were synthesized according to the literature procedure [35].

2.11.1. Synthesis of 4,8,14,18,23,26,28,31,32,35-deca-[Acylthioethoxy]-pillar[5]arene (**2**)

Potassium thioacetate (0.74 g, 6.47 mmol) with anhydrous DMF (12 mL) were placed in a round-bottom flask equipped with a magnetic stirrer. The solution was stirred until complete dissolution. Then pillar[5]arene **1** 0.4 g (0.32 mmol) was added in one portion. Then the reaction mixture was heated for 56 h at 90 °C in an argon atmosphere. After the reaction, the mixture was poured into distilled water. The precipitated beige precipitate was filtered off on a Schott filter and washed with distilled water. The organic phase was separated and evaporated to dryness on a rotary evaporator.

Yield: 0.46 g (88%), mp. 100 °C. ^{1}H NMR (CDCl$_3$): 2.38 s (30H, –CH$_3$), 3.30 t (20H, $^3J_{HH}$ = 5.9 Hz, –CH$_2$S–), 3.72 s (10H, –CH$_2$–), 3.93–4.19 m (20H, –OCH$_2$–), 6.79 s (10H, ArH). ^{13}C NMR (CDCl$_3$): 29.37; 29.57; 30.78; 67.55; 115.55; 128.53; 149.45; 195.34. IR (ν/CM^{-1}) 2932 (–C$_{Ph}$-H), 2868 (–CH$_2$–, C$_{Ph}$–O–CH$_2$), 1684 (C=O), 1496 (–CH$_2$–), 1465 (C$_{Ph}$-C$_{Ph}$), 1352 (–CH$_3$), 1204 (C$_{Ph}$–O–CH$_2$), 1102 (C$_{Ph}$–O–CH$_2$), 1027 (C$_{Ph}$–C$_{Ph}$), 879 (–C$_{Ph}$–H), 703 (C–S). MS (MALDI–TOF): calc. [M$^+$] m/z = 1632.1, found [M + Na]$^+$ m/z = 1654.6. Found (%): C, 54.98; H, 5.46; N, 19.95. Calc. for C$_{75}$H$_{90}$O$_{20}$S$_{10}$. (%):C, 55.19; H, 5.56; O, 19.64; S, 19.64.

2932 (–C$_{Ph}$–H), 2868 (–CH$_2$–, C$_{Ph}$–O–CH$_2$), 1684 (C=O), 1496 (–CH$_2$–), 1465 (C$_{Ph}$-C$_{Ph}$), 1352 (–CH$_3$), 1204 (C$_{Ph}$–O–CH$_2$), 1102 (C$_{Ph}$–O–CH$_2$), 1027 (C$_{Ph}$–C$_{Ph}$), 879 (–C$_{Ph}$–H), 703 (C–S).

2.11.2. Synthesis of 4,8,14,18,23,26,28,31,32,35-deca-[2-Mercaptoethoxy]-pillar[5]arene (**3**)

In a round-bottom flask equipped with a magnetic stirrer, pillar[5]arene **2** 0.46 g (0.28 mmol) was dissolved in anhydrous acetonitrile (23 mL). Then hydrazine hydrate 0.98 mL (31.46 mmol) was added dropwise, and a white precipitate formed. The reaction was carried out for a week at room temperature in an argon atmosphere. Then, the reaction mixture was filtered and washed with acetonitrile. The precipitate was dissolved in chloroform and evaporated on a rotary evaporator in an argon atmosphere. The resulting white powder is the target product.

Yield: 0.25 g (75%), mp. 158 °C. ^{1}H NMR (CDCl$_3$): 1.65 t (10H, $^3J_{HH}$ = 7.9 Hz, -SH), 2.78–2.83 m (20H, –CH$_2$S–), 3.80 s (10H, –CH$_2$–), 3.98 t (20H, $^3J_{HH}$ = 5.3 Hz, –OCH$_2$–), 6.78 s (10H, ArH). ^{13}C NMR (CDCl$_3$): 29.85; 29.90; 70.59; 115.80; 128.84; 149.85. IR (v, sm^{-1}) 2931 (–C$_{Ph}$–H), 2863 (–CH$_2$–, C$_{Ph}$–O–CH$_2$), 2552 (–SH–), 1496 (–CH$_2$–), 1463 (C$_{Ph}$–C$_{Ph}$), 1200 (C$_{Ph}$–O–CH$_2$), 1100 (C$_{Ph}$–O–CH$_2$); 1025 (C$_{Ph}$–C$_{Ph}$); 877 (–C$_{Ph}$–H); 702 (C–S). MS (MALDI–TOF): calc. [M$^+$] m/z = 1210.2, found [M + K + 2H]$^+$ m/z = 1252.1, [M + Na + H]$^+$ m/z = 1235.2. Found (%): C, 54.78; H, 5.98; S, 26.04, Calc. for C$_{55}$H$_{70}$O$_{10}$S$_{10}$. (%): C, 54.52; H, 5.82; O, 13.20; S, 26.46.

2.11.3. General Procedure for the Synthesis of **3n, 3/3S, 3/4S**

A total of 0.09 g (0.074 mmol) of pillar[5] arene **3** was dissolved in 6 mL of THF; then, 0.35 mL of polythiol (trimethylolpropane tris(3-mercaptopropionate) or pentaerythritol tetrakis(3-mercaptopropionate)) (or without for **3n**) dissolved in 6 mL of THF and 1.2 mL of H_2O_2 (30%) dissolved in 6 mL THF was added. The reaction proceeded at room temperature for 40 h. Then the reaction mixture was poured into water and centrifuged. The resulting precipitate was dissolved in THF. Then, the precipitate was dried in vacuum under reduced pressure. The target product is a white, stretching mass.

4,8,14,18,23,26,28,31,32,35-deca-[2-mercaptoethoxy]-pillar[5]arene (3)-based cross-linked supramolecular polymer (3n).

Yield: 0.08 g (89%), mp. 317 °C. IR (ν, sm^{-1}) 2917 (–C$_{Ph}$–H); 2859 (C$_{Ph}$–O–CH$_2$); 1496 (–CH$_2$–); 1463 (C$_{Ph}$–C$_{Ph}$); 1194 (C$_{Ph}$–O–CH$_2$); 1058 (C$_{Ph}$–O–CH$_2$); 772 (C$_{Ph}$–H); 698 (C–S).

Tetrablock co-monomer, based on 4,8,14,18,23,26,28,31,32,35-deca-[2-mercaptoethoxy]-pillar[5]arene (3) and trimethylolpropane tris(3-mercaptopropionate), (3/3S).

Yield: 0.36 g (70%). MS (MALDI-TOF): calc. [M + 5K + 4Na + 1 × 3S-11H]$^+$ m/z = 1886.7, [M + 5K + Na + 2 × 3S-12H]$^+$ m/z = 2212.1, [2M + 10K + 2 × 3S-18H]$^+$ m/z = 3589.6, [3M + Li + 3 × 3S-11H]$^+$ m/z = 4825.8, found [M + 5K + 4Na + 1 × 3S-11H]$^+$ m/z = 1887.4, [M + 5K + Na + 2 × 3S-12H]$^+$ m/z = 2212.4, [2M + 10K + 2 × 3S-18H]$^+$ m/z = 3590.2, [3M + Li + 3 × 3S-11H]$^+$ m/z = 4825.8.

Tetrablock co-monomer, based on 4,8,14,18,23,26,28,31,32,35-deca-[2-mercaptoethoxy]-pillar[5]arene (3) and trimethylolpropane tris(3-mercaptopropionate), (3/4S).

Yield: 0.41 g (76%). MS (MALDI-TOF): calc. [2M + 3K + 3Na + 2 × 4S-14H]$^+$ m/z = 3572, [2M + Na + 2 × 4S-1H]$^+$ m/z = 3413.5, found [2M + 3K + 3Na + 2 × 4S-14H]$^+$ m/z = 3572.0, [2M + Na + 2 × 4S-1H]$^+$ m/z = 3413.4.

2.11.4. General Procedure for the Synthesis of Cross-Linked Supramolecular Polymers **3/3Sn, 3/4Sn**

Freshly prepared **3/3S** and **3/4S** tetrablock co-monomers (m = 0.3 g) were dissolved in 10 mL of THF, sprayed onto the surface of a glass substrate, and dried in the presence of atmospheric oxygen at ambient temperature for 30–40 min. Next, the formed supramolecular polymeric film **3/3Sn, 3/4Sn** was used for further study.

Cross-linked supramolecular polymer, based on 4,8,14,18,23,26,28,31,32,35-deca-[2-mercaptoethoxy]-pillar[5]arene (3) and trimethylolpropane tris(3-mercaptopropionate), (3/3Sn).

Mp. 322 °C. IR (ν/CM^{-1}): 3450 (–C=O); 2958 (–CH$_2$–); 2941 (–C$_{Ph}$–H); 2907 (C$_{Ph}$–O–CH$_2$); 2568 (–S–H); 1729 (–C=O); 1499 (–CH$_2$–); 1471 (C$_{Ph}$–C$_{Ph}$); 1406 (–CH$_2$–C=O); 1233 (–CH$_2$–O–C=O); 1192 (C$_{Ph}$–O–CH$_2$); 1178 (C$_{Ph}$–O–CH$_2$); 1142 (–CH$_2$–O–C=O); 1048 (C$_{Ph}$–O–CH$_2$); 671 (C–S); 583 (–S–S–).

Cross-linked supramolecular polymer, based on 4,8,14,18,23,26,28,31,32,35-deca-[2-mercaptoethoxy]-pillar[5]arene (3) and pentaerythritol tetrakis(3-mercaptopropionate), (3/4Sn).

Mp. 324 °C. IR (ν, sm^{-1}) 3451 (–C=O); 2960 (–CH$_2$–); 2925 (–C$_{Ph}$–H); 1726 (–C=O); 1497 (–CH$_2$–); 1470 (C$_{Ph}$–C$_{Ph}$); 1407 (–CH$_2$–C=O); 1230 (–CH$_2$–O–C=O); 1209 (C$_{Ph}$–O–CH$_2$); 1173 (C$_{Ph}$–O–CH$_2$); 1127 (–CH$_2$–O–C=O); 663 (–C–S–); 595 (–S–S–).

Detailed information of physical-chemical characterization is presented in Electronic Supporting Information (ESI).

3. Results

3.1. Synthesis and Polymerization of Pillar[5]arene Containing Mercapto Groups

Self-healing is an attractive properties of materials, which is currently in demand [36]. The ability of a product to self-heal significantly increases its service life due to improved mechanical characteristics, surface renewal, and preservation of its integrity [37]. The design of such materials includes two main approaches to self-regeneration: the use of physical methods of cross-linking based on the mutual diffusion of individual parts and chemical methods of cross-linking using the formation of a reversible covalent bond [38]. In all of these approaches, macrocyclic compounds can be used. However, unlike self-healing through supramolecular interactions, chemical methods of self-healing provide higher mechanical strength and material stability [39].

Thus, to create self-healing materials based on pillar[5]arene, we chose to modify the macrocyclic platform with substituents able to form reversible covalent bonds. These interactions include the reversible formation of disulfide bonds, metal–ligand interactions, ionic interactions, and the formation of hydrogen bonds [40]. A common disadvantage of many such systems is the need for external influences, such as heating, UV irradiation, or pH changes through the addition of acid or alkali, which is necessary to initiate surface regeneration. The use of thiol/disulfide redox dynamic exchange reactions to form reversible disulfide bonds is the most accessible approach to date [40]. Although thiol/disulfide redox reactions can be accelerated in the presence of catalysts, they can also occur under ambient conditions (air temperature from 16 °C to 32 °C and relative humidity from 20% to 80%) with atmospheric oxygen as an external trigger [41].

To initiate thiol/disulfide cross-linking redox reactions in a polymeric material based on pillar[5]arene, the presence of thiol fragments in the structure of the macrocyclic platform is necessary. To this end, we developed an approach for introducing thiol fragments into the pillar[5]arene structure (Figure 1). Decabromoethoxy pillar[5]arene **1** was prepared according to the literature procedure [35] and reacted with potassium thioacetate in anhydrous DMF at 90 °C, whereupon macrocycle **2** was isolated by precipitation from water in 88% yield. Acetate fragments were cleaved with hydrazine hydrate in anhydrous acetonitrile [42] to give target macrocycle **3**, which was collected by filtration in 75% yield. Pillar[5]arene **3** was used without further purification in subsequent experiments (Figure 1, see SI, Figures S1–S13). The presence of free mercapto groups in macrocycle **3** was confirmed by one-dimensional ^{1}H NMR spectroscopy, where they are observed as a triplet of 10 SH-protons at δ = 1.65 ppm (see SI, Figure S2). It should be noted that the macrocycle **3** forms as a powder, which showed no oxidation over a month of storage under argon at room temperature. Macrocycle **3** was characterized using X-ray diffraction (Figure 1) from crystals grown from a CHCl$_3$–CH$_3$CN solvent mixture. The crystal habit of **3** is monoclinic, and the symmetry group is P 21/n.

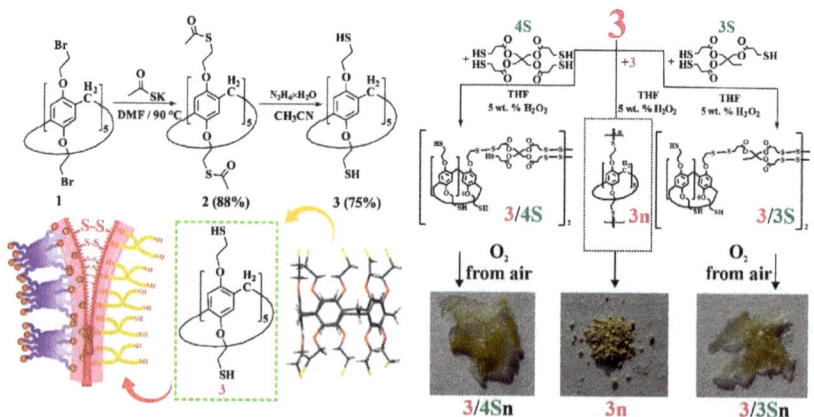

Figure 1. Synthesis of macrocycles **2** and **3**; X-ray lateral view of macrocycle **3** and sketch showing copolymerization of macrocycle **3** into **3n** and **3** with trimethylolpropane-tris(3-mercaptopropionate) **3S** and pentaerythritol-tetrakis(3-mercaptopropionate) **4S** in THF in the presence of 5 wt. % H$_2$O$_2$ and atmospheric oxygen.

Afterwards, we developed a procedure to prepare co-monomers and polymers using thiol/disulfide redox dynamic exchange reactions involving pillar[5]arene **3**. The polymerization proceeded under the action of 30% H$_2$O$_2$ in THF for 40 h at room temperature, and polymer **3n** was isolated in 88% yield (Figure 1) as a light-yellow powder (Figure 1). Polymer **3n** is practically insoluble in both polar and nonpolar solvents, and its decomposition onset temperature was 298 °C. Thus, it can be concluded that it is not suitable for the formation of self-healing films.

For the synthesis of self-healing films based on pillar[5]arene **3**, we chose to prepare cross-linked copolymers in the presence of low-molecular cross-linking agents: trimethylolpropane tris(3-mercaptopropionate) **3S** or pentaerythritol tetrakis(3-mercaptopropionate) **4S** (Figure 1). These commercially available compounds are used as gel formers [43] and polymer resin hardeners [44]. Thus, we carried out the selection of conditions for oxidative copolymerization to obtain copolymers in advance (see SI, Table S1).

Variation in the nature of the solvent (CH_3CN, DMF, THF) and oxidants (I_2, $FeCl_3$) does not lead to the formation of polymer products **3** with **3S** or **4S**. Analysis of the ^1H NMR spectra of the reaction mixtures showed the presence of the starting compounds and polymer **3n**. The use of H_2O_2 as an oxidizing agent and various ratios of reacting components led to the formation of comonomers **3/3S** and **3/4S**. The temperature regime of the syntheses varied in the range from 0–75 °C. However, the best results were achieved when the reaction was carried out at 25 °C for 24 h both for **3S** and **4S**. Thus, by reacting macrocycle **3** with 5-fold excesses of **3S** or **4S** in the presence of 30% H_2O_2 in THF for 24 h at room temperature, products **3/3S** and **3/4S** were isolated in 70% and 76% yields (see SI, Table S1).

The structure of the formed products **3/3S** and **3/4S** was studied by gel permeation chromatography (GPC) and MALDI mass spectrometry (see SI, Figures S7, S8, S29 and S30). We chose GPC as a convenient method to determine the relative molecular weight of **3/3S** and **3/4S**. Thus, GPC analysis of fractions of samples **3/3S** and **3/4S** freshly prepared in THF showed average mass values up to 3000 Da, which corresponds to diblock- or tetrablock-linked (Figure 1) fragments of macrocycle **3** with polythiols **3S** or **4S** (see SI, Figures S29 and S30). The formation of cross-linked tetrablock comonomers **3/3S** and **3/4S** can be explained by the polyfunctionality of the macrocyclic platform, the possible thermodynamic stability of the resulting tetrablock comonomers, and the low solubility of longer polymer units in THF [40].

The overall pattern of sequential fragmentation in the MALDI mass spectra of **3/3S** and **3/4S** tetrablock comonomers (see SI, Figures S7 and S8) also agrees with the GPC data, as the mass spectra of **3/3S** and **3/4S** contain peaks of molecular ions in the range from 1713 Da to 5219 Da, which corresponds to a possible crosslinking from two to six fragments of **3S**, **4S**, and macrocycle **3** (see SI, Figures S7 and S8).

The method of forming a film from a solution of co-monomers is a relatively simple process since the co-monomer is in a dissolved state. We used the method of spraying a solution of **3/3S** and **3/4S** in THF (1×10^{-3} M) [45] over the surface of a glass substrate. As a result, it was found that as the solvent evaporates, the solution passes into a gel-like state, followed by the formation of a film upon drying.

3.2. Interaction of Macrocycle 3 with the Antibiotic Moxifloxacin Hydrochloride

The films obtained on the basis of **3/3S** and **3/4S** contain fragments of pillar[5]arenes, which are capable of host–guest interactions with therapeutic drugs [26]. In this regard, we hypothesized that using an antimicrobial drug as a therapeutic agent would promote the formation of pillar[5]arene/drug complexes in the film structure to effectively suppress the development of bacteria. In addition, the presence of disulfide bonds (from **3/3S** and **3/4S**) in the film structure will contribute to the self-healing of the damaged surface under the action of atmospheric oxygen.

On account of the ability of macrocycle **3** to interact with antimicrobial drugs, benzalkonium chloride and moxifloxacin hydrochloride (moxi) were studied using UV–vis and NMR spectroscopy (see SI, Figures S25 and S26). The choice of substrates was determined by their use in medical practice as effective preparations to treat bacterial infections [46]. The studies were carried out in a mixture of $THF:CH_3OH = 100:1$. The choice of the solvent system was due to the good solubility of **3** in THF and its low solubility in solvents (water, alcohols), which dissolve binding substrates benzalkonium chloride and moxifloxacin hydrochloride.

It turned out that only when macrocycle **3** bound moxifloxacin hydrochloride were spectral changes at the absorption wavelength of moxifloxacin (λ = 340 nm) significant enough to establish quantitative binding characteristics (see SI, Figure S27). The association constant was determined based on spectrophotometric titration data. Thus, the concentration of moxi (1×10^{-5} M) was constant, while the concentration of macrocycle **3** (0–2.67×10^{-5} M) varied (see SI, Figure S27). Binding constants of **3**/moxi in THF:CH$_3$OH = 100:1 were calculated by UV–vis spectroscopy from the analysis of binding isotherms and were established on the binding model **3**/moxi = 2:1 using Bindfit [47], a statistical model widely used in supramolecular chemistry to determine the characteristics of intermolecular interactions [48]. To confirm the proposed stoichiometry, the titration data were also processed using a binding model with a host–guest ratio of 1:1 and 1:2. However, in this case, the constants were determined with a much larger error (see SI, Figure S28). The calculated logarithms of the association constants (logKa) for **3**/moxi were logK_{11} = 2.14 and logK_{12} = 6.20.

We chose 2D ^{1}H-^{1}H NOESY and 2D DOSY NMR (see SI) spectroscopy to confirm the formation of the **3**/moxi complex and set its structure. An analysis of the experimental data obtained using ^{1}H NMR spectroscopy did not make it possible to determine the nature of the interaction by changing the position of the host–guest chemical shifts. The chemical shifts of protons of moxi and **3** were broadened due to ongoing association processes (ESI). However, in the 2D ^{1}H-^{1}H NOESY NMR spectrum of associate **3**/moxi (2:1, C$_{moxi}$ = 5×10^{-3} M) in CHCl$_3$/CD$_3$OD cross peaks between protons of aromatic fragments (**H^a**) of macrocycle **3** and protons **H^{10}**, **H^{12}**, **H^{13}** of the octahydro-1H-pyrrolo[3,4-b]pyridine fragment were observed (see SI). Cross peaks are also observed between the protons of the cyclopropyl fragment of moxi (**H^4** and **H^5**) and the protons of the methylene bridges **H^b**. The 2:1 inclusion complex was supported by calculations (DFT/BLY3P/6-31G*) as shown in Figure 2.

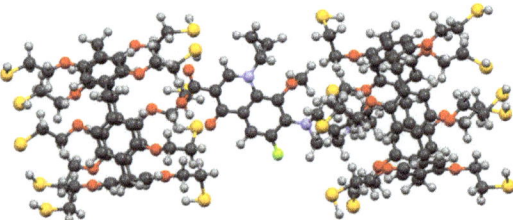

Figure 2. Geometry optimized structure of the **3**/moxi complex.

The formation of the **3**/moxi complex was additionally confirmed by 2D DOSY NMR spectroscopy. Diffusion coefficients of **3**, **3**/moxi, and moxi at 298 K (5×10^{-3} M) were determined. The 2D DOSY NMR spectrum of the **3**/moxi system shows the presence of signals of the complex lying on one straight line with one diffusion coefficient (D = 1.4×10^{-10} m^2 s^{-1}) (see SI). This value of the diffusion coefficient **3**/moxi is much lower than the self-diffusion coefficients of macrocycle **3** (D = 3.5×10^{-10} m^2 s^{-1}) and moxi (D = 4.7×10^{-10} m^2 s^{-1}) under the same conditions. The results obtained definitely indicate the formation of an associate **3**/moxi. The formation of the **3**/moxi complex is in good agreement with the literature data on the binding of pillar[5]arenes with fluoroquinolone derivatives [49].

It is also important to evaluate the possibility of the interaction of moxi with **3/3S** and **3/4S** tetrablock co-monomers, which are soluble in THF and contain fragments of pillar[5]arene. Experiments on the interaction of moxi with **3/3S** and **3/4S** were carried out in THF:CH$_3$OH (100:1). Highly sensitive fluorescence spectroscopy (Figure 3) was chosen as an effective method for detecting interactions between **3/3S**, **3/4S**, and the antibiotic moxi. The molecular weights of tetrablock co-monomers obtained by GPC (see SI, Figures S29 and S30) to calculate the concentrations of **3/3S** and **3/4S**.

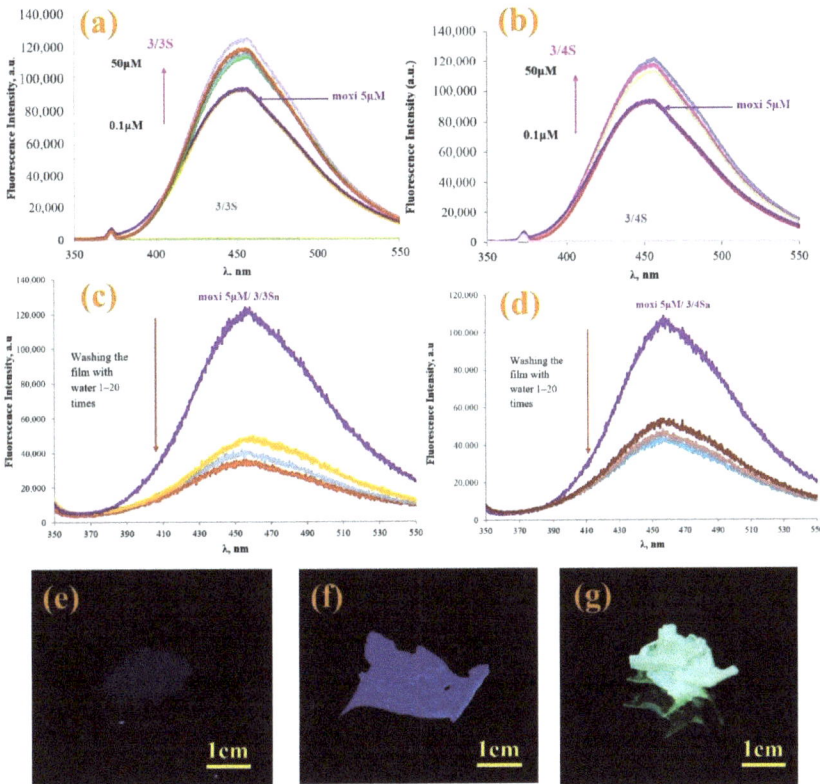

Figure 3. Fluorescence spectra of moxi (5×10^{-6} M) with various concentrations of: (**a**) **3/3S** (0–50 μM) and (**b**) **3/4S** (0–50 μM); (**c**) fluorescence spectra **3/3Sn**/moxi and (**d**) **3/4Sn**/moxi of the film before and after washing with distilled water; photographs of samples under UV irradiation at λ = 365 nm (**e**) **3n**; (**f**) **3/4Sn**; and (**g**) **3/4Sn** + moxi after washing with water (five times).

In the spectrum of moxi (5 μM), as **3/3S** (Figure 3a) and **3/4S** (Figure 3b) increases from 0.1 to 50 μM, the fluorescence also rises sequentially. The study was carried out at the emission wavelength of moxi (λ = 455 nm) in THF:CH$_3$OH (100:1). Thus, the data of fluorescence spectroscopy confirm the interaction of **3/3S** and **3/4S** with the antibiotic.

This increase in the intensity emission can be related with the antibiotic's inclusion into the cavity of pillar[5]arene, which is part of the tetrablock comonomer. The data obtained by 2D NOSY and DOSY NMR spectroscopy confirm this hypothesis.

For the **3/3S** and **3/4S**, the processes of self-association and aggregation in the presence of moxi were additionally studied by dynamic light scattering (DLS) in THF:CH$_3$OH (100:1) (see SI, Figures S15–S24, Table S2). It was shown that **3/3S** and **3/4S** do not form stable self-associates (see SI, Table S2) in the studied concentration range (1×10^{-3}–1×10^{-5} M). However, adding a 10-fold excess of moxi to the tetrablock co-monomers **3/3S** or **3/4S** reduces the polydispersity index to 0.28–0.34. Stabilization of the system is observed when the average particle size of 460 nm is reached in the case of the moxi/**3/4S** (see SI, Table S2) system and 620 nm for moxi/**3/3S** (see SI, Table S2). Separately, macrocycle **3** and moxi did not form stable associates in THF:CH$_3$OH (100:1) over the entire concentration range studied (1×10^{-3}–1×10^{-5} M). When the method of spraying a solution of moxi/**3/3S** and moxi/**3/4S** in THF (1×10^{-3} M) onto a glass substrate was used, the formation of a drug-loaded films was observed.

3.3. Formation of Supramolecular Polymer Networks

Thus, pillar[5]arene **3** and tetrablock comonomers **3/3S** and **3/4S** based on it were able to interact with the antimicrobial drug moxi and form host–guest complexes. As shown above, **3/3S** and **3/4S** are able to form films loaded with moxi. In this regard, it can be assumed that moxi can be placed in the structure of the film based on pillar[5]arene and contribute to the suppression of the development of pathogenic microorganisms as part of the polymer coating.

To confirm this hypothesis, the films obtained after spraying a solution of **3/3S** or **3/4S** in THF (1×10^{-3} M) on a glass substrate were additionally investigated by a number of physical methods. The resulting cross-linked copolymers based on macrocycle **3** and thiols **3S** or **4S** were transparent films (Figure 1) soluble in THF. However, after evaporation of the solvent, the film becomes insoluble in THF over ~10–20 min, apparently due to the formation of additional disulfide bonds under the action of atmospheric oxygen. To confirm the formation of additional disulfide bonds and cross-linking **3/3S**, **3/4S** into **3/3Sn**, and **3/4Sn** polymer network compositions (Figure 1) during film formation, the IR spectra of **3/3Sn** and **3/4Sn** were studied and compared to **3** and **3n**. The IR spectra of **3/3Sn** and **3/4Sn** (Figure 4) show characteristic bands for the structure of thiols **3S**, **4S** (1732 cm^{-1}), and macrocycle **3** (2960, 2925, 1470, 1209 cm^{-1}). The absence of free S-H bonds vibrations at $\nu = 2750$ cm^{-1} as well as the presence of SS bonds vibrations at $\nu = 640$ cm^{-1} [50] in the fingerprint region, which did not appear in the initial macrocycle **3**, confirm the formation of additional disulfide bridges [40] under the action of atmospheric oxygen when THF solutions of **3/3S** and **3/4S** comonomers are dried.

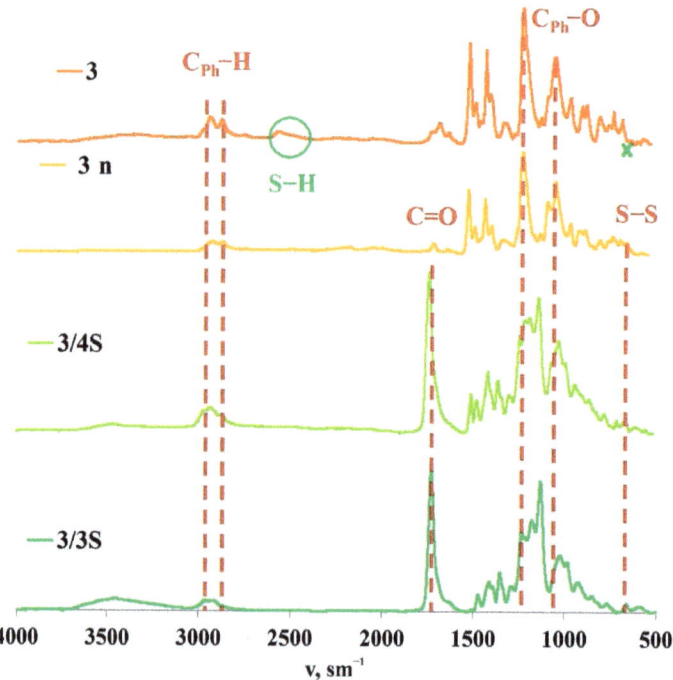

Figure 4. Attenuated total internal reflectance IR spectra of **3** and **3n** powders and **3/3Sn** and **3/4Sn** films.

Thermogravimetric analysis is widely used to assess the phase and thermal characteristics of polymeric materials [51], including self-regenerating ones [52] (see SI, Figures S31–S34). The DSC curve for macrocycle **3** includes three processes, one of which corresponds to the largest weight loss (34%), and three stages of decomposition on the TG curve. This exo-process (see SI, Figure S31) is observed over the temperature range of 261–285 °C, which is consistent with the oxidative S–S crosslinking that can occur in a sample with a change in temperature [53]. The TG curves of the **3/3Sn** and **3/4Sn** samples (see SI, Figures S33 and S34) contain four or five weight-loss steps. The first stage (up to 65 °C) corresponds to the removal of residual solvent. The second stage is different for each **3/3Sn** and **3/4Sn** sample and varies over the temperature range 137–211 °C. This stage is accompanied by a loss of up to 7% of the mass and is associated with the evaporation of water included in the structure of the film. In contrast to **3** alone, the DSC curve in the temperature range of 250–300 °C of the **3/3Sn** and **3/4Sn** (see SI, Figures S33 and S34) samples does not show an exo-process. This process corresponds to the main stage of weight loss. However, in a higher temperature range of 280–340 °C, in samples **3/3Sn** and **3/4Sn**, an endo-process is observed corresponding to the first stage of destruction (25–34% weight loss) and of the melting of substances **3/3Sn** and **3/4Sn**. The fourth step on the TG curve for the **3/3Sn** and **3/4Sn** samples corresponds to the weight loss of 13–17% and on the DSC curve in the temperature range of 347–374 °C corresponding of oxidative S–S cross-linking.

Thus, analysis of the TG-DSC (see SI, Figures S31–S34) data allows us to conclude that the **3/3Sn** and **3/4Sn** polymer structures are more thermally stable than macrocycle **3**, which is characterized by thermally sensitive S–S crosslinking processes. The melting temperature of free macrocycle **3** is 30–50 °C lower than that of **3/3Sn** and **3/4Sn**, which confirms the improved thermal characteristics of the obtained materials.

*3.4. Interaction of **3/3S**, **3/4S** and Supramolecular Copolymers **3/3Sn**, **3/4Sn** with the Antibiotic Moxifloxacin Hydrochloride*

Since **3/3Sn** and **3/4Sn** polymer films were formed from **3/3S** and **3/4S** tetrablock co-monomers, capable of interacting with moxi, it was necessary to investigate the ability to bind moxi in the structure of **3/3Sn** and **3/4Sn**. For this purpose, moxi/**3/3Sn**, moxi/**3/4Sn** polymer films were formed by spraying a solution of moxi/**3/3S** and moxi/**3/4S** (THF:CH$_3$OH = 100:1, 1×10^{-3} M) on to the glass substrate for 30 min (Figure 3f,g) and the fluorescence spectra of the resulting films recorded (Figure 3c,d). It is known that ionized forms of antibiotics of the fluoroquinolone series dissolve well in water, therefore, under real conditions, when the environmental humidity changes, the antibiotic can be removed from the polymer surface. In order to simulate this situation, the surface of moxi/**3/3Sn** and moxi/**3/4Sn** was repeatedly washed with distilled water. The number of washes varied from 1 to 20 times. According to fluorescence spectroscopy data, the intensity of moxi emission in the samples decreased during the first washing for moxi/**3/3Sn** by 67% (Figure 3c) and by 45% for moxi/**3/4Sn** (Figure 3d). Further washing did not lead to significant changes in the moxi fluorescence intensity in the **3/3Sn**, **3/4Sn** films.

Since the moxi/**3/4Sn** systems turned out to be the most monodisperse and resistant to washing, their morphology was studied using electron and atomic force microscopies (see SI, Figures S35–S43). According to scanning electron microscopy (SEM) data, the **3/4Sn** film is an irregular network polymer consisting of intertwining filaments with a thickness of 63 nm (Figure 5a). A similar morphology is confirmed by 3D images of the atomic force microscope (AFM) (Figure 5b). Also, according to transmission electron microscopy (TEM) (Figure 5c, see SI, Figures S35–S43), the formation of dendritic structures on the surface of interlacing threads is observed. The TEM images of moxi/**3/4Sn** show the formation of dendritic structures typical of polymer morphology with included spherical moxi particles on the surface (Figure 5d). The size of spherical particles moxi was 100 nm, and the thickness of the dendritic fragments corresponded to the thickness of the filaments and was 67 nm (Figure 5d, see SI, Figure S36).

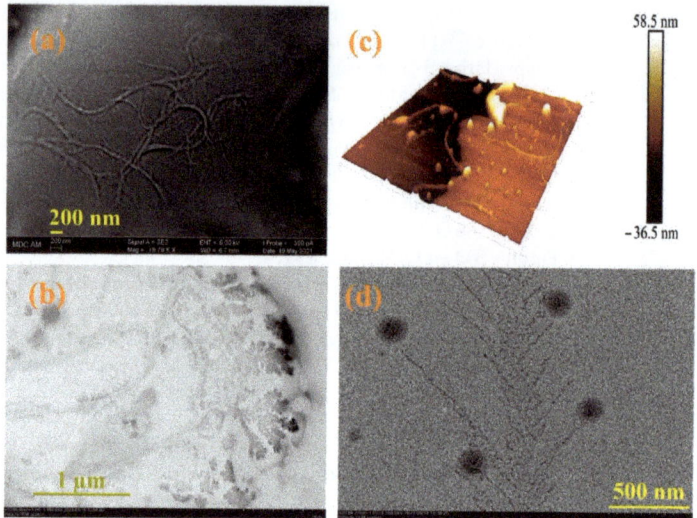

Figure 5. (**a**) SEM images of **3/4Sn** (1×10^{-5} M) after the solvent (THF:CH$_3$OH (100:1)) evaporation; (**b**) AFM images of a **3/4Sn** film (1×10^{-5} M) after the solvent (THF:CH$_3$OH (100:1)) evaporation; (**c**) TEM images of **3/4Sn** film (1×10^{-5} M) after the solvent (THF:CH$_3$OH (100:1)) evaporation; (**d**) TEM images of system **3/4S** (1×10^{-5} M)/**moxi** (1×10^{-4} M) after the solvent (THF:CH$_3$OH (100:1)) evaporation.

Thus, in the course of the studies, the formation of **3/3Sn**, **3/4Sn** films capable of binding antibacterial drugs of the fluoroquinolone series was demonstrated. The resulting **3/3Sn**, **3/4Sn** structures are built on the principle of the formation of dynamic -S-S- covalent bonds, which, due to thiol/disulfide redox dynamic exchange reactions, lead to self-healing of damaged surface areas. The mechanism of self-healing process is proposed to be through the formation of free sulfur radicals [54], which again form disulfide bonds under the action of external triggers (Figure 6a).

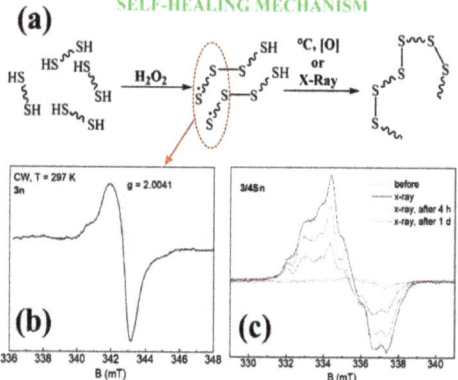

Figure 6. (**a**) Proposed mechanism of self-regeneration; (**b**) Stationary electron paramagnetic resonance spectrum of sample 3n (powder) at room temperature in the X-band continuous wave mode (9.6 GHz); (**c**) EPR spectra of **3/4Sn** films at T = 15 K before and after X-ray irradiation.

3.5. Study of the Process of Self-Regeneration of Films 3/3Sn, 3/4Sn

Electron paramagnetic resonance (EPR) was used to determine the presence of sulfur radicals in **3/3S**, **3/4S**, **3n**, **3/3Sn**, **3/4Sn** structures, since free radicals have a nonzero electronic (spin) magnetic moment, characterized by a quantum number of S = 1/2 [55]. Figure 6b shows the steady-state absorption spectrum for the **3n** powder sample at room temperature. Comparing the spectroscopic g-factor value with the literature data [54,55], it is obvious that sulfur radicals are present in the **3n** (see SI, Figures S44 and S45) powder sample, which confirms the hypothesis put forward.

No EPR signals in liquid **3/4S** samples with different THF concentrations were observed, which may indicate the instability of sulfur radicals in these solutions. The heating of **3n** powder also did not lead to the formation of new sulfur radicals. The shape of EPR spectrum has a weak (but visible) asymmetry, which indicates the anisotropy of the g-factor arising from the interaction of the electron shell of the radical with the surrounding electric field gradients of neighboring ions. The phase coherence time (T_M) of 578 ns is a rather short value for transverse relaxation compared to other stable radicals [56]. The asymmetry of the spectrum (anisotropy) and the short value of T_M (due to nuclear spin diffusion) are associated with the localization of the sulfur radical in the polymer structure.

Additionally, samples of **3/4Sn** and **3/3Sn** in the form of film structures (see SI, Figures S44 and S45) were studied. The EPR spectra show signals from sulfur radicals superimposed on a broader and structureless line. The formation of this line (underlayer signal) is possibly associated with a change in the structure or local environment of the sulfur radical, which leads to a strong (dipole-dipole) inhomogeneous broadening.

Since radiation exposure leads to the formation of stable free radicals, a sample of the **3/4Sn** film was undergone by X-ray irradiation (Figure 6c). To improve the signal-to-noise ratio (20 times), the experiments were carried out at 15 K. After irradiation, the spectrum is a sum of signals of different origins, which is possibly due to the presence of a hyperfine interaction of the paramagnetic center with the nuclei of hydrogen ^{1}H or sulfur ^{33}S or the formation of other types of sulfur radical. Figure 6c also shows the dynamics of the EPR spectra that decrease in intensity with time. This indicates the lower stability of these paramagnetic centers, in contrast to the radicals in the powder sample, the spectrum of which does not change with time.

The **3/3Sn** sample was also irradiated with an X-ray source for 1 h, but this procedure did not lead to the formation of additional EPR signals.

Modern microscopy methods are convenient and effective tools for dynamic study of the self-healing process [57]. Therefore, the process of self-healing of the formed films based on the **3/4Sn** system was qualitatively assessed using optical microscopy to monitor the healing of the cut surface under the action of atmospheric oxygen. Initially, the 3D surface of a **3/4Sn** film was created by SEM at low pressure. An uneven thickness distribution in the film upon drying in air was found (Figure 7a). Micrometer-sized surface scratches were then made on the surface using a micrometer blade (Figure 7b). The film was stored in the ambient atmosphere, and the damaged area was monitored using an optical microscope (Figure 7b, see SI, Figures S42 and S43). The healing response was clearly observed within 2 h at room temperature. The cut healed from the ends where the cut surfaces were closest to each other. In this case, the cut surface, according to SEM data, was an intergrowth of dendritic structures directed perpendicular to the cut wall (Figure 7c). This not only demonstrates the ability of the material to heal, but also opens up the possibility of creating a material with the function of self-regeneration under the action of biological substrates or in the body's environment.

Figure 7. (a) SEM images of **3/4Sn** at low pressure and topographic map of **3/4Sn** film; (b) optical microscope image of a **3/4Sn** film with surface disturbance over time (0–24 h); (c) SEM image of a section of a **3/4Sn** film.

3.6. Antibacterial Properties of Self-Regenerating Films **3/3Sn**, **3/4Sn**

Despite the ability of **3/4Sn** to self-heal and interact with moxi, the antibacterial drug may not be available for entry into the bacterial matrix. As a result, the moxi/**3/4Sn** system will not inhibit the formation of microbial biofilms.

It is worth noting that, according to the literature data [58], pillar[5]arenenes containing thioether fragments do not have pronounced cytotoxicity. This makes these compounds attractive for use in biomedical materials.

To test the ability of a moxi/**3/3Sn** and moxi/**3/4Sn** films to suppress the development of pathogenic microorganisms, we evaluated the formation of bacterial biofilms on the adhesive surfaces of slide chambers treated with moxi/**3/3Sn** and moxi/**3/4Sn**. Pathogenic microorganisms of the Gram-negative morphotype of the *Enterobacteriaceae* family [59] and Gram-positive bacteria *Staphylococcaceae* have serious impacts on human health. Methicillin-resistant strains of *Staphylococcus aureus* attract special attention in the clinic [60]. In this study, a clinical isolate of *Klebsiella pneumonia* belonging to the *Enterobacteriaceae* family and *Staphylococcus aureus* ATCC® 29213™ were selected as model pathogens capable of forming biofilms.

Modification of the surface of adhesive glasses with **3/4Sn**, **3/3Sn** and moxi/**3/4Sn**, moxi/**3/3Sn** led to a change in the thickness of microorganism biofilms (Figure 8b). On the surface coated with **3/4Sn** and **3/3Sn** films, in the case of *S. aureus* biofilms and biofilm of *K. pneumonia*, a slight increase in the total biomass of biofilm was observed compared to the untreated variant (Figure 8a). A significant scatter of data in this processing option can be associated with a different area of the modified surface. Addition of moxi into **3/4Sn** and **3/3Sn** films reduced the total biomass of biofilm of both *S. aureus* and *K. pneumoniae*.

Thus, moxi/**3/4Sn** and moxi/**3/3Sn** films were formed by sputtering solutions of tetrablock co-monomers in THF (1×10^{-3} M) in a chamber with an adhesive glass bottom. The resulting moxi/**3/4Sn** and moxi/**3/3Sn** systems were washed five times with distilled water to remove excess unbound moxi. It was shown that moxi/**3/4Sn** reduced the capacity of biofilms formed by *S. aureus* and *K. pneumoniae*, by 80% and 48%, respectively (Figure 8a). In the variant with the moxi/**3/3Sn** film (1×10^{-3} M), its application reduced the capacity of the biofilm formed by *S. aureus* and *K. pneumoniae*, by 77% and 43%, respectively (Figure 8a).

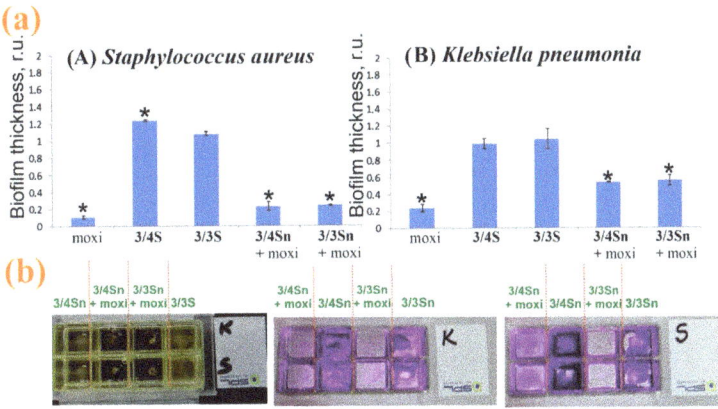

Figure 8. (a) Effect of pretreatment of the adhesive glass surface with **3/3Sn**, **3/4Sn** films and free moxi, as well as their composites with moxi on the ability to form *S. aureus* and *K. pneumoniae* biofilms. The power of the biofilm of microorganisms in the variant without pretreatment was taken as a unit. (b) Photographs of slide chambers with cultures of bacteria in the presence of **3/4Sn** and **3/3Sn** and moxi/**3/4Sn**, moxi/**3/3Sn**. *—$p \leq 0.05$ when compared with the variant without pre-treatment of the surface.

Additionally, the inhibition of pathogenic biofilm formation by *S. aureus* and *K. pneumoniae* in the presence of moxi was studied (Figure 8a). An analysis of experimental data showed that individual moxi suppresses the development of pathogenic biofilms more effectively. However, moxifloxacin hydrochloride is highly soluble in water, which makes moxi not applicable under changing environmental conditions. When humidity changes, moxi will be washed off the treated surface. Doping moxi into the composition of the polymer film makes it possible to keep it on the surface and create a concentration gradient near the biofilm. It should be noted that the efficiency of moxi in the composition of the polymer film remains as high as without it (Figure 8).

Drug release methods are as important as the encapsulation process. As a mechanism for releasing moxi from the moxi/**3/4Sn** complex, a guest exchange mechanisms can be assumed [61]. So, a molecule with a high affinity for the macrocyclic pillar[5]arene should be chosen as a new guest. As such molecules, some amino acids can be selected like arginine (Arg) and lysine (Lys). These amino acids are part of the human body proteins—elastin and collagen. Arginine (Arg) and lysine (Lys) have a large affinity for the carboxylated pillar[5]arene cavity [62]. The process of pathogenic biofilm formation on the surface of moxi/**3/4Sn** will gradually release the encapsulated moxi, thereby ensuring continuous maintenance of the concentration of the active form of moxifloxacin hydrochloride in the biofilm matrix and in close proximity to microbial cells.

4. Conclusions

A novel decasubstituted pillar[5]arene containing free mercapto groups, **3**, was synthesized and its structure determined by powder X-ray diffraction. Using UV–vis spectroscopy, the ability of pillar[5]arene **3** to interact with the antimicrobial drug moxifloxacin was shown. The association constant and stoichiometry of the **3**/moxi complex were calculated by UV–vis spectroscopy from the analysis of binding isotherms for model **3**/moxi = 2:1 ($logK_{11} = 2.14$ and $logK_{12} = 6.20$). The structure of the resulting complex was confirmed by 2D ^1H-^1H NOESY NMR spectroscopy. THF soluble tetrablock co-monomers **3/3S** and **3/4S** were isolated by thiol/disulfide redox reactions of **3** with trimethylolpropane tris(3-mercaptopropionate) **3S** or pentaerythritol tetrakis(3-mercaptopropionate) **4S**, the structure of which was studied by GPC and MALDI mass spectrometry. Spraying of **3/3S** and **3/4S** solutions in THF (1×10^{-3} M) on the surface of a glass substrate led to the formation of **3/3Sn** and **3/4Sn** polymer films. The formation of films occurs due to the formation of additional -S-S- bonds between tetrablock co-monomers **3/3S** and **3/4S**. They were characterized by TG-DSC analysis, FTIR spectroscopy, and their morphology studied by electron microscopy. Optical spectroscopy and EPR showed that the resulting **3/4Sn** film had the ability to self-heal under atmospheric oxygen. It was found that the **3/3Sn** and **3/4Sn** systems do not affect the formation of biofilms formed by *S. aureus* and *K. pneumoniae*. However, the introduction of the drug moxi into the composition of **3/3Sn** and **3/4Sn** films resulted in a noticeable inhibition of the formation of biofilms of these pathogenic microorganisms. The ability to retain an antimicrobial drug in **3/3Sn** and **3/4Sn** films after washing with water was shown by fluorescence spectroscopy. These results open up wide opportunities to develop new antibacterial polymeric materials with self-healing abilities that are resistant to external conditions.

Supplementary Materials: The following supporting information can be downloaded at: https://www.mdpi.com/article/10.3390/nano12091604/s1, Table S1: Variation of the conditions for the oxidative oligomerization of the macrocycle **3**, **3S**, and **4S**; Figure S1: ^{1}H NMR spectrum of 4,8,14,18,23,26,28,31,32,35—deca-[acylthioethoxy]-pillar[5]arene (**2**). CDCl$_3$, 298 K, 400 MHz; Figure S2: ^{1}H NMR spectrum of 4,8,14,18,23,26,28,31,32,35—deca-[2-mercaptoethoxy]-pillar[5]arene (**3**). CDCl$_3$, 298 K, 400 MHz; Figure S3: ^{13}C NMR spectrum of 4,8,14,18,23,26,28,31,32,35—deca-[acylthioethoxy]-pillar[5]arene (**2**). CDCl$_3$, 298 K, 400 MHz; Figure S4: ^{13}C NMR spectrum of 4,8,14,18,23,26,28,31,32,35—deca-[2-mercaptoethoxy]-pillar[5]arene (**3**). CDCl$_3$, 298 K, 400 MHz; Figure S5: Mass spectrum (MALDI-TOF, 4-nitroaniline matrix) of 4,8,14,18,23,26,28,31,32,35—deca-[acylthioethoxy]-pillar[5]arene (**2**); Figure S6: Mass spectrum (MALDI-TOF, 4-nitroaniline matrix) of 4,8,14,18,23,26,28,31,32,35—deca-[2-mercaptoethoxy]-pillar[5]arene (**3**); Figure S7: Mass spectrum (MALDI-TOF, 4-nitroaniline matrix) of tetrablock co-monomer, based on 4,8,14,18,23,26,28,31,32,35-deca-[2-mercaptoethoxy]-pillar[5]arene (**3**) and trimethylolpropane tris(3-mercaptopropionate), (**3/3S**); Figure S8: Mass spectrum (MALDI-TOF, 4-nitroaniline matrix) of tetrablock co-monomer, based on 4,8,14,18,23,26,28,31,32,35-deca-[2-mercaptoethoxy]-pillar[5]arene (**3**) and trimethylolpropane tris(3-mercaptopropionate), (**3/4S**); Figure S9: IR spectrum of 4,8,14,18,23,26,28,31,32,35—deca-[acylthioethoxy]-pillar[5]arene (**2**); Figure S10: IR spectrum of 4,8,14,18,23,26,28,31,32,35—deca-[2-mercaptoethoxy]-pillar[5]arene (**3**); Figure S11: IR spectrum of 4,8,14,18,23,26,28,31,32,35-deca-[2-mercaptoethoxy]-pillar[5]arene (**3**)-based supramolecular polymer (**3**)n; Figure S12: IR spectrum of tetrablock co-monomer, based on 4,8,14,18,23,26,28,31,32,35-deca-[2-mercaptoethoxy]-pillar[5]arene (**3**) and trimethylolpropane tris(3-mercaptopropionate), (**3/3S**); Figure S13: IR spectrum of tetrablock co-monomer, based on 4,8,14,18,23,26,28,31,32,35-deca-[2-mercaptoethoxy]-pillar[5]arene (**3**) and trimethylolpropane tris(3-mercaptopropionate), (**3/4S**); Figure S14: (a) The 2D 1H-1H NOESY NMR spectrum of the 3/moxi complex (2:1, 5×10^{-3} M;) in CHCl$_3$/CD$_3$OD = 100:1 at 25 °C; (b) 2D DOSY NMR 3/moxi complex in CHCl$_3$/CD$_3$OD = 100:1 at 25 °C (400 MHz, 298K); Figure S15: Size distribution of the particles by intensity for 3 (1×10^{-4} M) in solvent system THF:CH$_3$OH 100:1 (d = 845 ± 303 nm, PDI = 0.42 ± 0.06); Figure S16: Size distribution of the particles by intensity for **3/3S** (1×10^{-5} M) in solvent system THF:CH$_3$OH 100:1 (d = 782 ± 254 nm, PDI = 0.650 ± 0.174); Figure S17: Size distribution of the particles by intensity for **3/4S** (1×10^{-5} M) in solvent system THF:CH$_3$OH 100:1 (d = 640 ± 193 nm, PDI = 0.35 ± 0.02); Figure S18: Size distribution of the particles by intensity for moxi (1×10^{-4} M) in solvent system THF:CH$_3$OH 100:1 (d = 1269 ± 479 nm, PDI = 0.50 ± 0.14); Figure S19: Size distribution of the particles by intensity for BCl (1×10^{-4} M) in solvent system THF:CH$_3$OH 100:1 (d = 209 ± 21 nm, PDI = 0.24 ± 0.06); Figure S20: Size distribution of the particles by intensity for **3/3S** (1×10^{-5} M) + moxi (1×10^{-4} M) in solvent system THF:CH$_3$OH 100:1 (d = 617 ± 203 nm, PDI = 0.34 ± 0.05); Figure S21: Size distribution of the particles by intensity for **3/4S** (1×10^{-5} M) + moxi (1×10^{-4} M) in solvent system THF:CH$_3$OH 100:1 (d = 462 ± 13 nm, PDI = 0.28 ± 0.04); Figure S22: Size distribution of the particles by intensity for (4S)n (1×10^{-5} M) + moxi (1×10^{-4} M) in solvent system THF:CH$_3$OH 100:1 (d = 561 ± 7 nm, PDI = 0.24 ± 0.02); Figure S23: Size distribution of the particles by intensity for **3/3S** (1×10^{-5} M) + BCl (1×10^{-4} M) in solvent system THF:CH$_3$OH 100:1 (d = 556 ± 97 nm, PDI = 0.29 ± 0.05); Figure S24: Size distribution of the particles by intensity for **3/4S** (1×10^{-5} M) + BCl (1×10^{-4} M) in solvent system THF:CH$_3$OH 100:1 (d = 553 ± 140 nm, PDI = 0.36 ± 0.10); Table S2: Aggregation of thiols, cross-linked polymers and model polymers; Figure S25: Absorption spectra of macrocycle 3 (1×10^{-5} M) with a BCl solution (1

× 10^{-4} M) in the solvent system THF:CH$_3$OH = 100: 1; Figure S26: Absorption spectra of macrocycle **3** (1 × 10^{-5} M) with a moxi solution (1 × 10^{-4} M) in the solvent system THF:CH$_3$OH = 100: 1; Figure S27: Titration curve for the system macrocycle **3** (0–2.67 × 10^{-5} M)/moxi (1 × 10^{-5} M) in solvent system THF:CH$_3$OH = 100:1; Figure S28: Bindfit (Fit data to 1:1, 1:2 and 2:1 Host–Guest equilibria) screenshots taken from the summary window of the website supramolecular.org. This screenshots shows the raw data for UV–vis titration of 3 with moxi, the data fitted to 1:1 binding model (a), 1:2 binding model (b) and 2:1 binding model (c); Figure S29: GPC curves of products **3/3S** (eluent-THF, calibrated by PS standards); Figure S30: GPC curves of products **3/4S** (eluent-THF, calibrated by PS standards); Figure S31: TGA (green) and differential scanning calorimetry (DSC) (blue) curves of 3; Figure S32: TGA (green) and differential scanning calorimetry (DSC) (blue) curves of 3n; Figure S33: TGA (green) and differential scanning calorimetry (DSC) (blue) curves of **3/3Sn**; Figure S34: TGA (green) and differential scanning calorimetry (DSC) (blue) curves of **3/4Sn**; Figure S35: TEM image of **3/4S** (1 × 10^{-5} M) in the solvent system THF:CH$_3$OH (100:1) after the solvent evaporation; Figure S36: (a,b) TEM image of **3/4S** (1 × 10^{-5} M)/moxi (1 × 10^{-4} M) in the solvent system THF:CH$_3$OH (100:1) after the solvent evaporation; Figure S37: (a,b) SEM image of **3/4S** (1 × 10^{-5} M) in the solvent system THF:CH$_3$OH (100:1) after the solvent evaporation; Figure S38: SEM image of **3/4S** (1 × 10^{-5} M) at low pressure in the solvent system THF:CH$_3$OH (100:1) after the solvent evaporation; Figure S39: Three-dimensional model of **3/4S** (1 × 10^{-5} M) film from SEM images at low pressure; Figure S40: Three-dimensional model of **3/4S** (1 × 10^{-5} M)/moxi (10^{-4} M) film from SEM images at low pressure; Figure S41: AFM image of **3/4S** (1 × 10^{-5} M) film in the solvent system THF:CH$_3$OH (100:1) after the solvent evaporation; Figure S42: Optical microscope of **3/4Sn** film in the solvent system THF:CH$_3$OH (100:1) after the solvent evaporation with surface disturbance; Figure S43: Optical microscope of **3/4Sn** film after 120 min during the H$_2$O$_2$ surface treatment after 120 min; Figure S44: EPR spectra of 3n sample at room temperature in stationary (a) and in pulsed (b) modes of the X-band (9.6 GHz); Figure S45: EPR spectra of **3/4Sn** film and **3/3Sn** film at room temperature before irradiation with an X-ray source.

Author Contributions: Conceptualization, I.I.S.; writing—original draft preparation and methodology, D.N.S.; investigation and synthesis of target macrocycles, Y.I.A.; validation and visualization, O.A.M.; investigation and validation, V.A.N.; resources and software, R.E.T.; investigation and formal analysis, F.F.M.; formal analysis and original draft preparation, M.R.G.; formal analysis and investigation, P.V.Z.; resources and investigation, E.V.S.; conceptualization and investigation, E.A.S.; resources and investigation, A.V.G.; formal analysis and investigation, V.V.G.; software and methodology, D.R.I.; writing—review and editing and formal analysis, P.J.C. All authors have read and agreed to the published version of the manuscript.

Funding: The work was supported by Russian Science Foundation (№20-73-00161). Investigation of spatial structure of compounds by NMR spectroscopy was supported by the Kazan Federal University Strategic Academic Leadership Program ('PRIORITY-2030').

Data Availability Statement: Not applicable.

Acknowledgments: The authors are grateful to the staff of the Spectral-Analytical Center of Shared Facilities for Study of Structure, Composition, and Properties of Substances and Materials of the A.E. Arbuzov Institute of Organic and Physical Chemistry of the Kazan Scientific Center of the Russian Academy of Sciences for their research and assistance in discussion of the results.

Conflicts of Interest: The authors declare no conflict of interest.

References

1. Parsek, M.R.; Singh, P.K. Bacterial biofilms: An emerging link to disease pathogenesis. *Annu. Rev. Microbiol.* **2003**, *57*, 677–701. [CrossRef] [PubMed]
2. Verderosa, A.D.; Totsika, M.; Fairfull-Smith, K.E. Bacterial biofilm eradication agents: A current review. *Front. Chem.* **2019**, *7*, 824. [CrossRef] [PubMed]
3. Vestby, L.K.; Grønseth, T.; Simm, R.; Nesse, L.L. Bacterial biofilm and its role in the pathogenesis of disease. *Antibiotics* **2020**, *9*, 59. [CrossRef] [PubMed]
4. Scharnow, A.M.; Solinski, A.E.; Wuest, W.M. Targeting *S. mutans* biofilms: A perspective on preventing dental caries. *MedChemComm* **2019**, *10*, 1057–1067. [CrossRef] [PubMed]
5. Vishwakarma, V. Impact of environmental biofilms: Industrial components and its remediation. *J. Basic Microbiol.* **2020**, *60*, 198–206. [CrossRef] [PubMed]

6. Shen, N.; Cheng, E.; Whitley, J.W.; Horne, R.R.; Leigh, B.; Xu, L.; Jones, B.D.; Guymon, C.A.; Hansen, M.R. Photograftable zwitterionic coatings prevent *Staphylococcus aureus* and *Staphylococcus epidermidis* adhesion to PDMS surfaces. *ACS Appl. Bio Mater.* **2021**, *4*, 1283–1293. [CrossRef]
7. Berne, C.; Ellison, C.K.; Ducret, A.; Brun, Y.V. Bacterial adhesion at the single-cell level. *Nat. Rev. Microbiol.* **2018**, *16*, 616–627. [CrossRef]
8. Sapozhnikov, S.V.; Shtyrlin, N.V.; Kayumov, A.R.; Zamaldinova, A.E.; Iksanova, A.G.; Nikitina, E.V.; Krylova, E.S.; Grishaev, D.Y.; Balakin, K.V.; Shtyrlin, Y.G. New quaternary ammonium pyridoxine derivatives: Synthesis and antibacterial activity. *Med. Chem. Res.* **2017**, *26*, 3188–3202. [CrossRef]
9. Trizna, E.Y.; Yarullina, M.N.; Baidamshina, D.R.; Mironova, A.V.; Akhatova, F.S.; Rozhina, E.V.; Fakhrullin, R.F.; Khabibrakhmanova, A.M.; Kurbangalieva, A.R.; Bogachev, M.I.; et al. Bidirectional alterations in antibiotics susceptibility in *Staphylococcus aureus*—*Pseudomonas aeruginosa* dual-species biofilm. *Sci. Rep.* **2020**, *10*, 14849. [CrossRef]
10. Shtyrlin, N.V.; Sapozhnikov, S.V.; Galiullina, A.S.; Kayumov, A.R.; Bondar, O.V.; Mirchink, E.P.; Isakova, E.B.; Firsov, A.A.; Balakin, K.V.; Shtyrlin, Y.G. Synthesis and antibacterial activity of quaternary ammonium 4-deoxypyridoxine derivatives. *BioMed Res. Int.* **2016**, *2016*, 3864193. [CrossRef]
11. Boudarel, H.; Mathias, J.D.; Blaysat, B.; Grédiac, M. Towards standardized mechanical characterization of microbial biofilms: Analysis and critical review. *NPJ Biofilms Microbiomes* **2018**, *4*, 17. [CrossRef] [PubMed]
12. Si, Z.; Zheng, W.; Prananty, D.; Li, J.; Koh, C.H.; Kang, E.T.; Pethe, K.; Chan-Park, M.B. Polymers as advanced antibacterial and antibiofilm agents for direct and combination therapies. *Chem. Sci.* **2022**, *13*, 345–364. [CrossRef] [PubMed]
13. Chen, Y.; Sun, S.; Lu, D.; Shi, Y.; Yao, Y. Water-soluble supramolecular polymers constructed by macrocycle-based host-guest interactions. *Chin. Chem. Lett.* **2019**, *30*, 37–43. [CrossRef]
14. Dong, S.; Zheng, B.; Wang, F.; Huang, F. Supramolecular polymers constructed from macrocycle-based host–guest molecular recognition motifs. *Acc. Chem. Res.* **2014**, *47*, 1982–1994. [CrossRef]
15. Han, W.; Xiang, W.; Li, Q.; Zhang, H.; Yang, Y.; Shi, J.; Ji, Y.; Wang, S.; Ji, X.; Khashab, N.M.; et al. Water compatible supramolecular polymers: Recent progress. *Chem. Soc. Rev.* **2021**, *50*, 10025–10043. [CrossRef]
16. Correia, H.D.; Chowdhury, S.; Ramos, A.P.; Guy, L.; Demets, G.J.F.; Bucher, C. Dynamic supramolecular polymers built from cucurbit[n]urils and viologens. *Polym. Int.* **2019**, *68*, 572–588. [CrossRef]
17. Guo, D.S.; Liu, Y. Calixarene-based supramolecular polymerization in solution. *Chem. Soc. Rev.* **2012**, *41*, 5907–5921. [CrossRef]
18. Shurpik, D.N.; Padnya, P.L.; Stoikov, I.I.; Cragg, P.J. Antimicrobial activity of calixarenes and related macrocycles. *Molecules* **2020**, *25*, 5145. [CrossRef]
19. Gao, L.; Wang, H.; Zheng, B.; Huang, F. Combating antibiotic resistance: Current strategies for the discovery of novel antibacterial materials based on macrocycle supramolecular chemistry. *Giant* **2021**, *7*, 100066. [CrossRef]
20. Li, Z.; Yang, Y.W. Functional materials with pillarene struts. *Acc. Mater. Res.* **2021**, *2*, 292–305. [CrossRef]
21. Fa, S.; Kakuta, T.; Yamagishi, T.A.; Ogoshi, T. One-, two-, and three-dimensional supramolecular assemblies based on tubular and regular polygonal structures of pillar [n] arenes. *CCS Chem.* **2019**, *1*, 50–63. [CrossRef]
22. Shurpik, D.N.; Aleksandrova, Y.I.; Zelenikhin, P.V.; Subakaeva, E.V.; Cragg, P.J.; Stoikov, I.I. Towards new nanoporous biomaterials: Self-assembly of sulfopillar[5]arenes with vitamin D3 into supramolecular polymers. *Org. Biomol. Chem.* **2020**, *18*, 4210–4216. [CrossRef] [PubMed]
23. Shurpik, D.N.; Mostovaya, O.A.; Sevastyanov, D.A.; Lenina, O.A.; Sapunova, A.S.; Voloshina, A.D.; Petrov, K.A.; Kovyazina, I.V.; Cragg, P.J.; Stoikov, I.I. Supramolecular neuromuscular blocker inhibition by a pillar[5]arene through aqueous inclusion of rocuronium bromide. *Org. Biomol. Chem.* **2019**, *17*, 9951–9959. [CrossRef]
24. Shurpik, D.N.; Yakimova, L.S.; Gorbachuk, V.V.; Sevastyanov, D.A.; Padnya, P.L.; Bazanova, O.B.; Stoikov, I.I. Hybrid multicyclophanes based on thiacalix[4]arene and pillar[5]arene: Synthesis and influence on the formation of polyaniline. *Org. Chem. Front.* **2018**, *5*, 2780–2786. [CrossRef]
25. Antipin, I.S.; Alfimov, M.V.; Arslanov, V.V.; Burilov, V.A.; Vatsadze, S.Z.; Voloshin, Y.Z.; Volcho, K.P.; Gorbatchuk, V.V.; Gorbunova, Y.G.; Gromov, S.P.; et al. Functional supramolecular systems: Design and applications. *Russ. Chem. Rev.* **2021**, *90*, 895. [CrossRef]
26. Feng, W.; Jin, M.; Yang, K.; Pei, Y.; Pei, Z. Supramolecular delivery systems based on pillararenes. *Chem. Commun.* **2018**, *54*, 13626–13640. [CrossRef] [PubMed]
27. Li, M.H.; Lou, X.Y.; Yang, Y.W. Pillararene-based molecular-scale porous materials. *Chem. Commun.* **2021**, *57*, 13429–13447. [CrossRef] [PubMed]
28. Li, H.; Yang, Y.; Xu, F.; Liang, T.; Wen, H.; Tian, W. Pillararene-based supramolecular polymers. *Chem. Commun.* **2019**, *55*, 271–285. [CrossRef]
29. Yang, H.; Jin, L.; Zhao, D.; Lian, Z.; Appu, M.; Huang, J.; Zhang, Z. Antibacterial and Antibiofilm Formation Activities of Pyridinium-Based Cationic Pillar[5]arene Against *Pseudomonas aeruginosa*. *J. Agric. Food Chem.* **2021**, *69*, 4276–4283. [CrossRef]
30. Guo, S.; Huang, Q.; Chen, Y.; Wei, J.; Zheng, J.; Wang, L.; Wang, Y.; Wang, R. Synthesis and Bioactivity of Guanidinium-Functionalized Pillar[5]arene as a Biofilm Disruptor. *Angew. Chem. Int. Ed.* **2021**, *60*, 618–623. [CrossRef]
31. Sheldrick, G.M. SHELXT–Integrated space-group and crystal-structure determination. *Acta Crystallogr.* **2015**, *71*, 3–8. [CrossRef] [PubMed]
32. Sheldrick, G.M. A short history of SHELX. *Acta Crystallogr.* **2007**, *64*, 112–122. [CrossRef] [PubMed]

33. Macrae, C.F.; Edgington, P.R.; McCabe, P.; Pidcock, E.; Shields, G.P.; Taylor, R.; Towler, M.; Van De Streek, J. Mercury CSD 2.0—New features for the visualization and investigation of crystal structures. *J. Appl. Crystallogr.* **2006**, *39*, 453–457. [CrossRef]
34. Spartan '20, version 1.1.1; Wavefunction Inc.: Irvine, CA, USA, 2020.
35. Yao, Y.; Xue, M.; Chi, X.; Ma, Y.; He, J.; Abliz, Z.; Huang, F. A new water-soluble pillar[5]arene: Synthesis and application in the preparation of gold nanoparticles. *Chem. Commun.* **2012**, *48*, 6505. [CrossRef] [PubMed]
36. Pena-Francesch, A.; Jung, H.; Demirel, M.C.; Sitti, M. Biosynthetic self-healing materials for soft machines. *Nat. Mater.* **2020**, *19*, 1230–1235. [CrossRef] [PubMed]
37. Zhu, M.; Liu, J.; Gan, L.; Long, M. Research progress in bio-based self-healing materials. *Eur. Polym. J.* **2020**, *129*, 109651. [CrossRef]
38. Thakur, V.K.; Kessler, M.R. Self-healing polymer nanocomposite materials: A review. *Polymer* **2015**, *69*, 369–383. [CrossRef]
39. An, S.Y.; Arunbabu, D.; Noh, S.M.; Song, Y.K.; Oh, J.K. Recent strategies to develop self-healable crosslinked polymeric networks. *Chem. Commun.* **2015**, *51*, 13058–13070. [CrossRef]
40. Yoon, J.A.; Kamada, J.; Koynov, K.; Mohin, J.; Nicolaÿ, R.; Zhang, Y.; Balazs, A.C.; Kowalewski, T.; Matyjaszewski, K. Self-healing polymer films based on thiol–disulfide exchange reactions and self-healing kinetics measured using atomic force microscopy. *Macromolecules* **2012**, *45*, 142–149. [CrossRef]
41. Martin, R.; Rekondo, A.; De Luzuriaga, A.R.; Casuso, P.; Dupin, D.; Cabañero, G.; Grande, H.J.; Odriozola, I. Dynamic sulfur chemistry as a key tool in the design of self-healing polymers. *Smart Mater. Struct.* **2016**, *25*, 084017. [CrossRef]
42. Tyuftin, A.A.; Solovieva, S.E.; Murav'ev, A.A.; Polyantsev, F.M.; Latypov, S.K.; Antipin, I.S. Thiacalix[4]arenes with terminal thiol groups at the lower rim: Synthesis and structure. *Russ. Chem. Bull.* **2009**, *58*, 145–151. [CrossRef]
43. Naga, N.; Moriyama, K.; Furukawa, H. Synthesis and properties of multifunctional thiol crosslinked gels containing disulfide bond in the network structure. *J. Polym. Sci. A Polym. Chem.* **2017**, *55*, 3749–3756. [CrossRef]
44. Wang, Y.; Tang, L.; Li, Y.; Li, Q. Effects of networks composed of epoxy/dual thiol-curing agents on properties of shape memory polymers. *J. Appl. Polym. Sci.* **2022**, *139*, 51548. [CrossRef]
45. Felton, L.A. Mechanisms of polymeric film formation. *Int. J. Pharm.* **2013**, *457*, 423–427. [CrossRef] [PubMed]
46. Kowalski, R.P.; Romanowski, E.G.; Mah, F.S.; Yates, K.A.; Gordon, Y.J. Intracameral Vigamox® (moxifloxacin 0.5%) is non-toxic and effective in preventing endophthalmitis in a rabbit model. *Am. J. Ophthalmol.* **2005**, *140*, 497. [CrossRef]
47. Bindfit. Available online: http://supramolecular.org (accessed on 3 March 2022).
48. Hibbert, D.B.; Thordarson, P. The death of the Job plot, transparency, open science and online tools, uncertainty estimation methods and other developments in supramolecular chemistry data analysis. *Chem. Commun.* **2016**, *52*, 12792–12805. [CrossRef]
49. Barbera, L.; De Plano, L.M.; Franco, D.; Gattuso, G.; Guglielmino, S.P.; Lando, G.; Notti, A.; Parisi, M.F.; Pisagatti, I. Antiadhesive and antibacterial properties of pillar[5]arene-based multilayers. *Chem. Commun.* **2018**, *54*, 10203–10206. [CrossRef]
50. Cash, J.J.; Kubo, T.; Bapat, A.P.; Sumerlin, B.S. Room-temperature self-healing polymers based on dynamic-covalent boronic esters. *Macromolecules* **2015**, *48*, 2098–2106. [CrossRef]
51. Blaiszik, B.J.; Caruso, M.M.; McIlroy, D.A.; Moore, J.S.; White, S.R.; Sottos, N.R. Microcapsules filled with reactive solutions for self-healing materials. *Polymer* **2009**, *50*, 990–997. [CrossRef]
52. Ye, G.; Jiang, T. Preparation and Properties of Self-Healing Waterborne Polyurethane Based on Dynamic Disulfide Bond. *Polymers* **2021**, *13*, 2936. [CrossRef]
53. Zheng, S.; Brook, M.A. Reversible Redox Crosslinking of Thiopropylsilicones. *Macromol. Rapid Commun.* **2021**, *42*, 2000375. [CrossRef] [PubMed]
54. Nevejans, S.; Ballard, N.; Miranda, J.I.; Reck, B.; Asua, J.M. The underlying mechanisms for self-healing of poly(disulfide)s. *Phys. Chem. Chem. Phys.* **2016**, *18*, 27577–27583. [CrossRef] [PubMed]
55. Hernandez, M.; Grande, A.M.; Dierkes, W.; Bijleveld, J.; Van Der Zwaag, S.; García, S.J. Turning vulcanized natural rubber into a self-healing polymer: Effect of the disulfide/polysulfide ratio. *ACS Sustain. Chem. Eng.* **2016**, *4*, 5776–5784. [CrossRef]
56. Murzakhanov, F.F.; Grishin, P.O.; Goldberg, M.A.; Yavkin, B.V.; Mamin, G.V.; Orlinskii, S.B.; Komlev, V.S. Radiation-induced stable radicals in calcium phosphates: Results of multifrequency epr, ednmr, eseem, and endor studies. *Appl. Sci.* **2021**, *11*, 7727. [CrossRef]
57. Hu, J.; Mo, R.; Jiang, X.; Sheng, X.; Zhang, X. Structural design and antimicrobial properties of polypeptides and saccharide–polypeptide conjugates. *Polymer* **2019**, *183*, 1904683.
58. Shurpik, D.N.; Aleksandrova, Y.I.; Mostovaya, O.A.; Nazmutdinova, V.A.; Zelenikhin, P.V.; Subakaeva, E.V.; Mukhametzyanov, T.A.; Cragg, P.J.; Stoikov, I.I. Water-soluble pillar[5]arene sulfo-derivatives self-assemble into biocompatible nanosystems to stabilize therapeutic proteins. *Bioorg. Chem.* **2021**, *117*, 105415. [CrossRef]
59. Friedrich, A.W. Control of hospital acquired infections and antimicrobial resistance in Europe: The way to go. *Wien. Med. Wochenschr.* **2019**, *169*, 25–30. [CrossRef]
60. Yan, S.; Chen, S.; Gou, X.; Yang, J.; An, J.; Jin, X.; Yang, Y.-W.; Chen, L.; Gao, H. Biodegradable supramolecular materials based on cationic polyaspartamides and pillar[5]arene for targeting gram-positive bacteria and mitigating antimicrobial resistance. *Adv. Funct. Mater.* **2019**, *29*, 1904683. [CrossRef]
61. Pluth, M.D.; Raymond, K.N. Reversible guest exchange mechanisms in supramolecular host–guest assemblies. *Chem. Soc. Rev.* **2007**, *36*, 161–171. [CrossRef]
62. Bojtár, M.; Paudics, A.; Hessz, D.; Kubinyi, M.; Bitter, I. Amino acid recognition by fine tuning the association constants: Tailored naphthalimides in pillar[5]arene-based indicator displacement assays. *RSC Adv.* **2016**, *6*, 86269–86275. [CrossRef]

Article

Solid Lipid Nanoparticles Based on Monosubstituted Pillar[5]arenes: Chemoselective Synthesis of Macrocycles and Their Supramolecular Self-Assembly

Darya Filimonova [1], Anastasia Nazarova [1,*], Luidmila Yakimova [1,*] and Ivan Stoikov [1,2]

[1] A.M. Butlerov Chemistry Institute, Kazan Federal University, 18 Kremlyovskaya Str., 420008 Kazan, Russia
[2] Federal State Budgetary Scientific Institution «Federal Center for Toxicological, Radiation, and Biological Safety», Nauchny Gorodok-2, 420075 Kazan, Russia
* Correspondence: anas7tasia@gmail.com (A.N.); mila.yakimova@mail.ru (L.Y.); Tel.: +7-843-233-7241 (A.N. & L.Y.)

Citation: Filimonova, D.; Nazarova, A.; Yakimova, L.; Stoikov, I. Solid Lipid Nanoparticles Based on Monosubstituted Pillar[5]arenes: Chemoselective Synthesis of Macrocycles and Their Supramolecular Self-Assembly. *Nanomaterials* 2022, 12, 4266. https://doi.org/10.3390/nano12234266

Academic Editor: Eric Doris

Received: 7 November 2022
Accepted: 28 November 2022
Published: 30 November 2022

Publisher's Note: MDPI stays neutral with regard to jurisdictional claims in published maps and institutional affiliations.

Copyright: © 2022 by the authors. Licensee MDPI, Basel, Switzerland. This article is an open access article distributed under the terms and conditions of the Creative Commons Attribution (CC BY) license (https://creativecommons.org/licenses/by/4.0/).

Abstract: Novel monosubstituted pillar[5]arenes with one or two terminal carboxyl groups were synthesized by the reaction of succinic anhydride with pillar[5]arene derivative containing a diethylenetriamine function. The ability for non-covalent self-assembly in chloroform, dimethyl sulfoxide, as well as in tetrahydrofuran-water system was studied. The ability of the synthesized macrocycles to form different types of associates depending on the substituent nature was established. The formation of stable particles with average diameter of 192 nm in chloroform and of 439 nm in DMSO was shown for pillar[5]arene containing two carboxyl fragments. Solid lipid nanoparticles (SLN) based on monosubstituted pillar[5]arenes were synthesized by nanoprecipitation in THF-water system. Minor changes in the structure of the macrocycle substituent can dramatically influence the stability and shape of SLN (spherical and rod-like structures) accordingly to DLS and TEM. The presence of two carboxyl groups in the macrocycle substituent leads to the formation of stable spherical SLN with an average hydrodynamic diameter of 364–454 nm. Rod-like structures are formed by pillar[5]arene containing one carboxyl fragment, which diameter is about of 50–80 nm and length of 700–1000 nm. The synthesized stable SLN open up great prospects for their use as drug storage systems.

Keywords: macrocycles; pillar[5]arene; self-assembly; solid lipid nanoparticles

1. Introduction

Solid lipid nanoparticles (SLN) are an alternative generation of nanoparticles compared with well-known colloidal systems such as liposomes and polymeric micro- and nanoparticles [1–4]. They are applied in medical and pharmaceutical chemistry. SLN increase absorption and bioactivity, improve tissue distribution in the target organ and ensure controlled drug release [5–8]. Low toxicity, the capability to encapsulate hydrophilic and hydrophobic substances and the possibility of large-scale production are advantages of these systems. The stated above makes it possible to use SLN for the development of various pharmaceutical products [9–11]. Nowadays, the use of macrocyclic compounds as a lipid matrix is attracting more and more attention due to the possibility of molecules double encapsulation either in the host cavity or in the SLN matrix [12,13]. The ability to selective interaction with target cells through host-guest complexation will significantly improve targeted drug delivery. A number of works have proposed amphiphilic molecules, e.g., calix[n]arenes and cyclodextrins, as lipid analogues [14–18]. The use of a promising class of macrocycles, pillar[5]arenes, in the synthesis of SLN was shown for the first time in our research group [19–21].

The choice of this class of macrocyclic compounds was based on synthetic availability and possibility of regioselective functionalization [22–25]. Moreover, the macrocyclic cavity of pillar[5]arene can take an active part in the formation of host-guest complexes, and also makes it possible to obtain different types of associates (supramolecular polymers, micelles, vesicles, rotaxanes and pseudorotaxanes) [26–33].

The presented work is ongoing studies of the monosubstituted pillar[5]arene as a building block for nanoparticles [19–21]. We assumed that minor changes in the substituent structure of macrocycle can dramatically change the stability and shape of the SLN. We chose pillar[5]arenes containing one or two fragments of succinic acid to confirm this hypothesis. This research is presented an approach to the chemoselective synthesis of new monosubstituted pillar[5]arenes containing one or two carboxyl groups. The substituent effect on the aggregation ability of the macrocycle has also been evaluated. Solid lipid nanoparticles based on the obtained compounds were synthesized and the features of supramolecular self-assembly on the SLN size were studied.

2. Results and Discussion

2.1. Synthesis of Monosubstituted Pillar[5]arenes Containing Amide and Carboxyl Groups

Previously, our research group optimized acylation conditions with succinic anhydride of monosubstituted pillar[5]arenes containing N-propylamide and N-(aminobutyl)amide fragments [21]. Pillar[5]arene 1 containing one diethylenetriamine fragment, which has primary and secondary amino groups, was chosen as a starting amine for interaction with succinic anhydride. The compound 1 was obtained according to the literature [19] (Scheme 1).

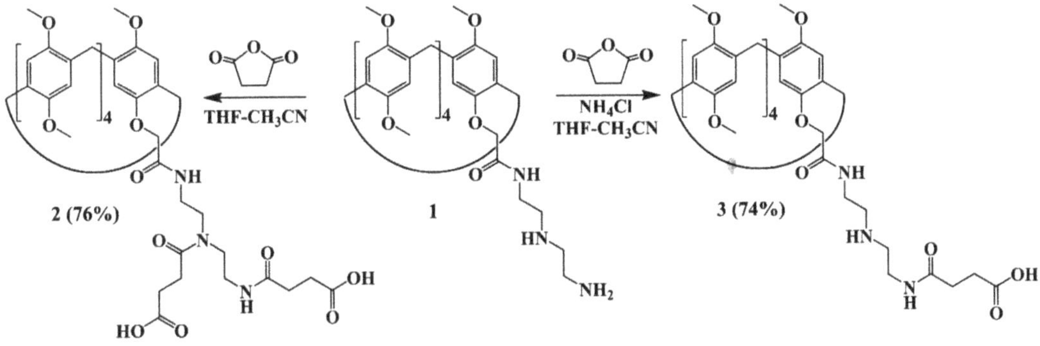

Scheme 1. Synthetic route for the preparation of pillar[5]arenes **2** and **3**: **1** is parent pillar[5]arene, **2** is monosubstituted pillar[5]arene containing two fragments of succinic acid, **3** is monosubstituted pillar[5]arene containing one fragment of succinic acid.

It is known that macrocycle **1** tends to form supramolecular polymers in proton-donor chloroform, which is not observed in proton-acceptor solvents such as dimethyl sulfoxide and acetonitrile [19]. Therefore, the reaction was carried out in a tetrahydrofuran-acetonitrile mixture to knock out the N-(aminoalkyl)amide substituent from the macrocyclic cavity. The use of acetonitrile led to acylation at both the primary and secondary amino groups in the case of macrocycle **1** containing diethylenetriamine fragment. It is resulting in the formation of monosubstituted pillar[5]arene **2** with two carboxyl groups (Scheme 1). The effect of acetonitrile leads to the knock out of the substituent from the cavity, as expected, whereby the secondary amino group becomes sterically accessible and reacts with the anhydride molecule. Monosubstituted pillar[5]arene **2** containing two fragments of succinic acid was obtained in 76% yield. The ^{1}H NMR spectrum is in good agreement with the structure of the synthesized macrocycle **2** (Figure 1).

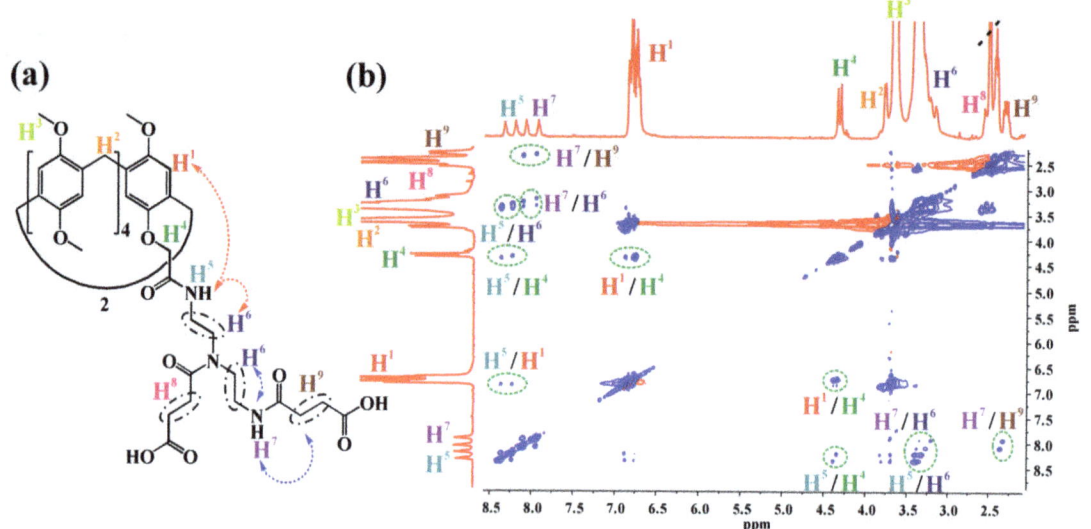

Figure 1. (a) Proposed intramolecular interactions for pillar[5]arene **2**; (b) ^{1}H-^{1}H NOESY NMR spectrum of pillar[5]arene **2** (DMSO-d_6, 298 K, 400 MHz). Green circles represent cross-peaks between protons. Blue intensity contours represent negative NOE's, and red intensity contours represent positive NOE's.

In the ^{1}H NMR spectrum of the compound **2** (Figure 1) the proton signals of aromatic fragments and methylene bridges as well as methoxy groups resonate as multiplets in the 6.70–6.83 ppm and 3.47–3.62 ppm, respectively. The H^4 proton signals of oxymethylene fragments split into two singlets (4.30 and 4.34 ppm). Similar splitting into two triplets is also characteristic for the protons of each amide group. Thus, the H^5 proton signals resonate as two triplets (8.19 and 8.32 ppm) with $^3J_{HH}$ 5.6 and 5.8 Hz. The H^7 proton signals of amide group also appear as two triplets with chemical shifts of 7.92 and 8.06 ppm and $^3J_{HH}$ 5.6 and 5.4 Hz, respectively. Such a signal-doubling is due to the coexistence of E/Z isomers, which appear by the tautomerism of the amide group. Therefore, the amide group either forms a hydrogen bond with the oxygen atom of the carbonyl fragment or it is in the shielding zone of the carbonyl group. This leads to a doubling of the proton signals for each amide fragment. The H^8 and H^9 proton signals of the methylene fragments resonate as a multiplet in the 2.26–2.56 ppm range. The H^6 methylene proton signals of diethylenetriamine fragment overlap with the residual proton signals of water and are also observed as a multiplet at 3.14–3.29 ppm.

It was decided to study the interaction of amine **1** with succinic anhydride in the presence of ammonium chloride to obtain monosubstituted pillar[5]arene containing only one fragment of succinic acid. We suppose that the addition of ammonium chloride to the reaction mixture leads to protonation of the secondary amino group. In view of this the reaction will proceed chemoselectively, and only the primary amino group will undergo acylation. Indeed, the addition of ammonium chloride to the reaction mixture made it possible to direct the reaction with macrocycle **1** only at the primary amino group. The compound **3** was obtained in 74% yield. Structure of all the synthesized compounds was confirmed by a number of physical methods (^{1}H NMR, ^{13}C NMR and IR spectroscopy). The composition was confirmed by mass spectrometry and elemental analysis (Figures S1–S8). It is worth noting that the melting point for macrocycle **2** is 104 °C and that is 120 °C for pillar[5]arene **3**. Apparently, the lower melting temperature for the compound **2** is caused by the lower packing density of molecules in space compared to the macrocycle **3**. This fact confirms that in the case of pillararene **3** the formation of supramolecular polymers is possible, which leads to denser packing.

2.2. Aggregation Properties of Monosubstituted Pillar[5]arenes Containing Carboxyl Groups

One of the important features of monosubstituted pillar[5]arenes is supramolecular self-assembly, which leads to the formation of various oligomeric and polymeric structures [34–39]. Therefore, the next stage of the work was studying the aggregation ability of **2** and **3**. It is known that monosubstituted pillar[5]arenes can form different typeCs of supramolecular architectures depending on the solvent used [40–42]. The macrocycle **2** contains a bulkier substituent including two fragments of succinic acid. In this regard, we hypothesized that formation of supramolecular polymers by pillar[5]arene **2** is impossible or unprofitable for steric reasons. We studied association properties (Figures S9–S12) of the obtained compounds **2** and **3** in chloroform and DMSO in the $1 \times 10^{-5} - 1 \times 10^{-3}$ M concentration range by dynamic light scattering (DLS) to confirm this hypothesis. Pillar[5]arene **2** at $C = 1 \times 10^{-3}$ M forms particles (Table 1) with hydrodynamic diameter (d) $d_1 = 289 \pm 9$ nm in DMSO. The polydispersity index (PDI) of this system is 0.44 ± 0.03. Macrocycle **3** forms systems similar in size and polydispersity index ($d_1 = 332 \pm 7$ nm, PDI = 0.39 ± 0.03) in the same conditions. A small fraction of larger particles with sizes $d_2 = 5050 \pm 254$ nm and $d_2 = 5150 \pm 128$ nm for pillar[5]arenes **2** and **3** respectively, can be additionally observed in both cases (Table 1). Decreasing the concentration of the pillar[5]arene **2** and **3** solutions in DMSO to $C = 1 \times 10^{-4}$ M leads to different effect on their self-assembly. A monodisperse system with PDI = 0.26 ± 0.03 and a particle diameter of 439 ± 59 nm was formed (Table 1) in the case of macrocycle **2**. There are no significant changes in the particle sizes and polydispersity index in the case of the compound **3** (Table 1).

Table 1. Physicochemical characteristic of particles formed by pillar[5]arenes **2** and **3** by intensity: d—hydrodynamic diameter by the DLS method, PDI—polydispersity index.

Macrocycles	Solvent	C, M	d, nm (S, %)	PDI
2	CHCl$_3$	1×10^{-5}	363 ± 50 (60%) 3636 ± 589 (40%)	0.63 ± 0.06
		1×10^{-4}	308 ± 34 (73%) 3519 ± 149 (27%)	0.46 ± 0.09
		1×10^{-3}	192 ± 20 (100%)	0.26 ± 0.01
	DMSO	1×10^{-5}	851 ± 99 (100%)	0.36 ± 0.04
		1×10^{-4}	439 ± 59 (100%)	0.26 ± 0.03
		1×10^{-3}	289 ± 9 (93%) 5050 ± 254 (7%)	0.44 ± 0.03
3	CHCl$_3$	1×10^{-5}	4226 ± 253 (66%) 346 ± 47 (34%)	0.78 ± 0.11
		1×10^{-4}	152 ± 46 (100%)	0.92 ± 0.09
		1×10^{-3}	155 ± 32 (100%)	0.85 ± 0.11
	DMSO	1×10^{-5}	367 ± 38 (77%) 4722 ± 115 (23%)	0.63 ± 0.12
		1×10^{-4}	344 ± 26 (84%) 4924 ± 84 (16%)	0.52 ± 0.05
		1×10^{-3}	332 ± 7 (94%) 5150 ± 128 (6%)	0.39 ± 0.03

The study of the aggregation properties of the obtained pillar[5]arenes **2** and **3** in chloroform (Figures S13–S16) showed that almost in the entire concentration range (from 1×10^{-5} to 1×10^{-3} M) macrocycles **2** and **3** form polydisperse systems (Table 1). The exception is the solution of the compound **2** at a 1×10^{-3} M concentration (d = 192 ± 20 nm, PDI = 0.26 ± 0.01). Apparently, the presence of two carboxyl functions in the structure of **2** leads to the fact that the pillar[5]arene **2** becomes similar to surfactant molecules in its aggregation properties. The macrocycle can be clearly distinguished between lipophilic (macrocyclic rim) and hydrophilic parts (substituent with two fragments of succinic acid). The aggregation of macrocycle **2** becomes possible only when a certain concentration is reached (1×10^{-3} M in CHCl$_3$ and 1×10^{-4} M in DMSO) due to the similar behavior to surfactants. The formation of polydisperse systems by the compound **3** in chloroform is probably caused by the formation of supramolecular polymers. The solution of pillararene **3** was studied by ^{1}H NMR spectroscopy (Figure 2) in two solvents (CDCl$_3$ and DMSO-d_6). It is worth noting that the chemical shifts of methylene proton signals of the succinic acid fragment (H^8 and H^9) practically do not change regardless of the solvent, whereas, chemical H^5 and H^6 proton signals of diethylenetriamine fragment undergo upfield shift for the solution of compound **3** in chloroform. It is due to their strong shielding by macrocyclic cavity. This indicates inclusion of the substituent in the macrocyclic cavity with the formation of supramolecular polymer (Figure 2b).

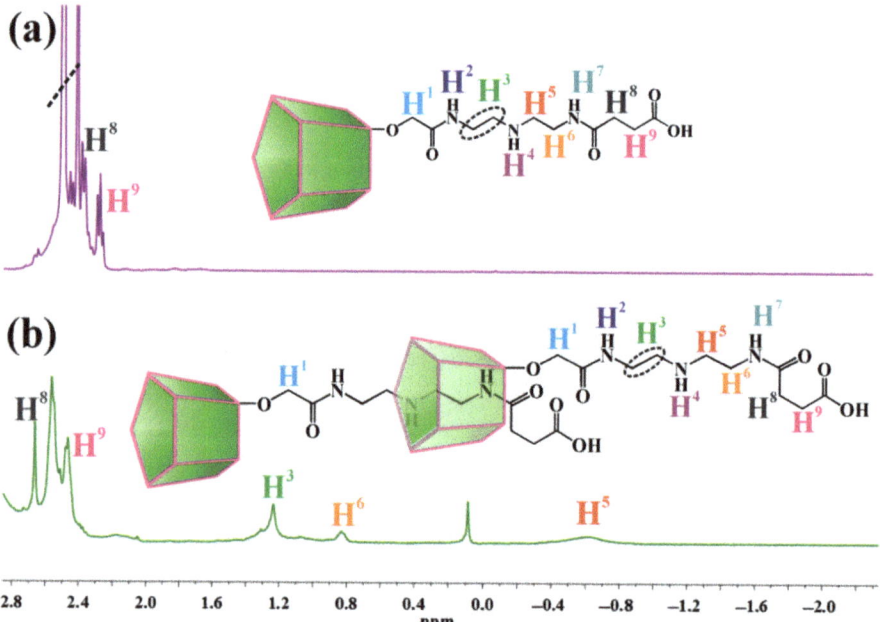

Figure 2. Fragments of ^{1}H NMR spectra of pillar[5]arene **3**: (**a**) DMSO-d_6, 400 MHz, 298 K; (**b**) CDCl$_3$, 400 MHz, 298 K. Black circles means H^3 proton signals.

Figure 3 shows the proposed self-assembly scheme of the macrocycles **2** and **3**.

The next stage of the work was a study of the noncovalent self-assembly patterns of pillar[5]arenes **2** and **3** with the formation of solid lipid nanoparticles (SLN-2 and SLN-3 respectively) in water (Table 2).

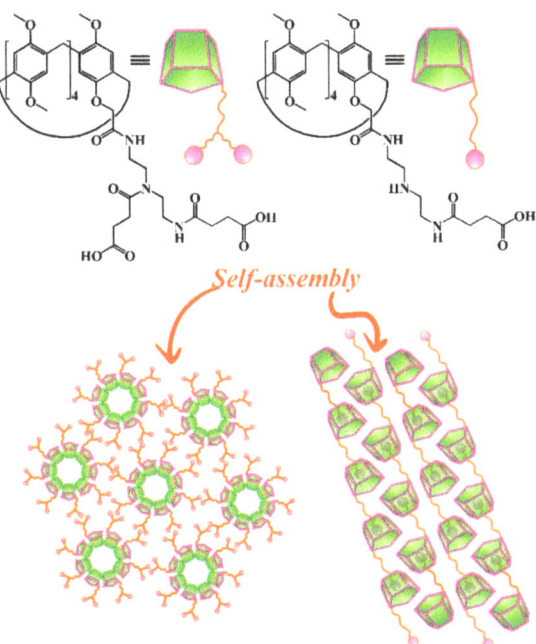

Figure 3. Schematic representation of self-assembly of pillar[5]arenes **2** (left) and **3** (right).

Table 2. Physicochemical characteristic of SLN formed by macrocycles **2** and **3** in water by intensity: d—hydrodynamic diameter by the dynamic light scattering (DLS) method, PDI—polydispersity index, ζ—zeta potential.

	C, M	d, nm	PDI	ζ, mV
SLN-2	3×10^{-6}	364 ± 39	0.24 ± 0.02	−25 ± 2
	3×10^{-5}	382 ± 8	0.21 ± 0.01	−33 ± 1
	3×10^{-4}	454 ± 19	0.18 ± 0.03	−33 ± 1
SLN-3	3×10^{-6}	582 ± 23	0.27 ± 0.01	−3 ± 0
	3×10^{-5}	444 ± 10	0.21 ± 0.01	−21 ± 1
	3×10^{-4}	209 ± 2	0.14 ± 0.01	−20 ± 1

Today, a number of works are devoted to the synthesis of different shaped SLN [20,43–46]. It has been shown that the shape of SLN can change depending on the nature and size of the substituents in the molecule [20,46], the interaction of the solute with the solvent [43], and the loading of SLN with some substance [45]. In this work, we studied the influence of substituent nature in the structure of monosubstituted pillar[5]arenes on the size and stability of formed particles. We obtained SLN in water by nanoprecipitation using the THF-water solvent system according to the literature method [47]. The synthesized nanoparticles were characterized by DLS and electrophoretic light scattering methods (Figures S17–S20). The initial concentration of pillar[5]arenes **2** and **3** was 3×10^{-4} M. The most stable SLN are formed at 3×10^{-4} M concentration (Table 2) for the macrocycle **2**, as indicated by the zeta potential value ($\zeta = -33$ mV). There is insignificant increase in the polydispersity index of the systems when the solution of the compound **2** is diluted, while the size of associates decreases. A different situation was observed for SLN formed by the pillar[5]arene **3** (Table 2). Both particle enlargement and an increase in the polydispersity index of the systems occur with the concentration decrease of the solutions. At the same time, the smallest aggregates size and PDI value (d = 209 ± 2 nm, PDI = 0.14 ± 0.01) are also characteristic for the highest concentration of 3×10^{-4} M as in the case of macrocycle **2**. Pillar[5]arene **2** containing a bulkier substituent, namely, two fragments of succinic acid, tends to form particles of larger size (d = 454 ± 19 nm) compared to the **3** (d = 209 ± 2 nm). The presence of two carboxyl groups in the macrocycle substituent leads to the formation of stable SLN over the studied concentration range within three orders of magnitude from 10^{-6} to 10^{-4} M, which is obviously due to the framing of particles by carboxylate groups. At the same time, the transition from the amido acid residue in the case of the compound **2** to the aminoamido acid fragment in the pillar[5]arene **3** leads to the formation of denser SLN by supramolecular self-assembly of the macrocycle into polymeric structures. It is in good agreement with its melting point and the size of the formed SLN. Measurement of ζ-potential allows to predict the stability of colloidal dispersions during storage. The use of the macrocycle **2** is preferable for obtaining more stable SLN because aggregation of charged particles with a high ζ-potential occurs to a lesser degree due to electrostatic repulsion. A similar pattern of ζ-potential changes has been described for non-macrocyclic surfactants [48].

The synthesized solid lipid nanoparticles were additionally studied by transmission electron microscopy (TEM). The formation of spherical nanosized aggregates with an average diameter of 200 nm was established for the SLN-2 (Figure 4a). Macrocycle **3** formed rod-like particles which diameter is about of 50–80 nm and length of 700–1000 nm (Figure 4b). The data obtained by TEM are in good agreement with the previously advanced hypothesis about the substituent effect on the self-assembly of the synthesized compounds **2** and **3**. The formulation of micelle-like structures is typical for pillar[5]arene **2** containing two carboxyl groups. Supramolecular polymers are formed in the case of macrocycle **3** containing one carboxyl function.

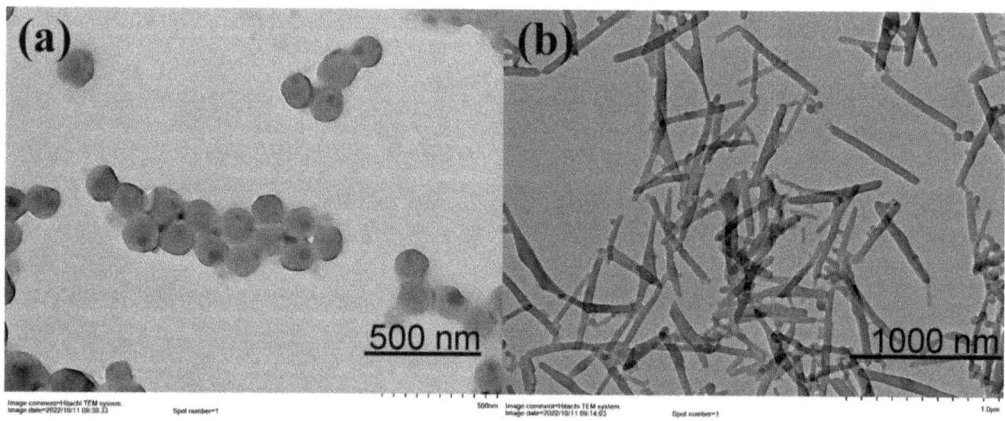

Figure 4. TEM images of: (**a**) SLN-2 (3×10^{-4} M) and (**b**) SLN-3 (3×10^{-4} M) prepared by nanoprecipitation using the THF-water solvent system.

3. Conclusions

Monosubstituted pillar[5]arenes containing one or two carboxyl groups were synthesized for the first time with good yields. The chemoselectivity of the process was controlled by adding ammonium chloride. The ability of the synthesized macrocycles to form different types of associates depending on the substituent nature was established. The formation of stable particles with average diameter of 192 nm in chloroform and of 439 nm in DMSO was shown for pillar[5]arene containing two carboxyl fragments. Monosubstituted pillar[5]arene containing one carboxyl function is prone to the supramolecular polymer formation in $CDCl_3$. Solid lipid nanoparticles (SLN) based on obtained macrocycles were synthesized and characterized by DLS and TEM. Minor changes in the structure of the macrocycle substituent can dramatically change the stability and shape of SLN (spherical and rod-like structures). The presence of two carboxyl groups in the macrocycle substituent leads to the formation of stable spherical SLN (364–454 nm) while rod-like structures are formed by pillar[5]arene containing one carboxyl fragment. The use of pillar[5]arene with two carboxyl functions is preferable to obtain more stable SLN. Synthesized stable SLN open up great prospects for their use as drug storage systems. This research could be regarded as a starting point for further investigation of the applicability of these materials.

Supplementary Materials: The following supporting information can be downloaded at: https://www.mdpi.com/article/10.3390/nano12234266/s1. Figures S1–S8: ^{1}H and ^{13}C NMR, IR and mass spectra of the compounds **2** and **3**; Figures S9–S12: DLS data of the compounds **2** and **3** in DMSO; Figures S13–S16: DLS data of the compounds **2** and **3** in CHCl$_3$; Figures S17 and S18: DLS data of SLN-2; Figures S19 and S20: DLS data of SLN-3.

Author Contributions: Conceptualization, writing—review and editing, supervision, I.S.; investigation, writing—original draft preparation, D.F.; investigation, writing—original draft preparation and visualization, A.N.; data curation, writing—review and editing, supervision, funding acquisition L.Y. All authors have read and agreed to the published version of the manuscript.

Funding: The work was supported by RFBR (Russian Foundation for Basic Research), project number 20-03-00816.

Institutional Review Board Statement: Not applicable.

Informed Consent Statement: Not applicable.

Data Availability Statement: The data presented in this study are available in Supplementary Materials.

Acknowledgments: The TEM images were recorded on the equipment of the Interdisciplinary Center for Analytical Microscopy of Kazan Federal University. Registration of mass spectra was carried out by the Kazan Federal University Strategic Academic Leadership Program ('PRIORITY-2030').

Conflicts of Interest: The authors declare no conflict of interest. The funders had no role in the design of the study; in the collection, analyses, or interpretation of data; in the writing of the manuscript; or in the decision to publish the results.

References

1. Cortesi, R.; Esposito, E.; Luca, G.; Nastruzzi, C. Production of lipospheres as carriers for bioactive compounds. *Biomaterials* **2002**, *23*, 2283–2294. [CrossRef] [PubMed]
2. Paliwal, R.; Rai, S.; Vaidya, B.; Khatri, K.; Goyal, A.K.; Mishra, N.; Mehta, A.; Vyas, S.P. Effect of lipid core material on characteristics of solid lipid nanoparticles designed for oral lymphatic delivery. *Nanomed. Nanotechnol. Biol. Med.* **2009**, *5*, 184–191. [CrossRef] [PubMed]
3. Mirchandani, Y.; Patravale, V.B.; Brijesh, S. Solid lipid nanoparticles for hydrophilic drugs. *J. Control. Release* **2021**, *335*, 457–464. [CrossRef] [PubMed]
4. Tenchov, R.; Bird, R.; Curtze, A.E.; Zhou, Q. Lipid nanoparticles-from liposomes to mRNA vaccine delivery, a landscape of research diversity and advancement. *ACS Nano* **2021**, *15*, 16982–17015. [CrossRef] [PubMed]
5. Andonova, V.; Peneva, P. Characterization methods for solid lipid nanoparticles (SLN) and nanostructured lipid carriers (NLC). *Curr. Pharm. Des.* **2017**, *23*, 6630–6642. [CrossRef] [PubMed]
6. Elbrink, K.; Van Hees, S.; Chamanza, R.; Roelant, D.; Loomans, T.; Holm, R.; Kiekens, F. Application of solid lipid nanoparticles as a long-term drug delivery platform for intramuscular and subcutaneous administration: In vitro and in vivo evaluation. *Eur. J. Pharm. Biopharm.* **2021**, *163*, 158–170. [CrossRef]
7. Mohamed, H.; Abdin, S.M.; Kamal, L.; Orive, G. Nanostructured lipid carriers for delivery of chemotherapeutics: A review. *Pharmaceutics* **2020**, *12*, 288. [CrossRef]
8. Kumar, R.; Singh, A.; Sharma, K.; Dhasmana, D.; Garg, N.; Siril, P.F. Preparation, characterization and in vitro cytotoxicity of Fenofibrate and Nabumetone loaded solid lipid nanoparticles. *Mater. Sci. Eng. C* **2020**, *106*, 110184. [CrossRef]
9. Mehnert, W.; Mäder, K. Solid lipid nanoparticles: Production, characterization and applications. *Adv. Drug Deliv. Rev.* **2012**, *64*, 83–101. [CrossRef]
10. Zur Mühlen, A.; Schwarz, C.; Mehnert, W. Solid lipid nanoparticles (SLN) for controlled drug delivery—Drug release and release mechanism. *Eur. J. Pharm. Biopharm.* **1998**, *45*, 149–155. [CrossRef]
11. Ghasemiyeh, P.; Mohammadi-Samani, S. Solid lipid nanoparticles and nanostructured lipid carriers as novel drug delivery systems: Applications, advantages and disadvantages. *Res. Pharm. Sci.* **2018**, *13*, 288–303. [CrossRef]

12. Burilov, V.A.; Artemenko, A.A.; Garipova, R.I.; Amirova, R.R.; Fatykhova, A.M.; Borisova, J.A.; Mironova, D.A.; Sultanova, E.D.; Evtugyn, V.G.; Solovieva, S.E.; et al. New calix[4]arene-fluoresceine conjugate by click approach-synthesis and preparation of photocatalytically active solid lipid nanoparticles. *Molecules* **2022**, *27*, 2436. [CrossRef]
13. Pojarova, M.; Ananchenko, G.S.; Udachin, K.A.; Daroszewska, M.; Perret, F.; Coleman, A.W.; Ripmeester, J.A. Solid lipid nanoparticles of p-hexanoyl calix[4]arene as a controlling agent in the photochemistry of a sunscreen blocker. *Chem. Mater.* **2006**, *18*, 5817–5819. [CrossRef]
14. Cirri, M.; Mennini, N.; Maestrelli, F.; Mura, P.; Ghelardini, C.; Mannelli, L.D.C. Development and in vivo evaluation of an innovative "Hydrochlorothiazide-in Cyclodextrins-in Solid Lipid Nanoparticles" formulation with sustained release and enhanced oral bioavailability for potential hypertension treatment in pediatrics. *Int. J. Pharm.* **2017**, *521*, 73–83. [CrossRef]
15. Helttunen, K.; Galán, A.; Ballester, P.; Bergenholtz, J.; Nissinen, M. Solid lipid nanoparticles from amphiphilic calixpyrroles. *J. Colloid Interface Sci.* **2016**, *464*, 59–65. [CrossRef]
16. Montasser, I.; Shahgaldian, P.; Perret, F.; Coleman, A.W. Solid lipid nanoparticle-based calix[n]arenes and calix-resorcinarenes as building blocks: Synthesis, formulation and characterization. *Int. J. Mol. Sci.* **2013**, *14*, 21899–21942. [CrossRef]
17. Pandey, A. Cyclodextrin-based nanoparticles for pharmaceutical applications: A review. *Environ. Chem. Lett.* **2021**, *19*, 4297–4310. [CrossRef]
18. Duan, Y.; Dhar, A.; Patel, C.; Khimani, M.; Neogi, S.; Sharma, P.; Siva Kumar, N.; Vekariya, R.L. A brief review on solid lipid nanoparticles: Part and parcel of contemporary drug delivery systems. *RSC Adv.* **2020**, *10*, 26777–26791. [CrossRef]
19. Yakimova, L.S.; Shurpik, D.N.; Guralnik, E.G.; Evtugyn, V.G.; Osin, Y.N.; Stoikov, I.I. Fluorescein-loaded solid lipid nanoparticles based on monoamine pillar[5]arene: Synthesis and interaction with DNA. *ChemNanoMat* **2018**, *4*, 919–923. [CrossRef]
20. Yakimova, L.S.; Guralnik, E.G.; Shurpik, D.N.; Evtugyn, V.G.; Osin, Y.N.; Subakaeva, E.V.; Sokolova, E.A.; Zelenikhin, P.V.; Stoikov, I.I. Morphology, structure and cytotoxicity of dye-loaded lipid nanoparticles based on monoamine pillar[5]arenes. *Mater. Chem. Front.* **2020**, *4*, 2962–2970. [CrossRef]
21. Nazarova, A.; Yakimova, L.; Filimonova, D.; Stoikov, I. Surfactant effect on the physicochemical characteristics of solid lipid nanoparticles based on pillar[5]arenes. *Int. J. Mol. Sci.* **2022**, *23*, 779. [CrossRef]
22. Demay-Drouhard, P.; Du, K.; Samanta, K.; Wan, X.; Yang, Y.; Srinivasan, R.; Sue, A.C.-H.; Zuihof, H. Functionalization at will of rim-differentiated pillar[5]arenes. *Org. Lett.* **2019**, *21*, 3976–3980. [CrossRef] [PubMed]
23. Nazarova, A.A.; Makhmutova, L.I.; Stoikov, I.I. Synthesis of pillar[5]arenes with a PH-containing fragment. *Russ. J. Gen. Chem.* **2017**, *87*, 1941–1945. [CrossRef]
24. Strutt, N.L.; Zhang, H.; Schneebeli, S.T.; Stoddart, J.F. Functionalizing pillar[n]arenes. *Acc. Chem. Res.* **2014**, *47*, 2631–2642. [CrossRef] [PubMed]
25. Nazarova, A.A.; Shibaeva, K.S.; Stoikov, I.I. Synthesis of the monosubstituted pillar[5]arenes with 1-aminophosphonate fragment. *Phosphorus Sulfur Silicon Relat. Elem.* **2016**, *191*, 1583–1584. [CrossRef]
26. Wei, T.-B.; Chen, J.-F.; Cheng, X.-B.; Li, H.; Han, B.-B.; Zhang, Y.-M.; Yao, H.; Lin, Q. A novel functionalized pillar[5]arene-based selective amino acid sensor for l-tryptophan. *Org. Chem. Front.* **2017**, *4*, 210–213. [CrossRef]
27. Lia, C. Pillararene-based supramolecular polymers: From molecular recognition to polymeric aggregates. *Chem. Commun.* **2014**, *50*, 12420–12433. [CrossRef]
28. Chen, J.-F.; Lin, Q.; Zhang, Y.-M.; Yao, H.; Wei, T.-B. Pillararene-based fluorescent chemosensors: Recent advances and perspectives. *Chem. Commun.* **2017**, *53*, 13296–13311. [CrossRef]
29. Hu, X.-Y.; Gao, L.; Mosel, S.; Ehlers, M.; Zellermann, E.; Jiang, H.; Knauer, S.K.; Wang, L.; Scmuck, C. From supramolecular vesicles to micelles: Controllable construction of tumor-targeting nanocarriers based on host–guest interaction between a pillar[5]arene-based prodrug and a RGD-sulfonate guest. *Small* **2018**, *14*, 1803952. [CrossRef]
30. Xia, D.; Wang, L.; Lv, X.; Chao, J.; Wei, X.; Wang, P. Dual-responsive [2]pseudorotaxane on the basis of a pH-sensitive pillar[5]arene and its application in the fabrication of metallosupramolecular polypseudorotaxane. *Macromolecules* **2018**, *51*, 2716–2722. [CrossRef]
31. Pearce, N.; Reynolds, K.E.A.; Kayal, S.; Sun, X.Z.; Davies, E.S.; Malagreca, F.; Schürmann, C.J.; Ito, S.; Yamano, A.; Argent, S.P.; et al. Selective photoinduced charge separation in perylenediimide-pillar[5]arene rotaxanes. *Nat. Commun.* **2022**, *13*, 415. [CrossRef]
32. Nazarova, A.A.; Yakimova, L.S.; Padnya, P.L.; Evtugyn, V.G.; Osin, Y.N.; Cragg, P.J.; Stoikov, I.I. Monosubstituted pillar[5]arene functionalized with (amino)phosphonate fragments are "smart" building blocks for constructing nanosized structures with some s-and p-metal cations in the organic phase. *New J. Chem.* **2019**, *43*, 14450–14458. [CrossRef]
33. Nazarova, A.; Khannanov, A.; Boldyrev, A.; Yakimova, L.; Stoikov, I. Self-assembling systems based on pillar[5]arenes and surfactants for encapsulation of diagnostic dye dapi. *Int. J. Mol. Sci.* **2021**, *22*, 6038. [CrossRef]
34. Zhang, H.; Liu, Z.; Zhao, Y. Pillararene-based self-assembled amphiphiles. *Chem. Soc. Rev.* **2018**, *47*, 5491–5528. [CrossRef]
35. Xu, L.; Wang, Z.; Wang, R.; Wang, L.; He, X.; Jiang, H.; Tang, H.; Cao, D.; Tang, Z. A conjugated polymeric supramolecular network with aggregation-induced emission enhancement: An efficient light-harvesting system with an ultrahigh antenna effect. *Angew. Chem.* **2020**, *59*, 9908–9913. [CrossRef]
36. Li, H.; Yang, Y.; Liang, T.; Wen, H.; Tian, W. Pillararene-based supramolecular polymers. *Chem. Commun.* **2019**, *55*, 271–285. [CrossRef]

37. Wang, P.; Liang, B.; Xia, D. A linear AIE supramolecular polymer based on a salicylaldehyde azine-containing pillararene and its reversible cross-linking by Cu^{II} and cyanide. *Inorg. Chem.* **2019**, *58*, 2252–2256. [CrossRef]
38. Ho, F.-C.; Huang, Y.-J.; Weng, C.-C.; Wu, C.-H.; Li, Y.-K.; Wu, J.I.; Lin, H.-C. Efficient FRET approaches toward copper(II) and cyanide detections via host-guest interactions of photo-switchable [2]pseudo-rotaxane polymers containing naphthalimide and merocyanine moieties. *ACS Appl. Mater. Interfaces* **2020**, *12*, 53257–53273. [CrossRef]
39. Lu, F.; Chen, Y.; Fu, B.; Chen, S.; Wang, L. Multistimuli responsive supramolecular polymer networks via host-guest complexation of pillararene-containing polymers and sulfonium salts. *Chin. Chem. Lett.* **2022**, *33*, 5111–5115. [CrossRef]
40. Strutt, N.L.; Zhang, H.; Giesener, M.A.; Lei, J.; Stoddart, J.F. A self-complexing and self-assembling pillar[5]arene. *Chem. Commun.* **2012**, *48*, 1647–1649. [CrossRef]
41. Zhu, H.; Li, Q.; Khalil-Cruz, L.E.; Khashab, N.M.; Yu, G.; Huang, F. Pillararene-based supramolecular systems for theranostics and bioapplications. *Sci. China Chem.* **2021**, *64*, 688–700. [CrossRef]
42. Liu, P.; Li, Z.; Shi, B.; Liu, J.; Zhu, H.; Huang, F. Formation of linear side-chain polypseudorotaxane with supramolecular polymer backbone through neutral halogen bonds and pillar[5]arene-based host–guest interactions. *Chem. Eur. J.* **2018**, *24*, 4264–4267. [CrossRef] [PubMed]
43. Dal Pizzol, C.; Filippin-Monteiro, F.B.; Restrepo, J.A.S.; Pittella, F.; Silva, A.H.; de Souza, P.A.; de Campos, A.M.; Creczynski-Pasa, T.B. Influence of surfactant and lipid type on the physicochemical properties and biocompatibility of solid lipid nanoparticles. *Int. J. Environ. Res. Public Health* **2014**, *11*, 8581–8596. [CrossRef] [PubMed]
44. Gordillo-Galeano, A.; Mora-Huertas, C.E. Solid lipid nanoparticles and nanostructured lipid carriers: A review emphasizing on particle structure and drug release. *Eur. J. Pharm. Biopharm.* **2018**, *133*, 285–308. [CrossRef]
45. Ledinski, M.; Marić, I.; Štefanić, P.P.; Ladan, I.; Mihalić, K.C.; Jurkin, T.; Gotić, M.; Urlić, I. Synthesis and in vitro characterization of ascorbyl palmitate-loaded solid lipid nanoparticles. *Polymers* **2022**, *14*, 1751. [CrossRef]
46. Cao, S.; Liu, X.; Li, X.; Lin, C.; Zhang, W.; Tan, C.H.; Liang, S.; Luo, B.; Xu, X.; Saw, P.E. Shape matters: Comprehensive analysis of star-shaped lipid nanoparticles. *Front. Pharmacol.* **2020**, *11*, 539. [CrossRef]
47. Galukhin, A.; Erokhin, A.; Imatdinov, I.; Osin, Y. Investigation of DNA binding abilities of solid lipid nanoparticles based on p-tert-butylthiacalix[4]arene platform. *RSC Adv.* **2015**, *5*, 33351–33355. [CrossRef]
48. Trotta, M.; Debernardi, F.; Caputo, O. Preparation of solid lipid nanoparticles by a solvent emulsification–diffusion technique. *Int. J. Pharm.* **2003**, *257*, 153–160. [CrossRef]

Article

New Approach to Non-Invasive Tumor Model Monitoring via Self-Assemble Iron Containing Protein Nanocompartments

Anna N. Gabashvili [1,2], Maria V. Efremova [3,4,5], Stepan S. Vodopyanov [1,6], Nelly S. Chmelyuk [1,7], Vera V. Oda [1], Viktoria A. Sarkisova [6,8], Maria K. Leonova [1], Alevtina S. Semkina [7,9], Anna V. Ivanova [1] and Maxim A. Abakumov [1,7,*]

1. Laboratory "Biomedical Nanomaterials", National University of Science and Technology "MISiS", Leninskiy Prospect, 4, 119049 Moscow, Russia; gabashvili.anna@gmail.com (A.N.G.); stepan.vodopianov@yandex.ru (S.S.V.); nellichmelyuk@yandex.ru (N.S.C.); oda-vera@mail.ru (V.V.O.); leonovamasha19@gmail.com (M.K.L.); super.fosforit@yandex.ru (A.V.I.)
2. Transplantation Immunology Laboratory, Biomedical Technology Department, National Medical Research Center for Hematology, Novy Zykovsky Drive, 4A, 125167 Moscow, Russia
3. Department of Chemistry and TUM School of Medicine, Technical University of Munich, Ismaninger Str. 22, 81675 Munich, Germany; m.efremova@tue.nl
4. Institute for Synthetic Biomedicine, Helmholtz Zentrum München GmbH, Ingolstaedter Landstr. 1, 85764 Neuherberg, Germany
5. Department of Applied Physics, Eindhoven University of Technology, Cascade P.O. Box 513, 5600 MB Eindhoven, The Netherlands
6. Biology Faculty, Lomonosov Moscow State University, Leninskiy Gory, 119234 Moscow, Russia; alice-lyddell@yandex.ru
7. Department of Medical Nanobiotechnology, Pirogov Russian National Research Medical University, Ostrovityanova St., 1, 117997 Moscow, Russia; alevtina.semkina@gmail.com
8. Cell Proliferation Laboratory, Engelhardt Institute of Molecular Biology, Vavilova Street, 32, 119991 Moscow, Russia
9. Department of Basic and Applied Neurobiology, Serbsky National Medical Research Center for Psychiatry and Narcology, Kropotkinskiy Per. 23, 119991 Moscow, Russia
* Correspondence: abakumov1988@gmail.com

Abstract: According to the World Health Organization, breast cancer is the most common oncological disease worldwide. There are multiple animal models for different types of breast carcinoma, allowing the research of tumor growth, metastasis, and angiogenesis. When studying these processes, it is crucial to visualize cancer cells for a prolonged time via a non-invasive method, for example, magnetic resonance imaging (MRI). In this study, we establish a new genetically encoded material based on *Quasibacillus thermotolerans* (*Q.thermotolerans*, *Qt*) encapsulin, stably expressed in mouse 4T1 breast carcinoma cells. The label consists of a protein shell containing an enzyme called ferroxidase. When adding Fe^{2+}, a ferroxidase oxidizes Fe^{2+} to Fe^{3+}, followed by iron oxide nanoparticles formation. Additionally, genes encoding mZip14 metal transporter, enhancing the iron transport, were inserted into the cells via lentiviral transduction. The expression of transgenic sequences does not affect cell viability, and the presence of magnetic nanoparticles formed inside encapsulins results in an increase in T2 relaxivity.

Keywords: encapsulins; magnetic resonance imaging; fluorescence; cell tracking

1. Introduction

Usually, tumor cells' in vivo visualization is carried out using optical methods, as well as MRI and positron emission tomography (PET) methods. For example, it is possible to perform two-color fluorescent imaging using tumor cells expressing red fluorescence protein (RFP) transplanted into transgenic mice expressing green fluorescence protein (GFP) [1]. In addition, there are two-color fluorescent cells with RFP expressed in the cytoplasm and GFP expressed in the nucleus [2]. Another example of fluorescent labels is quantum dots (QDs), which are inorganic fluorescent nanoparticles with great optical properties such as high quantum yield, high molar extinction coefficients, high resistance to photo bleaching and chemical decomposition, and restricted emission spectra, compared to organic dyes [3]. A common disadvantage of fluorescent labels, both exogenous labels and genetically encoded proteins, is a small penetration depth (about 2 mm), which makes their use only possible for surface tissues in small animals, for example, for subcutaneous tumor imaging in mice. In addition, a significant disadvantage of exogenous fluorescent labels is the decrease in the intensity of the fluorescent signal associated with cell proliferation.

Superparamagnetic iron oxide (SPIO) nanoparticles [4,5] are often considered for MRI tumor cells tracking. One of the drawbacks of SPIO nanoparticles is the negative contrast they produce that is often difficult to interpret. An alternative is to use gadolinium-based agents that give positive contrast on T2-weighted MRI scans. For example, in [6], rhodamine-conjugated gadolinium nanoparticles were utilized to label and track breast carcinoma cells in vivo. Cells labeled by such nanoparticles can be visualized using both MRI and optical imaging; however, the same problem as when using fluorescent labels remains—the intensity of the MR signal decreases over time due to cell proliferation. In addition, nanoparticles can be ejected by tumor cells and internalized by phagocytic cells in the tumor microenvironment, which degrades imaging accuracy.

Finally, more than a decade ago, PET was introduced to visualize malignant cells using the labels with various half-lives such as ^{18}F-FDG or ^{64}Cu-PTSM [7]. Their main disadvantages are toxicity and short half-life, which do not allow the long-term monitoring of tumor cells in vivo.

In light of the above, developing a stable and non-toxic label that allows the non-invasive tracking of tumor cells remains highly relevant. Encapsulins, protein nanocompartments with a high molecular weight, homologous in structure to viral capsids (icosahedral shells consisting of protomer proteins), were first discovered in 1994 in *Brevibacterium linens* [8]. Later, it was found that encapsulins occur in many other bacterial strains and archaea [9–11]. *Q. thermotolerans* encapsulin can accumulate more than 30,000 iron atoms inside its shell, which is an order of magnitude more than the amount of iron accumulated in ferritins. The diameter of the *Q.thermotolerans* encapsulin shell is large (42 nm) and consists of 240 protomers (32.2 kDa each). The accumulation of iron inside the encapsulin shell results from a catalytic activity of IMEF (Iron-Mineralizing Encapsulin-Associated Firmicute) cargo protein. It is assumed that, in wild-type *Q. thermotolerans*, encapsulins protect cells against oxidative stress by depositing iron inside the nanocompartments, thereby reducing the amount of free iron in the cytoplasm [12]. In this paper, we present *Q. thermotolerans* encapsulin-based genetic label, which is stably expressed in 4T1 mouse carcinoma cells.

To obtain a 4T1 cell line with a stable expression of *Q. thermotolerans* encapsulins genes (4T1-Qt), lentiviral transduction was performed using two viral vectors that carry genes encoding encapsulin shell and cargo protein (QtEnc^{FLAG}-QtIMEF), as well as genes encoding the divalent metal transporter (mZip14). We used ferrous ammonium sulfate (FAS) as a source of Fe^{2+}. The entire system works as follows: Fe^{2+} ions from FAS are transported into cells via mZip14 and enter the encapsulin nanocompartments. There, under the action of IMEF, Fe^{2+} is oxidized, resulting in iron oxide nanoparticles, which further allow 4T1-Qt cell tracking by MRI.

2. Materials and Methods

2.1. Cell Line

4T1 mammary mouse carcinoma cells were cultured in RPMI 1640 media (Gibco, Waltham, MA, USA) supplemented with antibiotics (100 U/mL penicillin, 100 mg/mL streptomycin, Gibco), GlutaMax Supplement (2 mM, Gibco), and 10% fetal bovine serum (HyClone, Cytiva, Washington, DC, USA). The cells were cultured under standard conditions (37 °C and 5% CO_2) in T-75 cultural flasks (Corning, New York, NY, USA) and used between 5 and 13 passages. Upon reaching high confluence, the cells were subcultured at a ratio of 1:4–1:6 following the standard trypsinization method.

2.2. Lentivirus Production and Lentiviral Transduction of Cells

HEK293T packaging cells were seeded in 6-well tissue culture plates (3.0×10^6 cells per plate) and cultured in complete DMEM medium (Gibco) at 37 °C and 5% CO_2. After 24 h of cultivation, lentiviral packaging plasmids (pRSV-Rev, pMDLg/pRRE, pCMV-VSV-G), plasmid-carrying encapsulin genes (pLCMV QtEnc^{FLAG}-QtIMEF), and plasmid encoding genes of iron transporter (pLCMV mZip14) were added to the cells in Opti-MEM medium. Individual lentiviral vector was constructed for QtEnc^{FLAG}-QtIMEF and mZip14. Then, 24 h after transfection, the medium was aspirated and replaced with DMEM supplemented with 2–5% FBS and antibiotics. Viruses were harvested 48 and 72 h post-transfection, loaded onto a 20 mL syringe, and filtered through a 0.45 µm syringe filter (Merck). Transduction 4T1 carcinoma cells with the lentiviral vectors was performed according to standard protocol in DMEM growth medium supplemented with 10% heat-inactivated FBS and polybrene (8 µg/mL, Sigma-Aldrich, Darmstadt, Germany). Lentiviruses were added to give a MOI of 4 for each virus; 48 h after transduction, the selection was started using puromycin (Thermo Fisher Scientific, Waltham, MA, USA) at a concentration of 5 µg/mL, and the medium with puromycin was changed to a new one every other day.

2.3. Reverse Transcription Polymerase Chain Reaction (RT-PCR)

The cells obtained after lentiviral infection were analyzed by reverse transcription polymerase chain reaction. Total RNA was extracted by Extract RNA reagent (Evrogen, Moscow, Russia) according to the manufacturer's protocol. RNA concentrations and quality were assessed by spectrophotometry. Then, cDNA was synthesized with Invitrogen SuperScript III Reverse Transcriptase (Thermo Fisher, Waltham, MA, USA) and oligo-DT and random primers. The cDNA and negative control (RNA with no reverse transcriptase added) were used for classical PCR with Taq polymerase (Fermentas, Waltham, MA, USA). The products of PCR were separated in a 1% agarose gel electrophoresis. The amplified fragments were identified by their length.

2.4. Western Blot Analysis

Western blot analysis was carried out as described earlier [13]. Briefly, 4T1-Qt cells were lysed using RIPA buffer, and the resulting lysate was precipitated (15 min, $14,000 \times g$). Sample buffer 5× was added to different amounts of cell lysate, heated at 95 °C, then cooled on ice. Samples were loaded onto a gel and electrophoresed (80 V for 25 min and 100 V for 1.5 h); then, the gel was transferred into a transfer buffer. The nitrocellulose membrane was activated and placed over the gel. The transfer was carried out in a chamber filled with transfer buffer for 1 h at 100 V. Then, the membrane was washed three times to remove transfer buffer residues in PBST. To prevent nonspecific binding, the membrane was incubated in a PBST solution with 5% non-fat milk for 2 h and washed again. The membrane was incubated with anti-DYKDDDDK Tag antibodies (1:1000, BioLegend, San Diego, CA, USA) for 2 h, followed by washing three times. After that, alkaline horseradish peroxidase conjugated secondary antibodies (1:1000, goat anti-mouse IgG, Santa Cruz Biotechnology,) were added. Clarity Max Western ECL Substrate kit (BioRad, Hercules, CA, USA) was used to reveal the result. The results were registered with the ChemidocMP Imaging system (BioRad, Hercules, CA, USA).

2.5. Immunofluorescence Staining

For direct immunofluorescence analysis, 4T1 and 4T1-Qt cells were seeded on a 35 mm µ-Dish with a polymer coverslip bottom for high-end microscopy (Ibidi, Grafelfing, Germany) in the amount of 5×10^4 cells/dish, cultured for 24 h in standard conditions, fixed by 4% formaldehyde in $1 \times$ PBS (Sigma-Aldrich, Darmstadt, Germany), and stained by Monoclonal Antibody to DYKDDDDK Tag (L5), Alexa Fluor 647 (1:250, BioLegend, San

Diego, CA, USA), according to the manufacturer's instructions. Nuclei were counterstained with DAPI (1:500, Sigma Aldrich, Darmstadt, Germany).

2.6. Laser Scanning Confocal Microscopy

Fluorescence confocal micrographs were captured with the Nikon Eclipse Ti2 microscope (Minato, Tokyo, Japan) equipped with ThorLabs laser (Newton, NJ, USA) and scanning systems, Nikon Apo 25×/1.10 water immersion objective lens (Minato, Tokyo, Japan), Nikon Plan Apo 10×/0.45 objective lens (Minato, Tokyo, Japan), and 405 and 642 lasers. Scanning was performed using the ThorImageLS software (Newton, NJ, USA), ImageJ2 FiJi was used to process the images.

2.7. MTS-Assay

Cytotoxicity FAS for 4T1 and 4T1-Qt cells were performed via CellTiter 96 AQueous One Solution Cell Proliferation Assay (Promega, Madison, WI, USA) according to manufacturer's protocol. 4T1 and 4T1-Qt cells were seeded at 8×10^3 cells/well in a 96-well culture plate in 100 µL of the medium per well. After 24 h of incubation, FAS (Sigma-Aldrich, Darmstadt, Germany) at different concentrations (4 mM, 2 mM, 1 mM, 0.5 mM, 0.25 mM, and 0.13 mM) was added to the cells. After 24 h incubation, the cells were washed with PBS, and a fresh growth medium with MTS reagent was added in each well. Non-treated cells by FAS were used as a positive control. The cells were incubated with MTS reagent for 4 h at 37 °C and 5% CO_2 in the humid atmosphere. The assay was conducted in three replicates. Optical density was measured using a Multiscan GO plate reader (Thermo Scientific, Waltham, MA, USA), λ = 490 nm.

Cell viability was calculated as:

$$\text{Cell viability (\%)} = (A_s - A_b)/(A_c - A_b) \times 100 \qquad (1)$$

where A_s—mean optical density in sample wells, A_b—mean optical density in blank wells, A_c—mean optical density in positive control wells.

2.8. Magnetic-Activated Cell Sorting (MACS)

The magnetic sorting of 4T1 and 4T1-Qt cells after 24 h of incubation with 2 mM FAS was performed using a magnetic separation kit (Miltenyi Biotech, Bergisch Gladbach, North Rhine-Westphalia, Germany). The cells were thoroughly washed from FAS with PBS and then removed from the plastic by trypsinization, precipitated (500 g, 5 min), and resuspended in 1.5 mL of PBS containing 2% FBS, and the number of cells was counted using an automatic cell counter. MS columns were placed in the magnetic field of an OctoMACS Separator and equilibrated with 0.5 mL PBS per column (Gibco, Waltham, MA, USA). The cell suspension was applied to magnetic columns, and free-passing cells were collected in a 15 mL tube. The columns were washed 3 times with 1 mL of PBS containing 2 % FBS, removed from the separator's magnetic field, and placed in a new 15 mL tube. Then, 1 mL of PBS with 2% FBS was added to the column, and the cells retained in the column were eluted using a plunger. Next, the cells were pelleted (500 g, 5 min), resuspended in a complete growth medium, and counted. The isolated cells were further cultured under standard conditions.

2.9. Prussian Blue Staining

4T1 and 4T1-Qt cells were seeded at 2×10^5 cells/dish in 35 mm Petri dishes and cultured for 24 h. FAS (4 mM, 2 mM, 1 mM, 0.5 mM, 0.25 mM, and 0.13 mM) was added to the cells, and they were incubated with FAS for 24 h. Afterward, the cells were thoroughly washed with PBS, fixed by 4% formaldehyde in PBS, and then stained using Iron Stain Kit (Sigma-Aldrich), according to the manufacturer's instructions. Images of 4T1 and 4T1-Qt cells were taken with an inverted microscope Primo Vert (Zeiss, Oberkochen, Germany).

2.10. Transmission Electron Microscopy (TEM)

4T1-Qt cells were seeded in µ-Slide 8 Well (Ibidi, Grafelfing, Germany) at a concentration of 4×10^4 cells/well and cultured in a growth medium (composition described above) supplemented with 2 mM FAS for 24 h. Afterward, the cells were thoroughly washed with PBS and fixed with 2% paraformaldehyde and 2.5% glutaraldehyde (Sigma-Aldrich, Darmstadt, Germany) in PBS, pH 7.4. Then, the cells were postfixed with 1% osmium tetroxide solution and dehydrated in ethanol of increasing (50%, 70%, 80%, and 95%) concentration. Finally, the cell sample was embedded in an Epoxy resin using an Epoxy embedding medium kit (Sigma Aldrich) according to the manufacturer's protocol. Ultrathin (70 nm) cell sections were obtained with an EM UC6 ultramicrotome (Leica, Wetzlar, Germany). Imaging was performed with a JEOL JEM 1400 electron microscope (Akishima, Tokyo, Japan).

2.11. Measurement of Intracellular Iron Content by Atomic Emission Spectroscopy (AES)

4T1-Q and 4T1 (control) cells were cultured in 6-well plates (5×10^5 cells/well). To investigate the iron accumulation in cells, the growth medium was supplemented with 2 mM, 1 mM, 0.5 mM, and 0.25 mM FAS for 24 h. Then, the cells were thoroughly washed with PBS, detached with 0.25% trypsin solution, precipitated by centrifugation, and counted. For each cell line, 1.6×10^6 cells were dissolved in 70 µL of concentrated nitric acid for 2 h at 60 °C. Iron concentration was determined using an Agilent 4200 MP-AES atomic emission spectrometer (Santa Clara, CA, USA).

2.12. MRI

For T2 relaxometry measurement, 4T1 and 4T1-Qt cells were incubated with 2 mM and 1 mM FAS for 24 h. Afterward, cells were washed 3 times with DPBS, detached with TripLE, and centrifuged at $500\times g$ for 5 min. The pellets (5×10^6 cells each) were resuspended in 200 µL DPBS and transferred to 500 µL PCR tubes. Cells were then spun down at $500\times g$ for 2 min and used for MRI. MRI images were acquired on ClinScan 7T system (Bruker Biospin, Billerica, MA, USA) in Spin Echo sequence with the following parameters: TR = 10,000, slice thickness 1.2 mm, FoV 84×120, base resolution 448×640, TE 8, 16, 24, ... , 256.

2.13. Animals and Tumor Model

All manipulations with experimental animals were approved by the local Ethical Committee of the Pirogov Russian National Research Medical University. Six- to eight-week-old female BALB/c mice were purchased from Andreevka Animal Center (Andreevka, Russia). Tumors were induced via subcutaneous injection of 1×10^6 cells (4T1 and 4T1-Qt. Tumor size and animal weight were monitored twice a week. The tumor volume (V) was calculated as V = a2/2 $\times$ b, where a is the smaller of the two orthogonal sizes (a and b), measured by caliper. Animals were euthanized when the tumors reached 500 mm^3 or body weight loss exceeded 10% with a lethal dose of isoflurane. For immunohistochemical analysis (12 days after 4T1 and 4T1-Qt cells implantation), isolated tumors were postfixed for 24 h in 4% paraformaldehyde solution in PBS, and 40–60 µm tumor slices were obtained using an HM 650v vibratome (Microm GmbH, Ettlingen, Germany). The tumor slices were stained by mono-clonal Antibody to DYKDDDDK Tag (L5) and Alexa Fluor 647 (1:100, BioLegend, San Diego, CA, USA), and nuclei were counterstained with DAPI (1:400, Sigma Aldrich, Darmstadt, Germany).

3. Results

3.1. Expression of Bacterial Genes in Eukaryotic Cells

We used two lentiviruses to obtain 4T1 cells with stable expression of *Q.thermotolerans* encapsulin-encoding genes: QtEnc^{FLAG}-QtIMEF-encoding encapsulin protomer and cargo protein, and mZip14-encoding iron transporter. The expression of encapsulin genes was validated via RT-PCR (Figure 1a). The protein expression of QtEncFLAG protomer was confirmed using Western blot analysis against FLAG-tag on the protomer protein. Figure 1b demonstrates a band at approximately 35 kDa. The signal level in Western blot increased with the amount of cell lysate loaded into the gel.

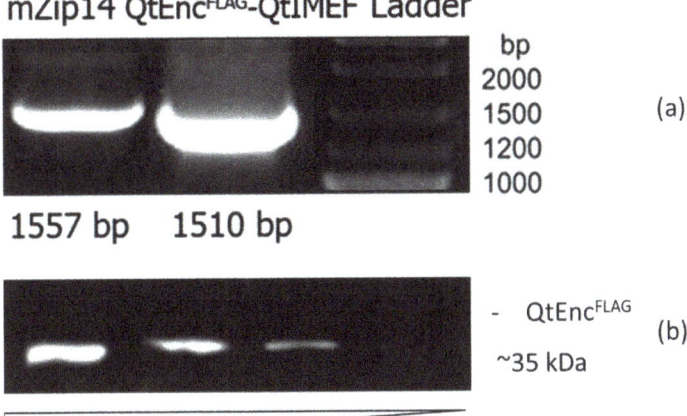

Figure 1. (a) RT-PCR analysis of 4T1-Qt cells; (b) Western blot analysis against FLAG-tag on *Q.thermotolerans* encapsulin protomer proteins in 4T1-Qt cells.

We also performed direct immunofluorescence staining of 4T1-Qt cells using primary labeled monoclonal anti-DYKDDDDK Tag antibodies that bind to the FLAG sequence coexpressed with QtEncFLAG, i.e., on encapsulin protomer proteins. Wild-type 4T1 cells were used as a control. The high-intensity red fluorescent signal from Alexa Fluor 647 label in 4T1-Qt cells (Figure 2a) is visible in the micrograph. In control 4T1 cells stained with anti-DYKDDDDK Tag antibodies, a red fluorescent signal was not detected (Figure 2b).

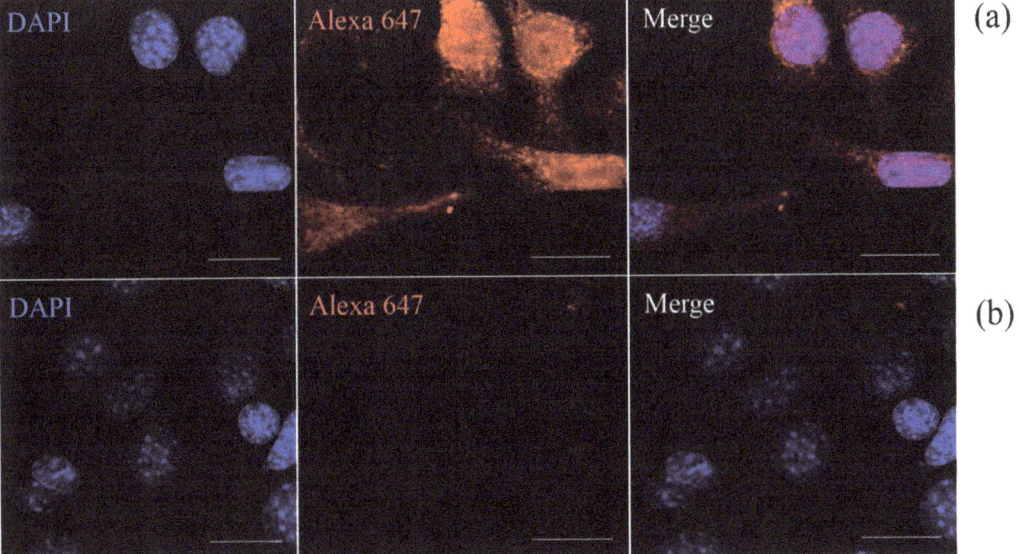

Figure 2. 4T1-Qt (a) and 4T1 (b) cells stained with Alexa Fluor 647 anti-DYKDDDDK Tag antibody (red fluorescence). Nuclei were counterstained with DAPI (blue fluorescence). Laser scanning confocal microscopy, scale bar 20 μm.

Confocal microscopy resolution is not enough to detect individual 42 nm diameter encapsulins, so this imaging method is relatively coarse. In addition, we used the TEM to visualize the iron oxide nanoparticles formed inside the encapsulins, allowing us to estimate the efficiency of iron biomineralization and storage.

3.2. Iron Biomineralization Inside the Encapsulins

We hypothesized that when FAS is added to wild-type 4T1 cells, excess iron may have a toxic effect due to ferritin overload and accumulation of free iron in the cytoplasm. In contrast, in genetically modified 4T1-Qt, part of the iron ions will be sequestered into encapsulins, which may reduce toxicity. From the data presented in Figure 3, it can be seen that our hypothesis was confirmed in the 0.25–4 mM FAS concentration range.

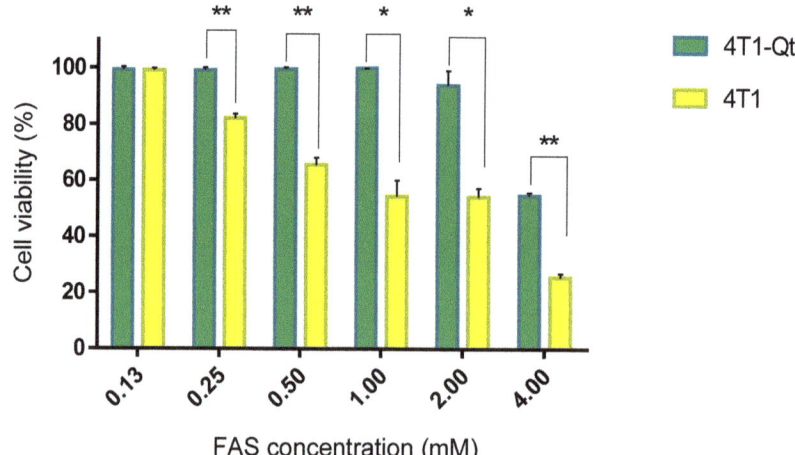

Figure 3. Cytotoxicity assay of various FAS concentrations in the growth medium for 4T1 and 4T1-Qt cells. The data are shown as the mean + S.D. of three independent experiments. p values were calculated using a one-tailed t-test, assuming unequal variances (** indicate p-value < 0.001, * indicate p-value < 0.05).

Knowing the toxicity of FAS for 4T1 and 4T1-Qt cells, we decided to qualitatively assess iron accumulation in cells via Prussian blue staining using the same concentrations of FAS as in the cytotoxicity assay. In a series of micrographs, the light blue staining of iron deposits in 4T1-Qt cells after 24 h incubation with FAS in the 0.25–2 mM concentration range is detected (Figure 4a, stained areas are indicated by black arrows). At 4 mM FAS, deviations in cell morphology and a decrease in the number of cells per well are seen, consistent with the toxicity assay data (only 54% of 4T1-Qt cells remained viable after 24 h incubation with 4 mM FAS). In contrast, no staining of control 4T1 cells was detectable after 24 h incubation with FAS at the same concentrations (Figure 4b). We can also observe the toxic effect of FAS on 4T1 cells at 0.25–4 mM concentration, in the form of a significant decrease in the density of the cell monolayer, the presence of cell debris, rounded cells, and cells with atypical morphology.

AES was used to quantify the iron accumulation in 4T1 and 4T1-Qt cells after the incubation with 0.25 mM, 0.5 mM, 1 mM, and 2 mM FAS (Figure 5).

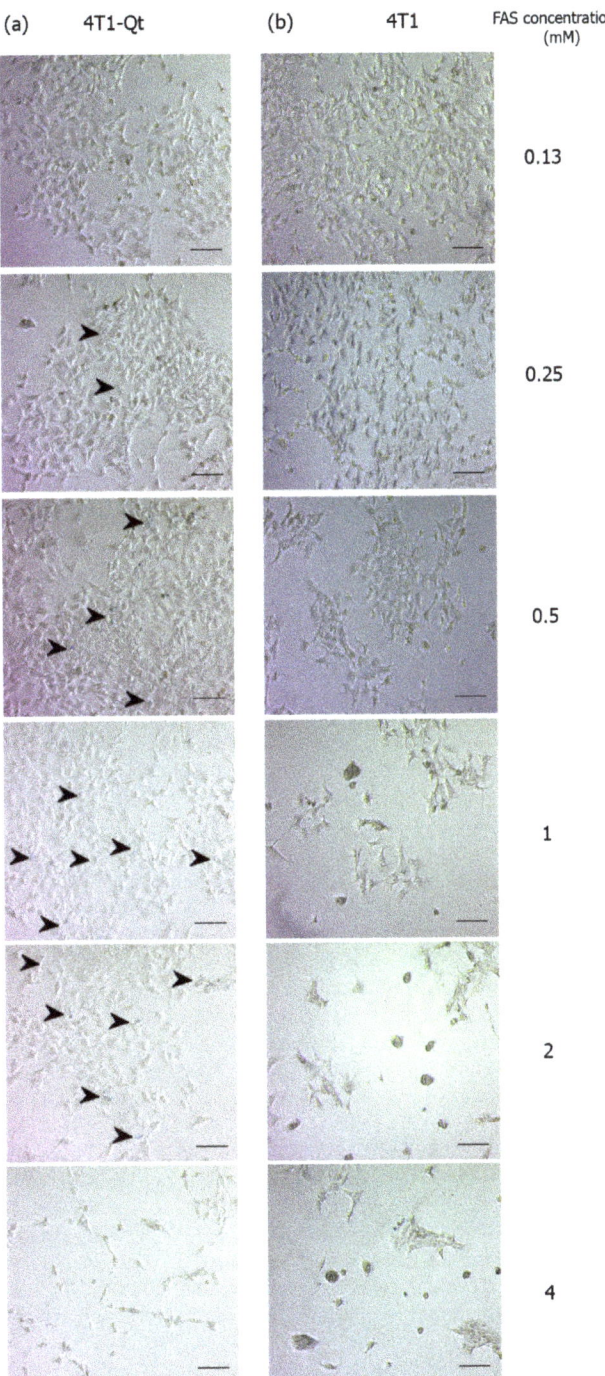

Figure 4. Prussian blue staining of 4T1-Qt (**a**) and 4T1 (**b**) cells after 24 h incubation with 0.13–4 mM FAS. Bright-field microscopy, scale bar 50 μm. Black arrows indicate iron deposits in 4T1-Qt cells.

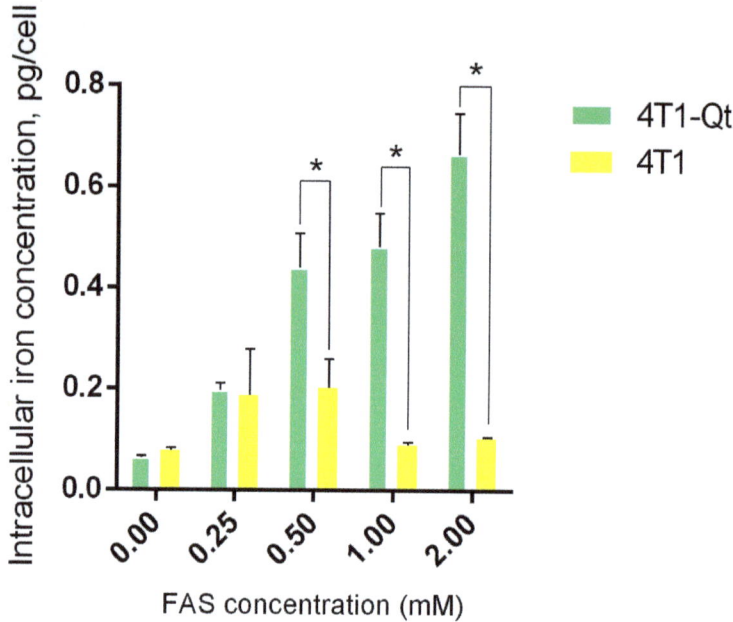

Figure 5. Cellular iron content in 4T1 and 4T1-Qt cells quantified by AES spectrometry. The data are shown as the mean + S.D of three independent experiments, p values were calculated using a one-tailed t-test, assuming unequal variances (* indicate p-value < 0.05).

It was found that iron accumulates in 4T1-Qt cells in a dose-dependent manner, and intracellular iron concentration increases proportionally to the amount of FAS added to the growth medium. The iron accumulation in 4T1-Qt cells is significantly higher than the same value in wild-type 4T1 cells.

Finally, we performed TEM imaging of encapsulins in 4T1-Qt cells after the incubation with 2 mM FAS for 24 h. We expected to find encapsulin-derived iron oxide nanoparticles within each cell. However, during TEM micrographs analysis, we noticed that the number of nanoparticles per cell is quite variable. For example, some of the cells contained electron-dense particles (Figure 6a); however, there were also cells without nanoparticles inside the encapsulin shells (Figure 6b). Both vectors used for transduction contained the same antibiotic resistance genes (puromycin). Thus, there were not only cells where both transgenic sequences survived during the selection, but also cells without mZip14. For MRI studies, such heterogeneity is unacceptable; therefore, it was necessary to obtain a more homogeneous cell line in terms of nanoparticle content.

3.3. Genetically Encoded Labels for MRI

To select the cells with high numbers of iron oxide-loaded encapsulins, MACS of 4T1-Qt cells preincubated with 2 mM FAS was performed. The magnetic sorting efficiency was about 2% (in other words, 2% of target cells were retained within the magnetic column). The sorted 4T1-Qt cells were then expanded and used to determine the T2 relaxation time in vitro. The same was conducted for the sorted 4T1-Qt cells after 24 h of incubation with 1 mM and 2 mM FAS. Wild-type 4T1 cells were used as a control (Figure 7).

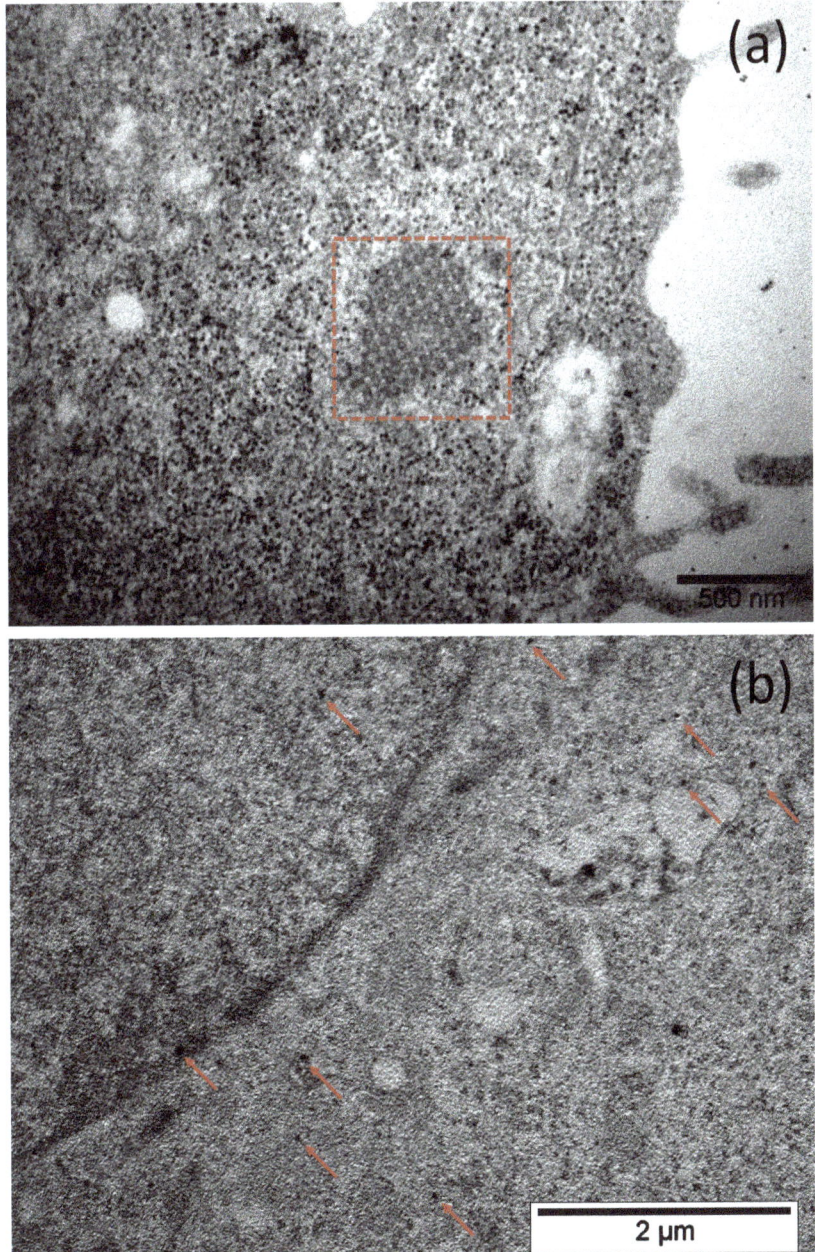

Figure 6. Bright-field TEM image of an ultrathin section of a 4T1-Qt cell, aggregated encapsulin shells in the cytoplasm are highlighted by a square (**a**), and red arrows indicate electron-dense nanoparticles in 4T1-Qt cell cytoplasm cell, scale bar 500 nm and 2 μm, respectively (**b**).

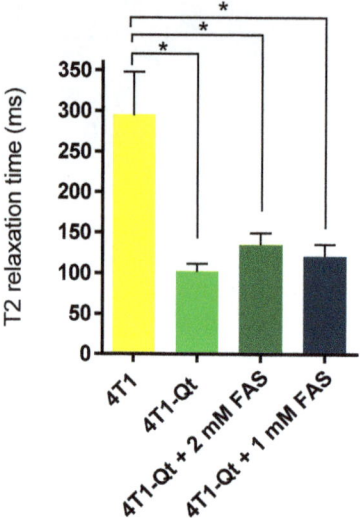

Figure 7. T2 relaxation time for 4T1 and 4T1-Qt cells. The data are shown as the mean + S.D. Statistical analysis was performed using an unpaired *t*-test (* corresponds to *p*-value < 0.05).

The relaxation time of wild-type 4T1 cells after their incubation with FAS was not measured. It is known that, in cells dying by ferroptosis, the intensity of the MR signal may increase [14,15], and this experiment was deemed inappropriate in our case (cf. FAS toxicity data presented above).

It is well-known that xenogeneic proteins expression can affect cell growth rate and their tumorigenic potential by suppression of the tumor growth via immune mechanisms [16]. For a comparative assessment of the growth dynamics of tumors obtained from 4T1 and 4T1-Qt cells, tumor volumes were measured on different days after subcutaneous tumor cells implantation. Transgenic and wild-type tumors transplanted into mice reached detectable size 3–5 days after implantation. Our experiments showed that the dynamics of 4T1 and 4T1-Qt tumor growth did not differ (Figure 8).

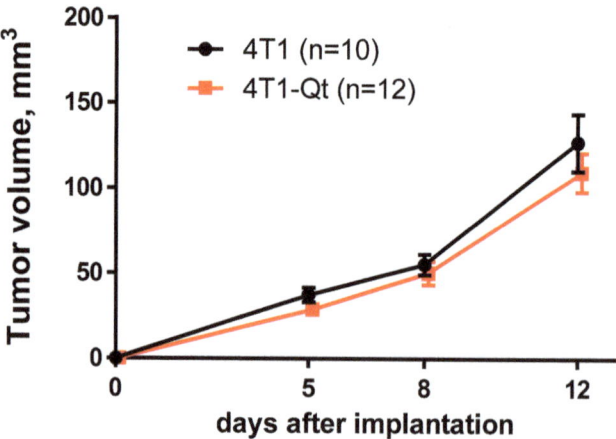

Figure 8. Tumor growth dynamics after subcutaneous implantation of wild-type 4T1 and 4T1-Qt cells. The data are shown as the mean + S.E.M.

This data indirectly confirm the absence of 4T1-Qt cells immunogenicity in immunocompetent mice.

We also provide direct immunofluorescence staining of 4T1- and 4T1-Qt-cell-induced tumor sections, using primary labeled antibodies to FLAG Tag sequence. Figure 9 shows confocal images of the tumor sections obtained 12 days after the injection of 4T1 and 4T1-Qt cells. Confocal microscopy images demonstrate a bright red fluorescent signal in 4T1-Qt cells induced tumor section. In control 4T1-cell-induced tumors stained with anti-FLAG Tag antibodies, a red fluorescent signal was not detected.

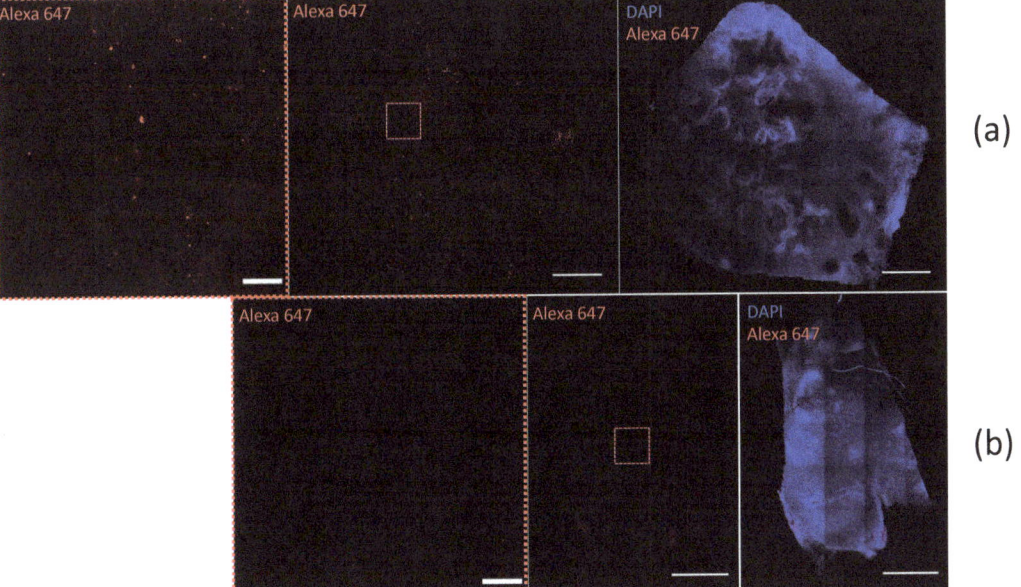

Figure 9. 4T1-Qt- (**a**) and 4T1- (**b**) cell-induced tumors sections stained with Alexa Fluor 647 anti-DYKDDDDK Tag antibody (red fluorescence). Nuclei were counterstained with DAPI (blue fluorescence). Laser scanning confocal microscopy, scale bar 100 and 1000 μm.

4. Discussion

The choice of the 4T1 cell line was based on the growth and metastasis of these cells in female BALB/c mice, closely mimicking stage IV human breast cancer. Tumors induced by subcutaneous injection of 4T1 cells in mice spontaneously metastasize to the lungs in almost 100% of cases, as well as to bones, lymph nodes, and, less often, to the brain [17,18]. We hope that the encapsulin-based label will allow us to study the growth and metastasis of 4T1-Qt tumors in vivo via MRI. Our previous study [13] has shown that the stable heterologous expression of *Myxococcus xanthus* encapsulin genes can be achieved in human mesenchymal stem cells (MSCs). The presence of transgenic sequences in cells did not reduce the rate of cell proliferation, and the T2 relaxation time of MSCs containing nanoparticles in encapsulins was significantly lower than the T2 relaxation time of intact MSCs. Another work [19] demonstrated that the encapsulin cargo system from *Q. thermotolerans* may be suitable for HepG2 hepatocellular carcinoma cells vizualisation. Thus, in the future, encapsulin-based genetic labels might be successfully used for MRI monitoring of eukaryotic cells.

Compared to our previous work, the label has been optimized; namely, two viral vectors instead of three are used to obtain a cell line stably expressing encapsulin genes. In addition, the efficiency of lentiviral transduction of malignant cells was higher compared to stem cells. Another significant improvement of this study is MACS, which is well-suited for tumor cells despite their low sorting efficiency. Due to the rapid proliferation of malignant cells, it is still possible to obtain a sufficiently large encapsulin-enriched cell fraction.

It is well-known that iron metabolism in tumor cells is often altered compared to non-malignant ones, which also applies to the 4T1 cell line. As in many carcinomas, 4T1 cells have an increased expression level of a type 1 transferrin receptor (TfR1) gene [20]. More importantly, these cells have a reduced expression of a gene encoding ferroportin (FPN) [21]—a transmembrane protein that transports iron ions from cells to the extracellular environment. As mentioned before, in *Q.thermotolerans*, encapsulins presumably perform a protective function by sequestering free cytoplasmic iron. Our data suggest that this protective mechanism can also be realized in eukaryotic cells with a heterologous expression of encapsulin genes.

AES data indicate that iron accumulation in 4T1-Qt cells is dose-dependent. The intracellular iron concentrations per cell for 4T1 and 4T1-Qt cells after the incubation with 2 mM FAS are almost an order of magnitude different (0.1 ± 0.05 pg/cell and 0.7 ± 0.2 pg/cell, respectively). For a 4T1-Qt line obtained in this study, the maximum concentration of FAS allowing us to reach the maximum concentration of intracellular iron while maintaining the cell viability is 2 mM. In wild-type 4T1 cells, iron is stored mostly in the protein ferritin. Each ferritin complex can accumulate about 4000 iron atoms, while in transgenic 4T1-Qt cells, iron storage also occurs inside the encapsulins, which can accumulate more than 30,000 iron atoms inside its shell. Thus, while intact cells die due to iron overload, 4T1-Qt cells remain viable because extra iron added in the cell culture medium is deposited inside encapsulins shells. That is why iron accumulation in 4T1-Qt cells is significantly higher than in wild-type 4T1 cells.

MRI data have shown lower T2 relaxation times for MACS-separated 4T1-Qt cells after 24 h incubation with FAS at 1 mM and 2 mM concentrations (120 ± 14 ms and 134 ± 15 ms, respectively) in comparison with 4T1 cells (294 ± 53 ms).

Finally, preliminary in vivo studies showed no significant differences in the dynamics of tumor growth, which indicates that the genetically encoded label does not affect cell proliferation in mouse breast carcinoma model, and the encapsulin encoding sequence is stably present in 4T1-Qt cells, even 12 days after implantation in mice.

5. Conclusions

In this work, we describe the 4T1 mouse carcinoma cell line stably expressing *Qt* encapsulin genes for the first time. The latter do not alter the viability and proliferation of 4T1-Qt cells; moreover, *Qt* encapsulins have a protective effect against high concentrations of iron ions, at the same time providing a dose-dependent iron accumulation in 4T1-Qt cells. Finally, in vitro MRI study showed a decrease in T2 relaxation time for magnetically sorted 4T1-Qt cells compared to wild-type 4T1 cells. We believe that this is an essential step towards the future in vivo MRI monitoring of malignant cells.

Author Contributions: Conceptualization, A.N.G. and M.A.A.; methodology, M.A.A. and A.N.G.; software, M.A.A., A.S.S. and N.S.C.; validation, M.V.E., A.N.G. and M.A.A.; formal analysis, A.N.G.; investigation, A.N.G., A.S.S., A.V.I., V.V.O., V.A.S., M.K.L. and S.S.V.; resources, V.A.S. and M.A.A.; data curation, A.N.G. and A.S.S.; writing—original draft preparation, A.N.G. and N.S.C.; writing—review and editing, M.V.E.; visualization, N.S.C. and S.S.V.; supervision, A.N.G.; project administration, A.N.G. and M.A.A.; funding acquisition, M.A.A. All authors have read and agreed to the published version of the manuscript.

Funding: This research was funded by RSF grant number 21-75-00096. M.V.E. gratefully acknowledges the support from the Humboldt Research Fellowship for Postdoctoral Researchers provided by the Alexander von Humboldt Foundation and the support from the Add-on Fellowship for Interdisciplinary Life Science provided by the Joachim Herz Foundation.

Institutional Review Board Statement: The animal study protocol was approved by the Institutional Ethics Committee of Pirogov Russian National Research Medical University # 20/2021 and #16/2021.

Data Availability Statement: All important data are included in the manuscript.

Acknowledgments: A.N.G. would like to thank Timur Nizamov for assistance in TEM images obtaining.

Conflicts of Interest: The authors declare no conflict of interest.

References

1. Hoffman, R.M. In Vivo Imaging with Fluorescent Proteins: The New Cell Biology. *Acta Histochem.* **2004**, *106*, 77–87. [CrossRef] [PubMed]
2. Jiang, P.; Yamauchi, K.; Yang, M.; Tsuji, K.; Xu, M.; Maitra, A.; Bouvet, M.; Hoffman, R.M. Tumor Cells Genetically Labeled with GFP in the Nucleus and RFP in the Cytoplasm for Imaging Cellular Dynamics. *Cell Cycle* **2006**, *5*, 1198–1201. [CrossRef] [PubMed]
3. Cai, W.; Chen, X. Preparation of Peptide-Conjugated Quantum Dots for Tumor Vasculature-Targeted Imaging. *Nat. Protoc.* **2008**, *3*, 89–96. [CrossRef] [PubMed]
4. Spira, D.; Bantleon, R.; Wolburg, H.; Schick, F.; Groezinger, G.; Wiskirchen, J.; Wiesinger, B. Labeling Human Melanoma Cells With SPIO: In Vitro Observations. *Mol. Imaging* **2016**, *15*, 1536012115624915. [CrossRef]
5. Wang, Z.; Cuschieri, A. Tumour Cell Labelling by Magnetic Nanoparticles with Determination of Intracellular Iron Content and Spatial Distribution of the Intracellular Iron. *Int. J. Mol. Sci.* **2013**, *14*, 9111–9125. [CrossRef]
6. Vuu, K.; Xie, J.; McDonald, M.A.; Bernardo, M.; Hunter, F.; Zhang, Y.; Li, K.; Bednarski, M.; Guccione, S. Gadolinium-Rhodamine Nanoparticles for Cell Labeling and Tracking via Magnetic Resonance and Optical Imaging. *Bioconjug. Chem.* **2005**, *16*, 995–999. [CrossRef]
7. Koike, C.; Watanabe, M.; Oku, N.; Tsukada, H.; Irimura, T.; Okada, S. Tumor Cells with Organ-Specific Metastatic Ability Show Distinctive Trafficking in Vivo: Analyses by Positron Emission Tomography and Bioimaging. *Cancer Res.* **1997**, *57*, 3612–3619.
8. Valdés-Stauber, N.; Scherer, S. Isolation and Characterization of Linocin M18, a Bacteriocin Produced by Brevibacterium Linens. *Appl. Environ. Microbiol.* **1994**, *60*, 3809–3814. [CrossRef]
9. Rosenkrands, I.; Rasmussen, P.B.; Carnio, M.; Jacobsen, S.; Theisen, M.; Andersen, P. Identification and Characterization of a 29-Kilodalton Protein from Mycobacterium Tuberculosis Culture Filtrate Recognized by Mouse Memory Effector Cells. *Infect. Immun.* **1998**, *66*, 2728–2735. [CrossRef]
10. Hicks, P.M.; Rinker, K.D.; Baker, J.R.; Kelly, R.M. Homomultimeric Protease in the Hyperthermophilic Bacterium Thermotoga Maritima Has Structural and Amino Acid Sequence Homology to Bacteriocins in Mesophilic Bacteria. *FEBS Lett.* **1998**, *440*, 393–398. [CrossRef]
11. Giessen, T.W.; Silver, P.A. Widespread Distribution of Encapsulin Nanocompartments Reveals Functional Diversity. *Nat. Microbiol.* **2017**, *2*, 17029. [CrossRef] [PubMed]
12. Giessen, T.W.; Orlando, B.J.; Verdegaal, A.A.; Chambers, M.G.; Gardener, J.; Bell, D.C.; Birrane, G.; Liao, M.; Silver, P.A. Large Protein Organelles Form a New Iron Sequestration System with High Storage Capacity. *Elife* **2019**, *8*, e46070. [CrossRef] [PubMed]
13. Gabashvili, A.N.; Vodopyanov, S.S.; Chmelyuk, N.S.; Sarkisova, V.A.; Fedotov, K.A.; Efremova, M.V.; Abakumov, M.A. Encapsulin Based Self-Assembling Iron-Containing Protein Nanoparticles for Stem Cells MRI Visualization. *Int. J. Mol. Sci.* **2021**, *22*, 12275. [CrossRef] [PubMed]
14. Luo, S.; Ma, D.; Wei, R.; Yao, W.; Pang, X.; Wang, Y.; Xu, X.; Wei, X.; Guo, Y.; Jiang, X.; et al. A Tumor Microenvironment Responsive Nanoplatform with Oxidative Stress Amplification for Effective MRI-Based Visual Tumor Ferroptosis. *Acta Biomater.* **2022**, *138*, 518–527. [CrossRef] [PubMed]
15. Yu, B.; Choi, B.; Li, W.; Kim, D.-H. Magnetic Field Boosted Ferroptosis-like Cell Death and Responsive MRI Using Hybrid Vesicles for Cancer Immunotherapy. *Nat. Commun.* **2020**, *11*, 3637. [CrossRef] [PubMed]
16. Vodopyanov, S.S.; Kunin, M.A.; Garanina, A.S.; Grinenko, N.F.; Vlasova, K.Y.; Mel'nikov, P.A.; Chekhonin, V.P.; Sukhinich, K.K.; Makarov, A.V.; Naumenko, V.A.; et al. Preparation and Testing of Cells Expressing Fluorescent Proteins for Intravital Imaging of Tumor Microenvironment. *Bull. Exp. Biol. Med.* **2019**, *167*, 123–130. [CrossRef]
17. Zhang, Y.; Zhang, N.; Hoffman, R.M.; Zhao, M. Surgically-Induced Multi-Organ Metastasis in an Orthotopic Syngeneic Imageable Model of 4T1 Murine Breast Cancer. *Anticancer Res.* **2015**, *35*, 4641–4646.
18. Guo, W.; Zhang, S.; Liu, S. Establishment of a Novel Orthotopic Model of Breast Cancer Metastasis to the Lung. *Oncol. Rep.* **2015**, *33*, 2992–2998. [CrossRef]
19. Efremova, M.V.; Bodea, S.-V.; Sigmund, F.; Semkina, A.; Westmeyer, G.G.; Abakumov, M.A. Genetically Encoded Self-Assembling Iron Oxide Nanoparticles as a Possible Platform for Cancer-Cell Tracking. *Pharmaceutics* **2021**, *13*, 397. [CrossRef]
20. Jiang, X.P.; Elliott, R.L.; Head, J.F. Manipulation of Iron Transporter Genes Results in the Suppression of Human and Mouse Mammary Adenocarcinomas. *Anticancer Res.* **2010**, *30*, 759–765.
21. Guo, W.; Zhang, S.; Chen, Y.; Zhang, D.; Yuan, L.; Cong, H.; Liu, S. An Important Role of the Hepcidin-Ferroportin Signaling in Affecting Tumor Growth and Metastasis. *Acta Biochim. Biophys. Sin.* **2015**, *47*, 703–715. [CrossRef] [PubMed]

Article

Heading toward Miniature Sensors: Electrical Conductance of Linearly Assembled Gold Nanorods

Marisa Hoffmann [1,2,3,†], Christine Alexandra Schedel [4,†], Martin Mayer [1], Christian Rossner [1,5], Marcus Scheele [4,*] and Andreas Fery [1,2,3,*]

1. Leibniz-Institut für Polymerforschung Dresden e.V., Institute of Physical Chemistry and Polymer Physics, Hohe Str. 6, 01069 Dresden, Germany
2. Physical Chemistry of Polymeric Materials, Technische Universität Dresden, Bergstr. 66, 01069 Dresden, Germany
3. Center for Advancing Electronics Dresden, Technische Universität Dresden, Helmholtzstr. 18, 01069 Dresden, Germany
4. Institute of Physical and Theoretical Chemistry, Eberhard Karls Universität Tübingen, Auf der Morgenstelle 18, 72076 Tübingen, Germany
5. Dresden Center for Intelligent Materials (DCIM), Technische Universität Dresden, 01069 Dresden, Germany
* Correspondence: marcus.scheele@uni-tuebingen.de (M.S.); fery@ipfdd.de (A.F.)
† These authors contributed equally to this work.

Citation: Hoffmann, M.; Schedel, C.A.; Mayer, M.; Rossner, C.; Scheele, M.; Fery, A. Heading toward Miniature Sensors: Electrical Conductance of Linearly Assembled Gold Nanorods. *Nanomaterials* 2023, *13*, 1466. https://doi.org/10.3390/nano13091466

Academic Editor: Pavel Padnya

Received: 28 March 2023
Revised: 17 April 2023
Accepted: 19 April 2023
Published: 25 April 2023

Copyright: © 2023 by the authors. Licensee MDPI, Basel, Switzerland. This article is an open access article distributed under the terms and conditions of the Creative Commons Attribution (CC BY) license (https://creativecommons.org/licenses/by/4.0/).

Abstract: Metal nanoparticles are increasingly used as key elements in the fabrication and processing of advanced electronic systems and devices. For future device integration, their charge transport properties are essential. This has been exploited, e.g., in the development of gold-nanoparticle-based conductive inks and chemiresistive sensors. Colloidal wires and metal nanoparticle lines can also be used as interconnection structures to build directional electrical circuits, e.g., for signal transduction. Our scalable bottom-up, template-assisted self-assembly creates gold-nanorod (AuNR) lines that feature comparably small widths, as well as good conductivity. However, the bottom-up approach poses the question about the consistency of charge transport properties between individual lines, as this approach leads to heterogeneities among those lines with regard to AuNR orientation, as well as line defects. Therefore, we test the conductance of the AuNR lines and identify requirements for a reliable performance. We reveal that multiple parallel AuNR lines (>11) are necessary to achieve predictable conductivity properties, defining the level of miniaturization possible in such a setup. With this system, even an active area of only 16 µm^2 shows a higher conductance (~10^{-5} S) than a monolayer of gold nanospheres with dithiolated-conjugated ligands and additionally features the advantage of anisotropic conductance.

Keywords: self-assembly; gold nanorods; anisotropic conductance

1. Introduction

Metal nanoparticles (NPs) are increasingly used as key elements in the fabrication and processing of advanced electronic systems and devices. At a comparably small size (e.g., >1.4 nm for gold [1]), an electronic band structure develops, and metal NPs become electrically conductive. In addition, gold NPs can serve as model systems in fundamental research [2] because of their precise shapes, chemical stability, ease of surface functionalization and processability.

In many applications, the conductivity of gold NP assemblies is crucial. It has been exploited, e.g., in the development of gold-NP-based conductive inks [3]. Assembled metal NPs and metal NP films can also be implemented into strain-sensitive [4] or resistive pressure-sensitive devices [5–7], which can be used, e.g., in healthcare [8]. Moreover, the fact that gold NPs possess a high surface-to-volume ratio proved them useful as sensing platforms to detect alcohols or neurotransmitters [9,10], solvent vapor [11–13], or electrochemical reactions [14] upon adsorption. Furthermore, arrays of parallel colloidal nanowires can be implemented to create surfaces with anisotropic conductance [15]. These could be used as interconnection structures to build electrical circuits, e.g., for signal transduction of directional mechanical events.

The step from single NPs to electronic components such as colloidal nanowires requires NP arrangement into tailored supracolloidal structures. The assembly of metal NPs into lines, as well as the optical properties of the resulting colloidal wires have been the subject of a plethora of investigations. Lines of plasmonic nanoparticles show optical effects, such as the coupling of plasmonic modes [16–19], and higher enhancement factors for surface-enhanced Raman spectroscopy (SERS) than single plasmonic NPs [20,21]. Whereas top-down methods can fabricate metal nanowires of arbitrary shape, large-scale fabrication is challenging, and lithographic methods are energy consuming and environmentally critical. In contrast, colloidal nanowire fabrication via self-assembly is scalable and has a reduced environmental footprint. Various bottom-up methods have been employed to assemble metal NPs into lines and stripes, including spin coating [20,22,23], dip coating [19,24], microfluidics [25] and capillary-assisted assembly [26]. The above-mentioned papers partially include testing the conductivity of the fabricated linearly assembled NPs. Studies on the conductivity mechanisms in three-dimensional networks of metal NPs [27] and in non-linear gold nanosphere chains [28] report that hopping dominates the charge transport at room temperature.

However, even for the fabrication of colloidal gold nanowires, top-down methods [17,26] are often used. Among the existing colloidal nanowires, there are only a few examples of linear gold nanorod (AuNR) assemblies. AuNR lines offer lower percolation thresholds for electrical conduction compared to colloidal nanowires composed of less anisotropic nanoparticles [29]. Therefore, they are well-suited for creating micron-sized surfaces with anisotropic conductivity. Most of the fabricated AuNR lines have widths in the (sub-)micrometer range, which is unfavorable for pushing the limit of device miniaturization. The printed stripes of poly [2-(3-thienyl)-ethyl-oxy-4-butylsulfonate]-functionalized AuNRs by Reiser et al. had low resistivities of 10^{-6} Ωm, but, on the other hand, widths of several hundred micrometers [3]. Rey et al. used a procedure of template-assisted capillary assembly of AuNRs by polydimethyl siloxane (PDMS) templates based on electron-beam lithography-made masters, which relies on lithographic processing steps. The resulting single AuNR lines had gaps of 5–7 nm between the AuNR tips and did not show any measurable conductance [26]. Despite the partial use of expensive methods, in none of these examples were the AuNR lines obtained with small dimensions while still maintaining acceptable conductivity values.

In contrast, our bottom-up, template-assisted, self-assembled AuNR lines feature comparably small widths [3,30] and better conductivities than other comparable assemblies, such as gold nanospheres with dithiolated-conjugated ligands [31]. Nevertheless, the bottom-up approach for preparing AuNR lines poses the question about the consistency of charge transport properties between individual lines, as this approach leads to heterogeneities among those lines with regard to AuNR orientation, as well as line defects. However, consistent charge transport properties are essential for future device integration. Therefore, the motivation of this work was to test the conductance of the AuNR lines and identify requirements for reliable performance. We revealed that multiple parallel AuNR lines (>11) are necessary to achieve predictable conductivity properties, defining the level of miniaturization possible in such a setup. Thereby, we set the foundation to use AuNR lines as resistance-based sensor wires or as anisotropically conducting surfaces in devices on the meso scale.

2. Materials and Methods

Materials. Cetyltrimethylammonium bromide (CTAB, 99%) was received from Merck chemicals. Cetyltrimethylammonium chloride (CTAC), $HAuCl_4 \cdot 3H_2O$ (99.9%), HBr (48% in water), silver nitrate ($AgNO_3$, 99.9999%), sodium borohydride ($NaBH_4$, 99%), and hydroquinone (99%) were purchased from Sigma-Aldrich. Photoresist maP-1215 and the developer maD-331/S were purchased from micro-resist technology. All chemicals were used as received. Purified water (Milli-Q-grade, 18.2 MΩ cm at 25 °C) was used as obtained from the purification system.

Gold Nanorod (AuNR) Synthesis. AuNRs were synthesized with minor modifications, as published elsewhere [32]. Briefly, seed particles were prepared by adding 3 mL of a freshly prepared 0.01 M $NaBH_4$ solution in a 47 mL mixture of 0.1 M CTAB and 0.25 mM $HAuCl_4$ under vigorous stirring at 40 °C. The solution was stirred rapidly for 2 min, followed by continued slow stirring at 32 °C for 30 min. A 1 L of 0.1 M CTAB solution was prepared and 5 mL of 0.1 M $HAuCl_4$ solution (fc.: 0.5 mM), 500 µL of 0.1 M HBr, and 4 mL of 0.1 M $AgNO_3$ were added (f.c.: 0.4 mM). Of the 0.1 M hydroquinone solution (f.c.: 5 mM), 50 mL was added as the reducing agent while stirring and 2 min later, 18 mL of the as-prepared seed solution was added and kept at 32 °C for at least 48 h.

Vis-NIR measurements. The extinction spectrum of the CTAC-stabilized AuNRs in aqueous solution were acquired with the spectrophotometer Cary 5000 (Agilent Technologies Deutschland GmbH, Germany). With the intensity of the extinction spectrum (Figure 1) at a wavelength of 400 nm (interband transition of gold [33], the concentration of the AuNR dispersion was calculated.

TEM measurements. Transmission electron microscopy (TEM) measurements were performed with a Libra200 (Zeiss, Germany) operated at an acceleration voltage of 200 kV. For TEM analysis, 1 mL as-synthesized nanoparticle solutions were concentrated to 50 µL via centrifugation, and washed twice to reduce the surfactant concentration below the critical micelle concentration (cmc; ~0.9 mM). Subsequently, 2–5 µL of these solutions were dried on a 400 mesh copper grid covered with a carbon support film. The geometric dimensions of over 250 AuNRs were determined by using ImageJ.

Template Fabrication. Wrinkled PDMS templates were fabricated as follows. Prepolymer and agent from the PDMS Sylgard 184 kit were mixed in a ratio of 10:1. It was hardened for 1 day at room temperature and subsequently cured at 80 °C for about 4 h. Stripes of 1.0 cm × 4.5 cm were cut from the cooled PDMS. To create wrinkles on the PDMS surface, a PDMS stripe was formed to 140% of its original length by a custom-made stretching device. After treatment with oxygen plasma (80 W, 0.2 mbar) for 5 min, the PDMS stripe was released to its original length. This procedure resulted in wrinkles with a wavelength of about 950 nm and a depth of ca. 220 nm.

Substrate Fabrication. Si/SiO_2(230 nm) wafers (15 × 15 mm) were used as the substrate, and gold electrodes of 80 µm width and 1.5 µm channel length (finger width 10 µm) were photolithographically prepared using a chromium adhesion layer. Photoresist maP-1215 was spin-coated on the wafer (3000 rpm; 30 s) and soft-baked on the hot plate (100 °C) for 60 s. A Karl Süss Mask Aligner MA/BA6 was used for light exposure and after development in maD-331/S, the substrates were metallized with 3 nm chromium and 30 nm gold in a PLS570 evaporator. The lift-off was done in acetone in an ultrasonic bath for two minutes.

Template-Assisted Self-Assembly of AuNRs. The substrates were cleaned by sonication in acetone and isopropanol, consecutively, blow-dried (air) and cleaned for 5 min in UV/ozone. The AuNRs were assembled from the aqueous solution. Of a 10 mg/mL solution with 1 mM CTAC, 2 µL was deposited on top of a substrate and a 12 × 12 mm PDMS template with the wrinkles being oriented perpendicular to the gold electrodes (Figure 2a) was left on top for 4 h, giving rise to the AuNR lines due to confinement assembly (see Figure 1c). Usually, a AuNR line has a width of 300 nm and is formed by the template with a periodicity of around 905 nm, see Figure 2a.

Electrical Measurements. Electrical measurements were executed at room temperature under nitrogen in a Lake Shore Cryotronics probe station CRX-6.5K with a Keithley 2634B System Source Meter, and the samples were contacted with tungsten two-point probes.

Scanning Electron Microscopy. Scanning electron micrographs were taken using a HITACHI model SU8030 at 30 kV.

3. Results and Discussion

3.1. Fabrication

Gold nanorods (AuNRs) with an aspect ratio of 7.3 were synthesized as previously described. Shortly, seeded growth was performed with tetrachloroauric acid in aqueous solution of Cetyltrimethylammoniumbromide (CTAB) with AgNO$_3$ and hydroquinone. Into this mixture, single crystalline gold seeds were injected rapidly while vortexing by hand, followed by overnight resting at 32 °C, exchange of the stabilizing surfactant to CTAC, and purification. Thorough purification and consistent behavior of the colloids would not have been ensured with CTAB due to crystallization [34]. The exchange of the surfactant from CTAB to CTAC, however, enabled easy handling of the colloidal AuNRs at room temperature.

The positions of the plasmonic modes of the AuNRs in the UV-vis spectrum (Figure 1a) at 505 nm (transversal) and at 1107 nm (longitudinal) correlate to the length of 118 ± 16 nm and width of 16 ± 1 nm derived from the TEM images (Figure 1b and Figure S1) [35,36].

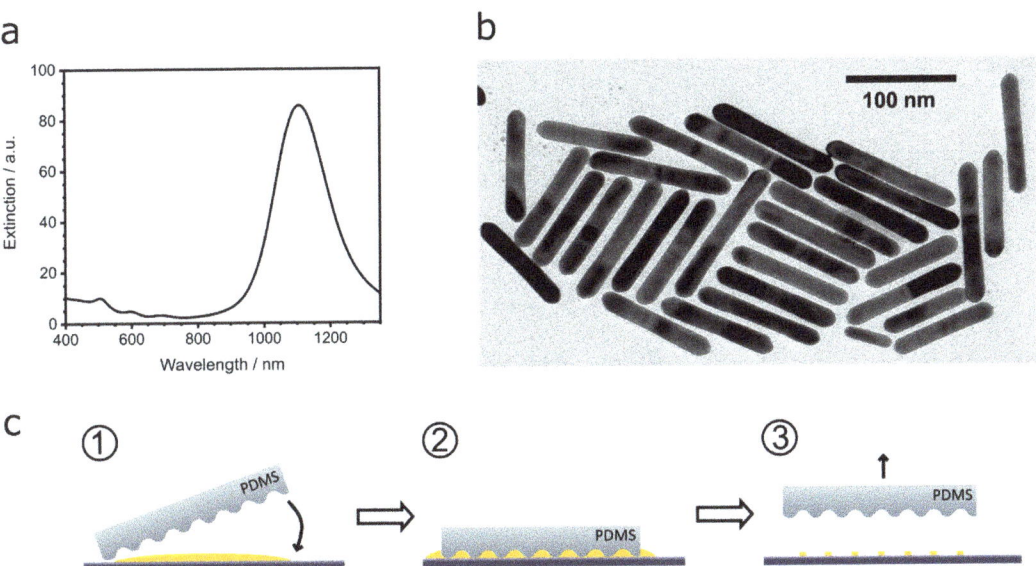

Figure 1. (**a**) Vis-NIR spectrum of the AuNRs with the transversal plasmon mode at 505 nm and the longitudinal plasmon mode at 1107 nm. (**b**) TEM image of AuNRs and (**c**) scheme of the template-assisted self-assembly of the AuNRs from aqueous solution on a Si/SiO$_2$ (230 nm) wafer with a wrinkled PDMS template.

To create AuNR lines, template-assisted self-assembly with wrinkled PDMS templates was used, since it allowed for easy and cost-effective assemblies on various substrates [37,38]. We chose an FET substrate consisting of a Si/SiO$_2$(230 nm) wafer with photolithographically deposited gold electrodes. The PDMS templates were fabricated by treating stretched PDMS stripes with oxygen plasma, followed by relaxation. This results in PDMS stripes with a wrinkled surface of sinusoidal shape. By changing the parameters of the plasma treatment, the geometrical dimensions of the wrinkles, periodicity and feature height can be tuned [39–41] and thereby, adjusted exactly to the required dimensions of the attempted application. AuNRs were assembled from aqueous solution into lines by confinement assembly (see Figure 1c). This is a template-assisted self-assembly method in which the colloids are confined between the PDMS template and the substrate [37]. The AuNR lines formed in the grooves of the wrinkles (Figure 1c). The PDMS templates had a periodicity of 950 nm and depth of 220 nm. These dimensions allowed for the flow of the AuNRs through the channels between the substrate and the template during the assembly process, but still provided enough confinement to result in narrow AuNR lines, with widths of 319 ± 139 nm (Figure 2a).

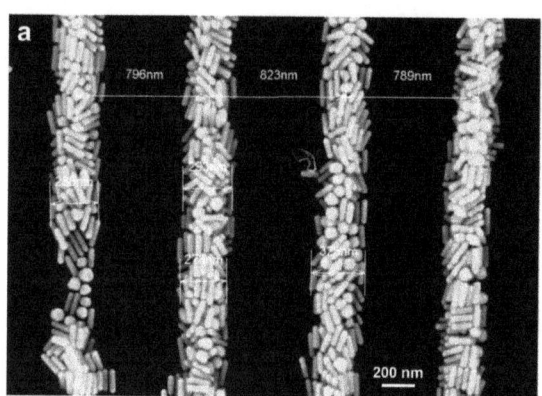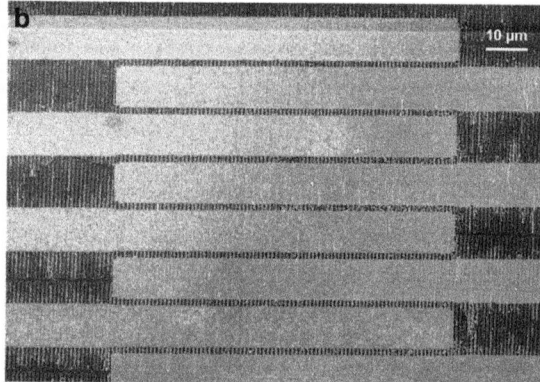

Figure 2. SEM images of (**a**) AuNR lines on Si/SiO$_2$ wafer and (**b**) on an FET substrate, with the AuNR lines perpendicular to the gold electrodes of the substrate.

3.2. Measurement Results

To measure the resistance of these AuNR lines, the electrode array of the substrate with parallel gold electrodes was used. The AuNR lines were assembled perpendicular to the electrodes, thereby connecting the electrodes (see Figure 2b). The center-to-center distance of the AuNR lines was about 905 ± 31 nm. The resistance was measured between each of the two parallel gold electrodes with a distance (channel length) of 1.5 µm.

The measured resistances were then correlated with SEM images of the AuNR lines. For each measured channel, the resistance and total number of continuous, electrode-connecting AuNR lines was noted as in the examples in Figure 3a,b. Any AuNR lines with gaps larger than 50 nm in between were not counted. This "coarse-grained" approach still does not rule out that there is no conductivity due to smaller gaps, as will become apparent in further discussion. As electronic conductivity decreases exponentially with the increasing spacing between the gold NPs [42–46], we do not expect charge to be transferred from one AuNR to another if the gap between them is 5–7 nm or larger [26].

The channel in Figure 3 (i) does not show a continuous AuNR line, and consequently, there is no conductance (R = 3.3 TΩ). In Figure 3 (ii) there is one continuous AuNR line visible and, thus, to that, the resistance measurement shows a much lower resistance of 79 GΩ. This is comparable with the resistances of 1 nm spaced gold nanowires, which have similar widths as this AuNR line [47]. In accordance with the literature of colloidal nanowires, the AuNR line shows an ohmic behavior at room temperature in this low-voltage regime [48,49]. The channel depicted in Figure 3 (iii) is connected by six AuNR lines, and a resistance of 7.4 MΩ was measured. For all the studied channels with a conductivity higher than the detection limit, the current-vs-voltage curves exhibited ohmic behavior, independent of the number of connecting AuNR lines. After acquiring the data for all channels, we correlated the number of apparently continuous AuNR lines with the corresponding resistances (see Figure 3c).

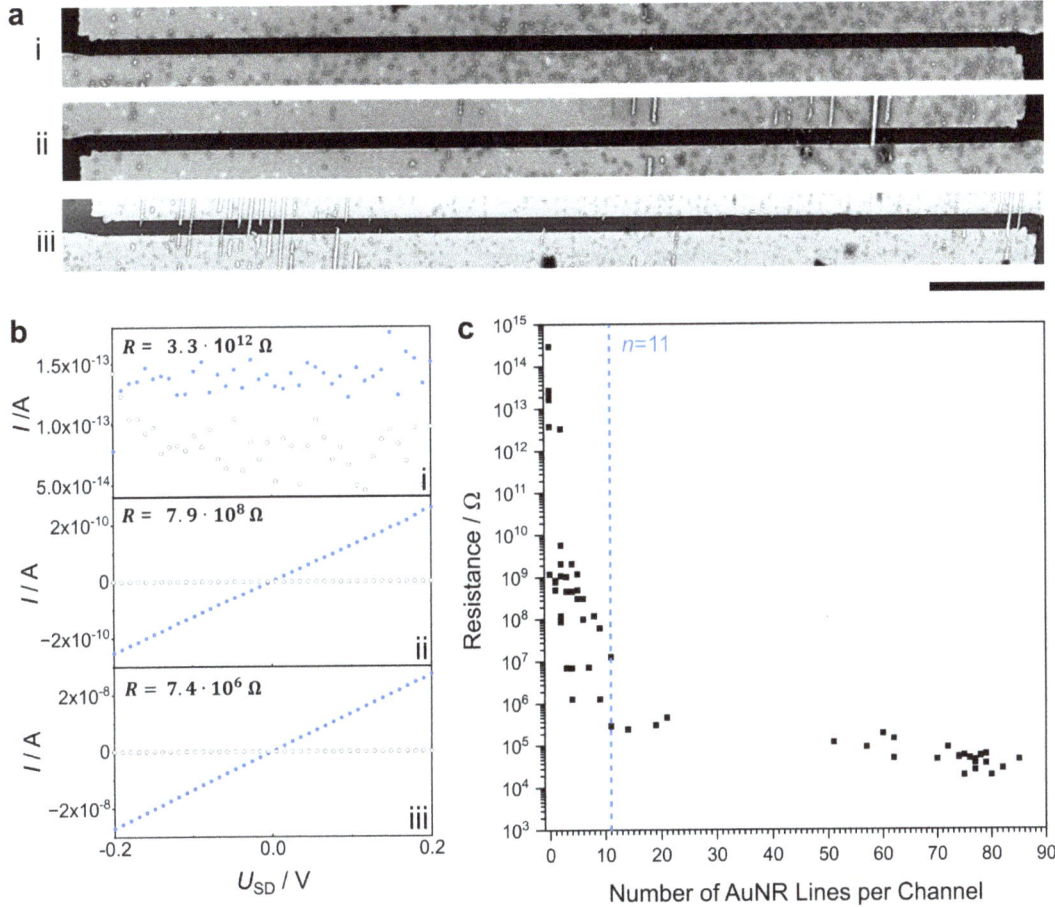

Figure 3. (a) SEM images, scale bar: 10 µm, and (b) the corresponding I–U plots obtained from source-drain measurements of the shown channels without a continuous AuNR line. (i) One AuNR line (ii) and six AuNR lines (iii), with blue = source-drain current, grey = leakage current, and (c) the plotted resistances for all channels.

3.3. Discussion

In Figure 3c, the measured resistances for each channel are plotted against the corresponding number of continuous AuNR lines, n. The more AuNR lines connect a pair of electrodes, the lower the measured resistance. The large scattering of the resistances for a lower n can be attributed to AuNR lines with strongly deviating resistance due to lower or a complete lack of conductance. If a channel has only a few conductive AuNR lines, its conductivity will be more severely affected by a low or non-conductive line, as is the case with many lines per channel. Unlike the scattering resistances for low numbers of AuNR lines, the total channel resistances converge for $n > 11$ and are constant within one order of magnitude (~10^5 Ω). Therefore, it can be concluded that more than 11 AuNR lines are needed for reliable conductance.

As for the non-conductive AuNR lines, we cannot rule out that some seemingly channel-bridging AuNRs still have charge-transport interrupting gaps which could not be detected by the coarse-grained SEM method. For the conducting AuNR lines, the reasons for deviating conductance can be manifold. Firstly, they can be attributed to the nonuniform arrangement of AuNRs in different AuNR lines. This causes a distribution of the number of charge–transport paths in the AuNR lines (corresponding to resistors connected in parallel) and of the number of resistive gaps within such a charge–transport path (corresponding to resistors connected in a series). Secondly, as we work with a CTAC concentration around the critical micelle concentration during the AuNR line assembly, the formation of the CTAC bilayers around the AuNRs could differ from one AuNR to another. The formation of the CTAC bilayers between the AuNRs is linked to the gap size and the latter one to the resistance of this gap. Assembled CTAC-stabilized AuNRs were shown to have a minimum distance of 3.4 nm, which corresponds to the thickness of a shared interdigitated CTA^+ bilayer [50]. The thickness of CTA^+ multilayers can also be significantly larger (i.e., 9 nm) as Sau and Murphy [50] reported, thus giving rise to a larger gap resistance, contributing to an increased total resistance of the AuNR line.

To attempt a more detailed account of the consistency of the resistances, we consider the AuNR lines as ohmic resistors R_i connected in parallel. The measured total resistance R_{total} of each channel then results as:

$$\frac{1}{R_{total}} = \sum_{i=1}^{n} \frac{1}{R_i}, \qquad (1)$$

with R_i being the resistance of an individual AuNR line in one channel. We cannot know the resistance of every single AuNR line, but we can model the resistance with the assumption that every AuNR line in a considered channel has the same resistance, R_{single}. Although this assumption may have weaknesses (as can be seen from the comparably large scattering of resistances in Figure 3c), it helps in assessing the resistance measurement data. Assuming all $R_i = R_{single}$, Equation (1) yields

$$\frac{1}{R_{total}} = n \cdot \frac{1}{R_{single}}, \qquad (2)$$

with n as the total number of continuous AuNR lines within the relevant channel. Figure S2 shows that the R_{single} values are roughly constant for $n > 11$, and therefore supports the assumption of our model of uniform AuNR lines acting as ohmic resistors connected in parallel. To calculate a general R_{single}, we can use a linear regression for the total channel resistances, R_{total} versus $\frac{1}{n}$, with $R_{single,\,fitted}$ as the slope of the linear graph:

$$R_{total} = R_{single,\,fitted}(n > 11) \cdot \frac{1}{n}. \qquad (3)$$

With a coefficient of determination of 0.81, this linear regression gives a slope, i.e., $R_{single,\,fitted}$ of $5.3 \cdot 10^6 \, \Omega$ (see Figure S3). This single AuNR line resistance is in the range of the lowest resistances measured for similarly sized gold nanowires, which were about 1 nm apart from each other [47]. This shows that despite the surfactant-induced gaps, the charge transport along the AuNR lines works remarkably well. Inserted into the plot of the total channel resistances versus the number of AuNR lines per channel (Figure 4), the modeled total channel resistance, $R_{total,model}$, fits the measured data well for the ohmic-resistors regime for $n > 11$ and even for the smallest values of the total channel resistances for a lower n.

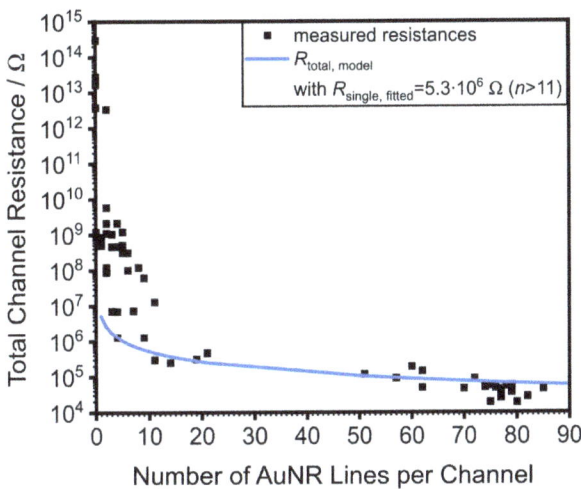

Figure 4. Conductivity measurements: measured and modeled total channel resistances.

Our conductivity measurement results mirror the heterogeneities of the AuNR arrangement between different AuNR lines, which results in strongly fluctuating resistances for small numbers of AuNR lines. However, by connecting electrodes with several ($n > 11$) AuNR lines, this heterogeneity does not negatively impair the consistency of conductivity measurements. Hence, the assembly of multiple conductive supracolloidal lines offers a suited approach to mitigate inconsistency in the transport behavior of this promising class of mesoscale electronic materials. This approach takes up an area of 10.8 µm × 1.5 µm for 12 parallelly aligned AuNR lines with a center-to-center distance of about 950 nm. Such a system still shows higher conductance ($\sim 10^{-5}$ S) than a monolayer of gold nanospheres with dithiolated-conjugated ligands ($>10^{-7}$ S) [31], and additionally features the advantage of anisotropic conductance.

4. Conclusions

In summary, we successfully fabricated AuNR lines via template-assisted self-assembly and characterized their conductance. By using bottom-up fabricated PDMS templates and wet-chemically synthesized AuNRs for the confinement assembly, the whole process of the linear assembly did not require expensive equipment. Another advantage is the possibility to print these AuNR lines on a plethora of materials [37,38], including heat-sensitive polymer films, since our fabrication process does not include sintering. Additionally, our structures feature comparably low dimensions in terms of the AuNR line width [3,30]. We observed a dependence of the conductance on the number of channel-bridging AuNR lines. For more than 11 AuNRs per channel, the single-line resistances approached a unified behavior, described by the ohmic model of uniform resistors connected in parallel. The results demonstrate that consistent conductivity properties can be reached if several supracolloidal wires are employed, even if their conductivity properties fluctuate strongly among the individual lines. This is especially applicable to the development of sensors based on surfaces with anisotropic resistance properties. With our approach, the active areas can be as small as 16 µm², but also as large as cm² [18], depending on the intended application. In this regard, the up-scalable fabrication and integration of our AuNR lines into robust technical processes is promising for future device integration [51].

Supplementary Materials: The supporting information can be downloaded at: https://www.mdpi.com/article/10.3390/nano13091466/s1, Figure S1: Histograms of the AuNRs' dimensions, their width and length, derived from TEM measurements; Figure S2: Conductivity measurements: Calculated mean resistance for single AuNR lines; Figure S3: Linear fit of the measured R_total and model for R_total by using the R_(single, fitted) derived from the linear regression; Figure S4: Conductivity measurements: (a) calculated conductances per AuNR line and (b) total channel conductances G, measured and modeled values; Figure S5: The measured channel resistances for $n = 1$, $n = 2$, $n = 3$ and $n = 4$ exemplarily illustrate the scattering of the total channel resistances R_total for small n.

Author Contributions: Conceptualization, M.H., C.A.S., M.M., C.R., M.S. and A.F.; methodology, M.H. and C.A.S.; formal analysis, M.H., C.A.S.; investigation, M.H., C.A.S., M.M.; resources, M.S. and A.F.; data curation, M.H. and C.A.S.; writing—original draft preparation, M.H. and C.R.; writing—review and editing, M.H., C.A.S., M.M., C.R., M.S. and A.F.; visualization, M.H. and C.A.S.; supervision, M.S. and A.F.; project administration, M.S. and A.F.; funding acquisition, M.S. and A.F. All authors have read and agreed to the published version of the manuscript.

Funding: The studies were performed within the LaSensA project carried out under the M-ERA.NET 2 scheme European Union's Horizon 2020 research and innovation program (grant no. 685451), and were co-funded by the Research Council of Lithuania (LMTLT), agreement no. S-M-ERA.NET-21-2, the National Science Centre of Poland, project no. 2020/02/Y/ST5/00086, and the Saxon State Ministry for Science, Culture and Tourism (Germany), grant no. 100577922, as well as by the tax funds on the basis of the budget passed by the Saxon state parliament. We gratefully acknowledge financial support from the Deutsche Forschungsgemeinschaft (DFG) within RTG2767, project no. 451785257. C.A.S. gratefully acknowledges funding by the DFG under grant SCHE1905/9-1. M.M. gratefully acknowledges funding by the Deutsche Forschungsgemeinschaft (DFG, German Research Foundation, 453211202). C.R. gratefully acknowledges receipt of a Liebig fellowship (Fonds der Chemischen Industrie). M.S. gratefully acknowledges funding by the European Research Council (ERC) under the European Union's Horizon 2020 research and innovation program (grant agreement no. 802822), as well as by the DFG under grant SCHE1905/9-1.

Data Availability Statement: The data presented in this study are available on request from the corresponding authors.

Acknowledgments: We would like to thank Elke Nadler for SEM measurements, Andre Maier for the optical lithography and for SEM measurements, and Daniel Schletz for valuable discussion.

Conflicts of Interest: The authors declare no conflict of interest.

References

1. Schmid, G.; Corain, B. Nanoparticulated Gold: Syntheses, Structures, Electronics, and Reactivities. *EurJIC* **2003**, *2003*, 3081–3098. [CrossRef]
2. Maurer, J.H.; Gonzalez-Garcia, L.; Reiser, B.; Kanelidis, I.; Kraus, T. Templated Self-Assembly of Ultrathin Gold Nanowires by Nanoimprinting for Transparent Flexible Electronics. *Nano Lett.* **2016**, *16*, 2921–2925. [CrossRef] [PubMed]
3. Reiser, B.; Gonzalez-Garcia, L.; Kanelidis, I.; Maurer, J.H.M.; Kraus, T. Gold nanorods with conjugated polymer ligands: Sintering-free conductive inks for printed electronics. *Chem. Sci.* **2016**, *7*, 4190–4196. [CrossRef] [PubMed]
4. Jia, P.; Kong, D.; Ebendorff-Heidepriem, H. Flexible Plasmonic Tapes with Nanohole and Nanoparticle Arrays for Refractometric and Strain Sensing. *ACS Appl. Nano Mater.* **2020**, *3*, 8242–8246. [CrossRef]
5. Schlicke, H.; Rebber, M.; Kunze, S.; Vossmeyer, T. Resistive pressure sensors based on freestanding membranes of gold nanoparticles. *Nanoscale* **2016**, *8*, 183–186. [CrossRef]
6. Schlicke, H.; Kunze, S.; Rebber, M.; Schulz, N.; Riekeberg, S.; Trieu, H.K.; Vossmeyer, T. Cross-Linked Gold Nanoparticle Composite Membranes as Highly Sensitive Pressure Sensors. *Adv. Funct. Mater.* **2020**, *30*, 2003381. [CrossRef]
7. Su, Y.S.; Yang, W.R.; Jheng, W.W.; Kuo, W.; Tzeng, S.D.; Yasuda, K.; Song, J.M. Optimization of Piezoresistive Strain Sensors Based on Gold Nanoparticle Deposits on PDMS Substrates for Highly Sensitive Human Pulse Sensing. *Nanomaterials* **2022**, *12*, 2312. [CrossRef]
8. Ketelsen, B.; Yesilmen, M.; Schlicke, H.; Noei, H.; Su, C.H.; Liao, Y.C.; Vossmeyer, T. Fabrication of Strain Gauges via Contact Printing: A Simple Route to Healthcare Sensors Based on Cross-Linked Gold Nanoparticles. *ACS Appl. Mater. Interfaces* **2018**, *10*, 37374–37385. [CrossRef]
9. Lin, H.Y.; Chen, H.A.; Lin, H.N. Fabrication of a single metal nanowire connected with dissimilar metal electrodes and its application to chemical sensing. *Anal. Chem.* **2008**, *80*, 1937–1941. [CrossRef]
10. Hsu, M.S.; Chen, Y.L.; Lee, C.Y.; Chiu, H.T. Gold nanostructures on flexible substrates as electrochemical dopamine sensors. *ACS Appl. Mater. Interfaces* **2012**, *4*, 5570–5575. [CrossRef]
11. Joseph, Y.; Krasteva, N.; Besnard, I.; Guse, B.; Rosenberger, M.; Wild, U.; Knop-Gericke, A.; Schlogl, R.; Krustev, R.; Yasuda, A.; et al. Gold-nanoparticle/organic linker films: Self-assembly, electronic and structural characterisation, composition and vapour sensitivity. *Faraday Discuss.* **2004**, *125*, 77–97, discussion 99–116. [CrossRef] [PubMed]

12. Ahn, H.; Chandekar, A.; Kang, B.; Sung, C.; Whitten, J.E. Electrical Conductivity and Vapor-Sensing Properties of ω-(3-Thienyl)alkanethiol-Protected Gold Nanoparticle Films. *Chem. Mater.* **2004**, *16*, 3274–3278. [CrossRef]
13. Milyutin, Y.; Abud-Hawa, M.; Kloper-Weidenfeld, V.; Mansour, E.; Broza, Y.Y.; Shani, G.; Haick, H. Fabricating and printing chemiresistors based on monolayer-capped metal nanoparticles. *Nat. Protoc.* **2021**, *16*, 2968–2990. [CrossRef] [PubMed]
14. MacKenzie, R.; Fraschina, C.; Dielacher, B.; Sannomiya, T.; Dahlin, A.B.; Voros, J. Simultaneous electrical and plasmonic monitoring of potential induced ion adsorption on metal nanowire arrays. *Nanoscale* **2013**, *5*, 4966–4975. [CrossRef] [PubMed]
15. Chen, S.; Pan, Q.; Wu, T.; Xie, H.; Xue, T.; Su, M.; Song, Y. Printing nanoparticle-based isotropic/anisotropic networks for directional electrical circuits. *Nanoscale* **2022**, *14*, 14956–14961. [CrossRef]
16. Barrow, S.J.; Funston, A.M.; Gomez, D.E.; Davis, T.J.; Mulvaney, P. Surface plasmon resonances in strongly coupled gold nanosphere chains from monomer to hexamer. *Nano Lett.* **2011**, *11*, 4180–4187. [CrossRef]
17. Slaughter, L.S.; Willingham, B.A.; Chang, W.S.; Chester, M.H.; Ogden, N.; Link, S. Toward plasmonic polymers. *Nano Lett.* **2012**, *12*, 3967–3972. [CrossRef]
18. Hanske, C.; Tebbe, M.; Kuttner, C.; Bieber, V.; Tsukruk, V.V.; Chanana, M.; Konig, T.A.; Fery, A. Strongly coupled plasmonic modes on macroscopic areas via template-assisted colloidal self-assembly. *Nano Lett.* **2014**, *14*, 6863–6871. [CrossRef]
19. Tebbe, M.; Mayer, M.; Glatz, B.A.; Hanske, C.; Probst, P.T.; Muller, M.B.; Karg, M.; Chanana, M.; Konig, T.A.; Kuttner, C.; et al. Optically anisotropic substrates via wrinkle-assisted convective assembly of gold nanorods on macroscopic areas. *Faraday Discuss.* **2015**, *181*, 243–260. [CrossRef]
20. Mueller, M.; Tebbe, M.; Andreeva, D.V.; Karg, M.; Alvarez Puebla, R.A.; Pazos Perez, N.; Fery, A. Large-area organization of pNIPAM-coated nanostars as SERS platforms for polycyclic aromatic hydrocarbons sensing in gas phase. *Langmuir* **2012**, *28*, 9168–9173. [CrossRef]
21. Pazos-Pérez, N.; Ni, W.; Schweikart, A.; Alvarez-Puebla, R.A.; Fery, A.; Liz-Marzán, L.M. Highly uniform SERS substrates formed by wrinkle-confined drying of gold colloids. *Chem. Sci.* **2010**, *1*, 174. [CrossRef]
22. Horváth, B.; Křivová, B.; Bolat, S.; Schift, H. Fabrication of Large Area Sub-200 nm Conducting Electrode Arrays by Self-Confinement of Spincoated Metal Nanoparticle Inks. *Adv. Mater. Technol.* **2019**, *4*, 1800652. [CrossRef]
23. Steiner, A.M.; Mayer, M.; Seuss, M.; Nikolov, S.; Harris, K.D.; Alexeev, A.; Kuttner, C.; Konig, T.A.F.; Fery, A. Macroscopic Strain-Induced Transition from Quasi-infinite Gold Nanoparticle Chains to Defined Plasmonic Oligomers. *ACS Nano* **2017**, *11*, 8871–8880. [CrossRef] [PubMed]
24. Hyun, D.C.; Moon, G.D.; Cho, E.C.; Jeong, U. Repeated Transfer of Colloidal Patterns by Using Reversible Buckling Process. *Adv. Funct. Mater.* **2009**, *19*, 2155–2162. [CrossRef]
25. Tadimety, A.; Kready, K.M.; Chorsi, H.T.; Zhang, L.; Palinski, T.J.; Zhang, J.X.J. Nanowrinkled thin films for nanorod assembly in microfluidics. *Microfluid. Nanofluid.* **2019**, *23*, 17. [CrossRef]
26. Rey, A.; Billardon, G.; Lortscher, E.; Moth-Poulsen, K.; Stuhr-Hansen, N.; Wolf, H.; Bjornholm, T.; Stemmer, A.; Riel, H. Deterministic assembly of linear gold nanorod chains as a platform for nanoscale applications. *Nanoscale* **2013**, *5*, 8680–8688. [CrossRef] [PubMed]
27. Franke, M.E.; Koplin, T.J.; Simon, U. Metal and metal oxide nanoparticles in chemiresistors: Does the nanoscale matter? *Small* **2006**, *2*, 36–50. [CrossRef] [PubMed]
28. Bayrak, T.; Martinez-Reyes, A.; Arce, D.D.R.; Kelling, J.; Samano, E.C.; Erbe, A. Fabrication and temperature-dependent electrical characterization of a C-shape nanowire patterned by a DNA origami. *Sci. Rep.* **2021**, *11*, 1922. [CrossRef]
29. Mutiso, R.M.; Winey, K.I. Electrical percolation in quasi-two-dimensional metal nanowire networks for transparent conductors. *Phys. Rev. E Stat. Nonlinear Soft Matter. Phys.* **2013**, *88*, 032134. [CrossRef]
30. Ahn, B.Y.; Lorang, D.J.; Lewis, J.A. Transparent conductive grids via direct writing of silver nanoparticle inks. *Nanoscale* **2011**, *3*, 2700–2702. [CrossRef]
31. Liao, J.; Mangold, M.A.; Grunder, S.; Mayor, M.; Schönenberger, C.; Calame, M. Interlinking Au nanoparticles in 2D arrays via conjugated dithiolated molecules. *New J. Phys.* **2008**, *10*, 065019. [CrossRef]
32. Schnepf, M.J.; Mayer, M.; Kuttner, C.; Tebbe, M.; Wolf, D.; Dulle, M.; Altantzis, T.; Formanek, P.; Forster, S.; Bals, S.; et al. Nanorattles with tailored electric field enhancement. *Nanoscale* **2017**, *9*, 9376–9385. [CrossRef] [PubMed]
33. Hendel, T.; Wuithschick, M.; Kettemann, F.; Birnbaum, A.; Rademann, K.; Polte, J. In situ determination of colloidal gold concentrations with UV-vis spectroscopy: Limitations and perspectives. *Anal. Chem.* **2014**, *86*, 11115–11124. [CrossRef] [PubMed]
34. Yamamoto, T.; Yagi, Y.; Hatakeyama, T.; Wakabayashi, T.; Kamiyama, T.; Suzuki, H. Metastable and stable phase diagrams and thermodynamic properties of the cetyltrimethylammonium bromide (CTAB)/water binary system. *Colloids Surf. A Physicochem. Eng. Asp.* **2021**, *625*, 126859. [CrossRef]
35. Link, S.; Mohamed, M.B.; El-Sayed, M.A. Simulation of the Optical Absorption Spectra of Gold Nanorods as a Function of Their Aspect Ratio and the Effect of the Medium Dielectric Constant. *J. Phys. Chem. B* **1999**, *103*, 3073–3077, Correction in *J. Phys. Chem. B* **2005**, *109*, 10531–10532. [CrossRef]
36. Brioude, A.; Jiang, X.C.; Pileni, M.P. Optical properties of gold nanorods: DDA simulations supported by experiments. *J. Phys. Chem. B* **2005**, *109*, 13138–13142. [CrossRef]
37. Schweikart, A.; Fortini, A.; Wittemann, A.; Schmidt, M.; Fery, A. Nanoparticle assembly by confinement in wrinkles: Experiment and simulations. *Soft Matter* **2010**, *6*, 5860. [CrossRef]

38. Schletz, D.; Schultz, J.; Potapov, P.L.; Steiner, A.M.; Krehl, J.; König, T.A.F.; Mayer, M.; Lubk, A.; Fery, A. Exploiting Combinatorics to Investigate Plasmonic Properties in Heterogeneous Ag–Au Nanosphere Chain Assemblies. *Adv. Opt. Mater.* **2021**, *9*, 2001983. [CrossRef]
39. Chiche, A.; Stafford, C.M.; Cabral, J.T. Complex micropatterning of periodic structures on elastomeric surfaces. *Soft Matter* **2008**, *4*, 2360. [CrossRef]
40. Claussen, K.U.; Tebbe, M.; Giesa, R.; Schweikart, A.; Fery, A.; Schmidt, H.-W. Towards tailored topography: Facile preparation of surface-wrinkled gradient poly(dimethyl siloxane) with continuously changing wavelength. *RSC Adv.* **2012**, *2*, 10185. [CrossRef]
41. Glatz, B.A.; Fery, A. The influence of plasma treatment on the elasticity of the in situ oxidized gradient layer in PDMS: Towards crack-free wrinkling. *Soft Matter* **2018**, *15*, 65–72. [CrossRef] [PubMed]
42. Zamborini, F.P.; Leopold, M.C.; Hicks, J.F.; Kulesza, P.J.; Malik, M.A.; Murray, R.W. Electron hopping conductivity and vapor sensing properties of flexible network polymer films of metal nanoparticles. *J. Am. Chem. Soc.* **2002**, *124*, 8958–8964. [CrossRef] [PubMed]
43. Parthasarathy, R.; Lin, X.-M.; Jaeger, H.M. Electronic Transport in Metal Nanocrystal Arrays: The Effect of Structural Disorder on Scaling Behavior. *Phys. Rev. Lett.* **2001**, *87*, 186807. [CrossRef]
44. Wuelfing, W.P.; Murray, R.W. Electron Hopping through Films of Arenethiolate Monolayer-Protected Gold Clusters. *J. Phys. Chem. B* **2002**, *106*, 3139–3145. [CrossRef]
45. Wessels, J.M.; Nothofer, H.G.; Ford, W.E.; von Wrochem, F.; Scholz, F.; Vossmeyer, T.; Schroedter, A.; Weller, H.; Yasuda, A. Optical and electrical properties of three-dimensional interlinked gold nanoparticle assemblies. *J. Am. Chem. Soc.* **2004**, *126*, 3349–3356. [CrossRef]
46. Liljeroth, P.; Vanmaekelbergh, D.; Ruiz, V.; Kontturi, K.; Jiang, H.; Kauppinen, E.; Quinn, B.M. Electron transport in two-dimensional arrays of gold nanocrystals investigated by scanning electrochemical microscopy. *J. Am. Chem. Soc.* **2004**, *126*, 7126–7132. [CrossRef]
47. Park, H.; Lim, A.K.L.; Alivisatos, A.P.; Park, J.; McEuen, P.L. Fabrication of metallic electrodes with nanometer separation by electromigration. *Appl. Phys. Lett.* **1999**, *75*, 301–303. [CrossRef]
48. Pearson, A.C.; Liu, J.; Pound, E.; Uprety, B.; Woolley, A.T.; Davis, R.C.; Harb, J.N. DNA origami metallized site specifically to form electrically conductive nanowires. *J. Phys. Chem. B* **2012**, *116*, 10551–10560. [CrossRef]
49. Teschome, B.; Facsko, S.; Schonherr, T.; Kerbusch, J.; Keller, A.; Erbe, A. Temperature-Dependent Charge Transport through Individually Contacted DNA Origami-Based Au Nanowires. *Langmuir* **2016**, *32*, 10159–10165. [CrossRef]
50. Sau, T.K.; Murphy, C.J. Self-assembly patterns formed upon solvent evaporation of aqueous cetyltrimethylammonium bromide-coated gold nanoparticles of various shapes. *Langmuir* **2005**, *21*, 2923–2929. [CrossRef]
51. Wang, R.; Zimmermann, P.; Schletz, D.; Hoffmann, M.; Probst, P.; Fery, A.; Nagel, J.; Rossner, C. Nano meets macro: Furnishing the surface of polymer molds with gold-nanoparticle arrays. *Nano Select.* **2022**, *3*, 1502–1508. [CrossRef]

Disclaimer/Publisher's Note: The statements, opinions and data contained in all publications are solely those of the individual author(s) and contributor(s) and not of MDPI and/or the editor(s). MDPI and/or the editor(s) disclaim responsibility for any injury to people or property resulting from any ideas, methods, instructions or products referred to in the content.

Article

Light-Induced Clusterization of Gold Nanoparticles: A New Photo-Triggered Antibacterial against *E. coli* Proliferation

Angela Candreva [1,2], Renata De Rose [1], Ida Daniela Perrotta [3], Alexa Guglielmelli [2,4] and Massimo La Deda [1,2,*]

1. Department of Chemistry and Chemical Technologies, University of Calabria, 87036 Rende, Italy
2. CNR-NANOTEC, Institute of Nanotechnology U.O.S, Cosenza, 87036 Rende, Italy
3. Department of Biology, Ecology and Earth Sciences, Centre for Microscopy and Microanalysis (CM2), University of Calabria, 87036 Rende, Italy
4. Department of Physics, NLHT-Lab, University of Calabria, 87036 Rende, Italy
* Correspondence: massimo.ladeda@unical.it

Abstract: Metallic nanoparticles show plasmon resonance phenomena when irradiated with electromagnetic radiation of a suitable wavelength, whose value depends on their composition, size, and shape. The damping of the surface electron oscillation causes a release of heat, which causes a large increase in local temperature. Furthermore, this increase is enhanced when nanoparticle aggregation phenomena occur. Local temperature increase is extensively exploited in photothermal therapy, where light is used to induce cellular damage. To activate the plasmon in the visible range, we synthesized 50 nm diameter spherical gold nanoparticles (AuNP) coated with polyethylene glycol and administered them to an *E. coli* culture. The experiments were carried out, at different gold nanoparticle concentrations, in the dark and under irradiation. In both cases, the nanoparticles penetrated the bacterial wall, but a different toxic effect was observed; while in the dark we observed an inhibition of bacterial growth of 46%, at the same concentration, under irradiation, we observed a bactericidal effect (99% growth inhibition). Photothermal measurements and SEM observations allowed us to conclude that the extraordinary effect is due to the formation, at low concentrations, of a light-induced cluster of gold nanoparticles, which does not form in the absence of bacteria, leading us to the conclusion that the bacterium wall catalyzes the formation of these clusters which are ultimately responsible for the significant increase in the measured temperature and cause of the bactericidal effect. This photothermal effect is achieved by low-power irradiation and only in the presence of the pathogen: in its absence, the lack of gold nanoparticles clustering does not lead to any phototoxic effect. Therefore, it may represent a proof of concept of an innovative nanoscale pathogen responsive system against bacterial infections.

Keywords: gold nanoparticles; light-induced clusterization; pathogen responsive; *E. coli* infection

Citation: Candreva, A.; De Rose, R.; Perrotta, I.D.; Guglielmelli, A.; La Deda, M. Light-Induced Clusterization of Gold Nanoparticles: A New Photo-Triggered Antibacterial against *E. coli* Proliferation. *Nanomaterials* **2023**, *13*, 746. https://doi.org/10.3390/nano13040746

Academic Editor: Pavel Padnya

Received: 30 January 2023
Revised: 13 February 2023
Accepted: 14 February 2023
Published: 16 February 2023

Copyright: © 2023 by the authors. Licensee MDPI, Basel, Switzerland. This article is an open access article distributed under the terms and conditions of the Creative Commons Attribution (CC BY) license (https://creativecommons.org/licenses/by/4.0/).

1. Introduction

Gold nanoparticles are widely used in the manufacture of numerous products, from electronics to biomedical devices [1–9]. In particular, they have raised high interest in cell biology and biomedicine due to their unique chemical, optical and electronic properties that result from their minute size [10–13]. Interestingly, when light interacts with gold nanoparticles, it is both scattered and absorbed causing a surface plasmon excitation and the resulting spectral shape of the absorbed radiation depends on many factors such as the size, shape, composition, and environment of the nanoparticles [14–20]. Over the years, scientists around the world have been experimenting with the synthesis of metal nanoparticles of different shapes and sizes. This is because the properties of nanoparticles depend on their nanostructure. By changing the nanostructural properties, the plasmonic properties change accordingly. Great interest has been aroused in the gold nanoparticles, which have a plasmon band that falls in the visible range, but can also move towards the wavelengths of NIR [21–23]. In addition, the surface of gold nanoparticles can be easily functionalized with amino groups or thiol groups [24], and still biomolecules such as DNA and proteins are used as covering agents of gold nanoparticles [25]. The ease of characterization of nanoparticles has also contributed to increased interest in the "nano world" [26,27].

A key aspect in uncovering more new biomedical applications of gold nanoparticles is their cellular uptake [28,29]. Despite numerous studies in this field, the current understanding of the factors influencing the cellular internalization of nanoparticles remains very limited. It is accepted that commonly used gold nanoparticles are able to cross cell membranes, usually via endocytic pathways; however, the effectiveness of absorption depends on the charge, as well as the size, shape, and surface chemistry of the nanoparticles [29–38]. Bulk gold is considered to be biologically inert; on the other hand, at nanoscale size, gold has different attributes due its surface plasmon resonance excitation features [39]. It is important to underline that, as the nanotechnology field continues to develop, several studies in understanding the size- and shape-dependent toxicity of gold nanomaterials are being carried out. These studies are increasingly emphasizing how different morphologies of the nanoparticles have a different impact on organisms and on the consequent applications [40–43]. Several studies have been conducted on clusterization of nanoparticles during their interaction with living cells. The result is that the nanoparticle cluster greatly enhances diagnostic sensitivity and therapeutic effectiveness compared to individual nanoparticles [44].

Gold nanoparticles (AuNP) are appealing photothermal candidates because they show efficient local heating upon excitation of surface plasmon oscillations [45–47]. The strong absorption, efficient light/heat conversion, and high photostability, contribute to arousing increasing interest in the photothermal applications of gold nanoparticles that permit a directional control of the incident radiation on the administration region of the these phototransducers, resulting in localized heat transfer to the surrounding environment [28,48–50]. When discussing the photothermal activity of gold nanoparticles, several important parameters are implicitly considered: the wavelength of the laser, that should be matched with the peak of the plasmonic band of the used nanoparticles [51], the power of the laser, and the nanoparticle's size. In particular, the higher the power of the laser, the higher the temperature increase [5,46,52,53]. Furthermore, by increasing the size of the nanoparticles, the temperature increases as well [17]. However, a high-power laser can itself cause damage to the cellular environment, and it is not possible to administer nanoparticles with a large size [54].

Herein, we show the interaction between gold nanoparticles and bacterial populations [55–57], i.e., *Escherichia coli* [58–61], in dark or light conditions [62–64]. In fact, recently, gold nanoparticles have also gained interest for their antibacterial properties against different microorganisms [65]. We prepared 50 nm-diameter gold nanospheres, covered with thiolate polyethylene glycol (PEG-SH), well known to not have significant antibacterial effects [36]. From an accurate study performed by the use of electron microscopies, it was possible to observe the uptake of nanoparticles within bacterial cells as well as the pleomorphism (rough surface) and shrinkage in size because of increased cell death [61,66–70] An antibacterial test, performed by varying AuNP concentration (from 0.26 to 3.54 µg/mL), clearly indicates that *E. coli* exposure to gold nanoparticles inhibits bacterial growth, and that this inhibition, directly proportional to the nanoparticle concentration, is mainly observed in light conditions. We administered the light to the bacteria simultaneously with the dosage of the nanoparticles, and this caused the fast aggregation of nanoparticles, induced by light [71], on the bacterial surface. According to our previous studies [17,18], the clusters have a greater photothermal effect than single nanoparticles, and this was confirmed by the total bacterial inhibition growth. Since this clustering occurs only in the presence of bacteria, this system may represent a proof-of-concept of an innovative nanoscale pathogen responsive system against bacterial infections. To the best of our knowledge, we are proposing the first example of a system able to activate only in the presence of the bacterial cells, and to have a bactericidal effect even at low concentrations. In particular, it is important to underline that in this work several limitations have been overcome, in fact, a low-power laser and small-sized gold nanoparticles have been employed obtaining, as the result, a complete inhibition of bacterial *E. coli* growth.

2. Materials and Methods

Chemicals. All chemicals were purchased from Sigma-Aldrich (Schnelldorf, Germany) (highest purity grade available) and used as received. Tetrachloroauric acid trihydrate ($HAuCl_4 \cdot H_2O$, $\geq$99.9%), sodium citrate $C_6H_5Na_3O_7 \cdot 2H_2O$ (99%), Milli-Q water (resistivity 18.2 M$\Omega \cdot$cm at 25 °C) were used in all experiments. All glassware was washed with aqua regia, rinsed with water, sonicated threefold for 3 min with Milli-Q water, and dried before use.

Sterile tissue culture plates of polystyrene, 6-well, 35mm, non-treated Biofil. *E. coli* (DSM 1576 Medium 1) from DSMZ (German Collection of Microorganisms and Cell Cultures, Braunschweig).

Exponential growth phase. *E. coli* solution was diluted (1:100 v/v) in 20 mL of fresh DMS Medium 1 to restart the cell cycle and after 3 h of incubation at 37 °C the cells were synchronized at the log phase of the growth curve, featured with the optical density at 600 nm of 0.4–0.6. At this OD value, the cells divide and the growth rate is constant and the cells are in exponential growth phase.

Gold spheres covered with sodium citrate (AuNS@Citrate). Seed solution. A water solution of sodium citrate (6.0 E-2 M) was stirred vigorously and heated until the boiling temperature. At this point, 1 mL of $HAuCl_4$ 0.025 E-3 M was added. Immediately, the reaction was cooled to 90 °C. The decrease in the temperature provides the inhibition of a new nucleation, favoring the consequent overgrowth of the seeds. Sodium citrate has two functions: it reduces Au (III) to Au (0) and coats the seed to prevent their aggregation. Within 10 min, the color of the solution changed from yellow to bluish grey and then to soft pink. *Growth solution*. After temperature stabilized to 90 °C, 1 mL of sodium citrate (0.060 E-2 M) and 1 mL of a $HAuCl_4$ solution (0.025 E-3 M) were consecutively added. By repeating this process various times, it was possible to obtain citrate-coated gold nanospheres with an increasing diameter [17,72]. To accelerate the process, the sample was diluted by extracting a 55 mL aliquot and adding 53 mL of hot distilled water to it, followed by 2 mL of 0.06 E-3 M sodium citrate water solution. When the temperature was stabilized at 90 °C, 1 mL of $HAuCl_4$ was added. By repeating the process six times, nanospheres with 50 nm diameter were obtained, characterized by UV–Vis absorption spectroscopy and TEM.

Gold spheres covered with thiolate polyethylene glycol (AuNS@PEG-SH). A water solution of PEG-SH was prepared (30 mg in 1 mL) and added to 25 mL of 5 E-4 M water-dispersed AuNS@Citrate. The sample was left under stirring overnight, and then purified by unlinked PEG-SH (three centrifuge cycles, 600 rpm). The solid residue was dissolved in water (Figure 1) [15].

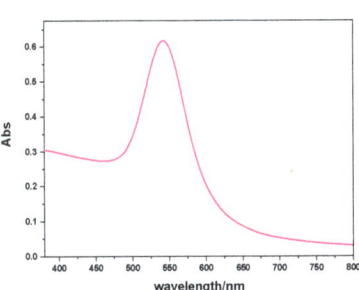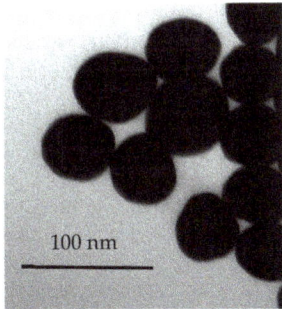

Figure 1. UV–Vis extinction spectrum and TEM image of AuNS@PEG-SH.

Preparation of microorganism suspension. The antibacterial potential of synthesized AuNPs with and without irradiation was tested against one human pathogen bacterial strain: *E. coli* (DSM 1576 Medium 1), as a model for Gram-negative bacteria. The lyophilized microorganisms were pre-cultured aerobically, with shaking in 50 mL of DSM Medium 1 for *E. coli* for 24 h at 37 °C, and were maintained on nutrient agar NA slants (0.5% beef extract, 1% peptone, 0.5% NaCl, and 1.5% agar). The organisms were stored at 4 °C and subcultured at regular intervals of 30 days to maintain the cell viability. The bacterium was transferred from stored slants at 4 °C to 10 mL of nutrient broth (meat extract, peptone, NaCl), and cultivated overnight at 37 °C. Considering that, for *E. coli*, an OD600 nm of 0.1 corresponds to a concentration of approx. 10^8 cells/mL, for experiments, the bacterial cultures were diluted in sterile PBS (phosphate-buffered saline) to obtain a microorganism suspension of about 10^4 cells/mL. To evaluate the antibacterial performances of gold nanoparticles against *E. coli* and of their photothermal bactericidal property, we used the growth inhibition assay. This method determines the number of viable bacteria, colony-forming units (CFU/mL) after 24 h of contact between microorganism and AuNPs, in the dark and irradiated. The bacterial suspension of 2×10^4 cells, in the exponential phase of growth, were added to the gold nanoparticles diluted in the growth DSM Medium 1 (1:500) to reach four final gold nanoparticle concentrations: 0.26, 0.39, 1.56, 3.54 µg/mL, in a final volume of 2ml, and placed in petri wells of 35 mm. A duplicate of these samples was irradiated with a green light source for 5 min. In addition, the bacterial strain sample without nanoparticles was used as negative control. All petri wells were incubated at 35 °C for 24 h with 95% humidity and were shaken at 200 rpm on a stirrer plate (Orbital Shaker, PSU-10i, Grant-bio). Cell division occurs slowly because the growth medium is very diluted.

All samples were collected and 100 µL of appropriate dilution was spread over the surface of the nutritive agar using a sterile bent plastic rod. After incubation at 35 °C for 24 h, the number of CFU was evaluated. Each experiment was performed in duplicate and repeated 3 times. The inhibitory effect was calculated using the following formula:

$$\text{Percent inhibition} = 1 - T/C \times 100$$

where T is the CFU/mL of the test sample after 24 h, and C is the CFU/mL of the control after 24 h [73,74].

Electron microscopy (TEM and SEM), dynamic light scattering (DLS), photophysical and photothermal measurements

Transmission electron microscopy (TEM). The size and shape of gold nanoparticles were characterized using a transmission electron microscope (TEM). Samples for TEM were prepared by depositing a drop of a diluted solution on formvar/carbon-film-coated 300 mesh copper grids for 15 min. This process was followed by the removal of extra solution using blotting paper. After that, the grids were allowed to dry prior to measurement. The analysis was carried out on a JEOL JEM-1400 Plus transmission electron microscope at an operating voltage of 80 kV [75].

Scanning electron microscopy (SEM). SEM analysis was carried out to investigate the uptake and effects of AuNP on bacterial cell morphology. Briefly, samples were fixed with 3% glutaraldehyde for 2 h and dehydrated with a graded series of ethanol solutions (50%, 60%, 70%, 80%, 90%, 99%, and anhydrous ethanol) for 10 min each. Prior to observation, specimens were coated with graphene films and finally viewed under a scanning electron microscope ZEISS Crossbeam 350 operating at 10.00 kV. Both secondary electrons (SE) and backscattered electrons (BSE) images were simultaneously acquired and compared. Acquisition by SE revealed topographic information with excellent resolution. With the use of BSE, the inorganic AuNP and the organic structures were distinguished by virtue of their different atomic number. This atomic number sensitivity creates contrast in the image where the inorganic NPs (with high atomic number) appear as bright spots, while the organic structures (bacterial cells) with low atomic number appear darker. This enables the ready visualization of inorganic NPs and their location both outside and/or inside the cells.

The analysis of interactions of gold nanoparticles with bacteria and their morphological effects requires high resolution imaging techniques due the extremely small size of the plasmonic metal nanoparticles. Nanoparticles are not individually distinguishable with conventional optical microscopy, since their size is below the resolution limit. Thanks to its high resolution, transmission electron microscopy (TEM) has proven to be a powerful tool that allows observation of nanoparticles inside the cells, and is widely used for the analysis of nanoparticle uptake and relationships with cell and tissue components [76]. However, sample preparation can be a rather challenging, laborious, and/or time-consuming process, and the obtained information for each sample is limited to the thickness of the cell slices [77]. A new generation of high-resolution SEM (HRSEM) provides less limitation, with a final image resolution better than 1 nm, that makes it possible to analyze the interactions/uptake of metallic nanoparticles by cells. HRSEM benefits from the rapid (but accurate) method for the sample preparation and the higher depth-of-field imaging, thereby providing detailed information on the 3D morphological organization of cells. Using this technique, it is possible to visualize the interaction of NPs with the cell membrane and map their 3D distribution.

Dynamic light scattering (DLS). Size distribution of nanoparticles was measured by dynamic light scattering, by using a Zetasizer Nano S from Malvern Instruments (632.8 nm, 4 mW HeNe gas laser, avalanche photodiode detector, 175° detection). The measurements were performed in triplicate at 25 °C. AuNS@PEG in water and AuNS@PEG in PBS were characterized. The results are showed in Figure S1 (Supplementary Material).

Photophysical characterization. Perkin Elmer Lambda 900 spectrophotometer was employed to obtain the absorption spectra [78–80]. An amount of 3 mL of the nanoparticle dispersion was transferred from the reaction flask to a quartz cuvette to carry out the measurement.

Photothermal characterization. Solutions were irradiated, within a customized thermo-optical setup, by using a CW laser source (gem532; Laser Quantum, Stockport, UK), emitting at 532 nm in the high-absorption plasmonic band of the investigated AuNPs. The laser beam acted perpendicularly (from the top) to the air/solution interface, using three mirrors, in the central part of a quartz cuvette. A high-resolution thermal camera was used to map and quantify the temperature increase of the AuNPs solutions under top-pumping laser excitation. The IR thermoimages were recorded by ThermoCamera FLIR (A655sc), providing thermal images with 640 × 480 pixels, with an accuracy of ±2 °C.

3. Results

The extinction spectrum in Figure 1 shows a 540 nm band due to the plasmonic resonance of 50 nm- diameter AuNS@PEG-SH, while the TEM image confirms the shape and size of synthetized gold nanoparticles.

AuNS@PEG-SH were administered to *E. coli* culture, at different concentrations, in dark or light conditions. The results of cell inhibition growth are shown in Table 1. In Table S1 (Supplementary Material), for more clarity, the results of the original experiments with counting colonies are also reported.

According to these results, in the dark, the administration of gold nanoparticles to *E. coli* has no detectable effect at low concentrations (0.26 and 0.39 µg/mL), while at the concentration of 1.56 µg/mL there is a growth inhibition of 15%, which reaches a greater bacteriostatic effect at the concentration value of 3.54 µg/mL, causing a growth inhibition of 46%. These results show that the MIC (minimum inhibitory concentration) of AuNS@PEG-SH in reference to *E. coli* culture is in the range 0.39–1.56 µg/µL. This behavior is due to the toxic effect exerted by the gold nanoparticles: due to their nano size, they penetrate cells, causing cytotoxic damage [81–83]. According to SEM images (Figure 2A,B), SE and BSE paired images demonstrate the uptake of AuNPs in non-irradiated *E. coli* cells.

Table 1. *E. coli* CFU/mL exposed to different concentrations of AuNP, with and without irradiation, and the percentage of growth inhibition after 24 h of growth in incubator. No changes were recorded for the control (*E. coli* CFU/mL 2.1×10^8 with and without irradiation). The data are reported as the average of three determinations made in duplicate ± standard deviation.

	Dark Condition		Under Irradiation	
AuNP Concentration	*E. coli* (CFU/mL)	% Growth Inhibition	*E. coli* (CFU/mL)	% Growth Inhibition
0.26 µg/mL	$2.1 \times 10^8 \pm 0.179 \times 10^8$	0	$2.12 \times 10^8 \pm 0.133 \times 10^8$	0
0.39 µg/mL	$2.08 \times 10^8 \pm 0.227 \times 10^8$	0	$2.08 \times 10^8 \pm 0.232 \times 10^8$	0
1.56 µg/mL	$1.8 \times 10^8 \pm 0.145 \times 10^8$	−15%	$0.977 \times 10^8 \pm 0.117 \times 10^8$	−53%
3.54 µg/mL	$1.23 \times 10^8 \pm 0.232 \times 10^8$	−46%	$2.1 \times 10^6 \pm 0.219 \times 10^6$	−99%

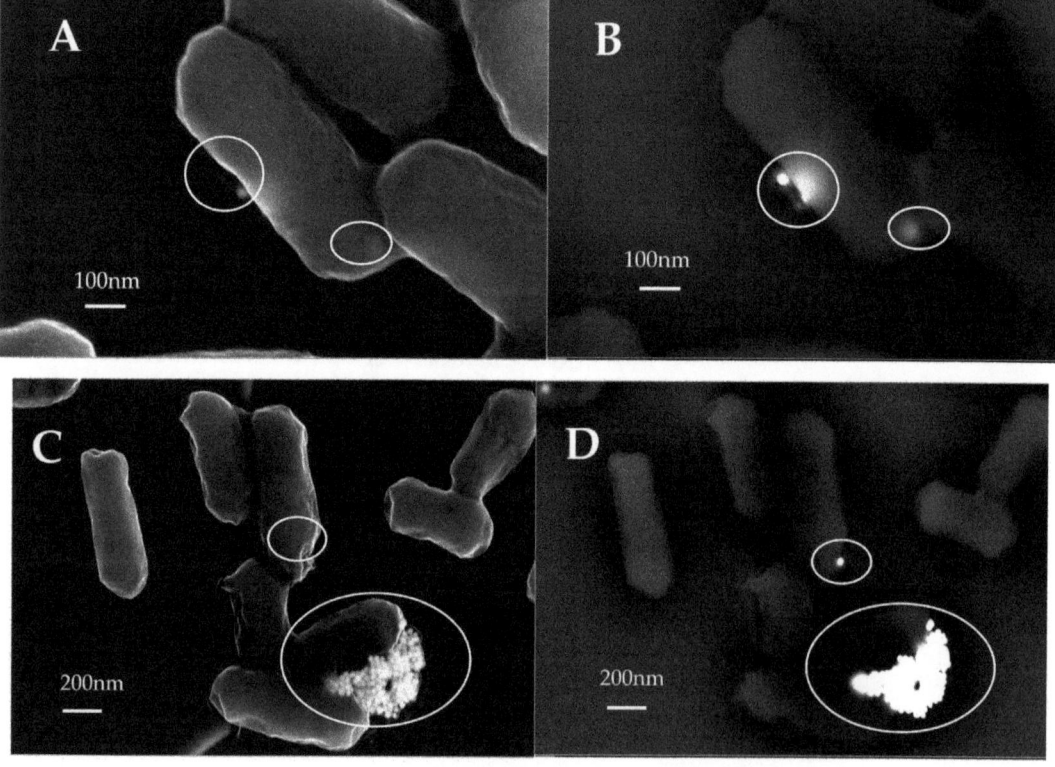

Figure 2. Secondary electrons (on the left) and backscattered electrons (on the right) paired images demonstrating the uptake of AuNP (3.54 µg/mL) in both non-irradiated (**A**,**B**) and irradiated (**C**,**D**) *E. coli* cells. Irradiated *E. coli* cells show greater pleomorphism (rough surface) and shrinkage in size because of increased cell death. Large aggregates of AuNP adhere onto the cell surface.

To activate plasmon resonance, we irradiated the bacterial cultures by using a 532 nm laser source (i.e., at a wavelength matching the nanoparticles plasmonic band) with a soft power of 60 mW. The results reported in Table 1 show that the growth inhibition is equal to 53%, while at the same concentration in the dark the value was 15%; it increases as the concentration of nanoparticles increases, reaching the remarkable value of 99% when the concentration of gold nanoparticles is equal to 3.54 µg/mL (while this value measured in the dark was 46%). This exceptional inhibition value, that corresponds to a bactericidal effect, is attributed to the cytotoxic effect, due the gold nanoparticles' uptake, combined with the photothermal one (Figure 2C,D). According to the literature, a bactericidal effect is obtained when the radiation induces a temperature increase over 60 °C [83,84]. By examining the SEM images of Figure 2C,D, a formation of gold nanoparticles clusters is clearly visible, which we assume to be responsible for the bactericidal photothermal effect.

Under the same conditions, an aqueous solution of AuNS@PEG-SH at a concentration of 3.54 µg/mL (i.e., the concentration inducing a bactericidal effect) was irradiated, achieving a temperature increase of 2 °C (Figure S2 in Supplementary Information). This slight increase in temperature, due to the irradiation of the gold nanoparticles, remained modest also by increasing the concentration of the nanoparticles: in a saturated solution, the maximum measured value was 11.6 °C (Figure S3 in Supplementary Information). Interestingly, in these aqueous solutions of AuNS@PEG-SH, cluster formation has never been observed. The only way to measure a noticeable rise in temperature was to increase the laser power (1500 mW). Indeed, by irradiating the gold nanoparticle solution with the concentration showing a bactericidal effect (i.e., 3.54 µg/mL), we measured a temperature increase of 26 °C (Figure S4 in Supplementary Information), while by irradiating the saturated solution, it showed a temperature increase of 35.4 °C (Figure S5 in Supplementary Information); in the latter case, we observed the formation of nanoparticle clusters. In fact, within seconds, a layer of gold appeared on the surface of the colloidal solution under irradiation. Upon interrupting the irradiation and shaking the sample, the layer of gold disappeared; this phenomenon, induced by high-power laser irradiation, is widely reported in the literature [44,71,85,86].

4. Discussion

The photothermal effect was studied both in the dispersion of nanoparticles alone (at two different concentrations, i.e., 3.54 µg/mL and saturated solution) and in presence of bacteria (at the 3.54 µg/mL concentration of nanoparticles). In the colloidal dispersion of nanoparticles alone, at the 3.54 µg/mL concentration, a relevant photothermal effect was not observed, while this effect was detected in the saturated solution, where clusters formation was observed. This leads to the conclusion that the clusters are responsible for the photothermal effect.

In the bacterial cultures that showed significant growth inhibition, clusters of gold nanoparticles were formed even at low concentrations (3.54 µg/mL) and with a low-power laser (60 mW). Since these clusters are formed, at this concentration, only in the presence of bacterial cells, we suppose that it is the bacterial wall that favors the aggregation of gold nanoparticles, once irradiated. It is precisely this cluster the effective photothermal tool, determining an increase of the local temperature, responsible of the observed bacterial growth inhibition and to the final cell death, as displayed in Figure 2C,D, where irradiated *E. coli* cells exhibit greater pleomorphism (rough surface) and shrink in size caused by cell death.

To obtain a photothermal effect, mediated by gold nanoparticles, it is necessary to have two requirements: a high-power light source and nanoparticles of considerable size. These two necessary aspects have disadvantages: the use of high-power lasers can itself be harmful to the cellular environment in many ways; the use of large nanoparticles makes administration difficult. In this work we have tried to overcome these two limitations by using a low-power laser and small size gold nanoparticles. A complete inhibition of bacterial growth was observed, due to the fact that the low-power laser induced the formation of gold nanoparticle clusters of such dimensions that are able to induce, under irradiation, a temperature increase capable of obtaining a bactericidal effect [17]. This fact, which under the same experimental conditions was not observed in the absence of bacteria, is due to a sort of catalytic effect exerted by the bacterial wall in the formation of the clusters, as can be clearly seen from the SEM images.

5. Conclusions

Local temperature increase is extensively exploited in photothermal therapy, where light is used to induce cellular damage. Gold nanoparticles, showing plasmon resonance phenomena when irradiated with electromagnetic radiation, are able to induce relevant temperature rises mainly dependent on their size.

In this work we present a proof-of-concept nano-sized system for a photothermal treatment of an *E. coli* culture. We synthesized 50 nm-diameter gold nanospheres covered with thiolate polyethylene glycol, AuNS@PEG-SH, and administered these nanoparticles in water solutions at different concentrations, to a bacterial culture, observing cell growth in the dark or under light irradiation.

Previously, we have measured the temperature values of the AuNP solutions under high-power (1500 Mw) laser irradiation, measuring a temperature value of 54.4 °C, while, as expected, low-power laser source (60 Mw) causes a negligible temperature rise; in both cases the values depend directly on the concentration of the solutions.

Bacterial treatments with AuNS@PEG-SH aqueous solutions show no growth inhibition at low concentration, while at 1.56 μg/mL and 3.54 μg/mL, we observe, in the dark, a bacteriostatic effect (46% of growth inhibition at a concentration of 3.54 μg/mL). SEM images show an uptake of nanoparticles in the bacteria cells, responsible for the observed bacterial inhibition growth.

To irradiate the bacteria, it was preferred to use a soft-power laser (i.e., 60 mW) in order to have no harmful effects due to the power of the source alone. Under these conditions, we observed an AuNP concentration-dependent inhibition of the bacterial growth, which at the concentration of 3.54 μg/mL leads to a bactericidal effect. To explain this surprising effect, we collected SEM images of the bacterial culture which was administered the AuNP solution with a concentration of 3.54 μg/mL, and irradiated with the low-power laser. We observed the formation of nanoparticle clusters on the bacterial wall. These clusters are not formed in the absence of bacteria except using high power lasers. These clusters are responsible for the low-power-induced photothermal effect, and their formation is catalyzed by the bacterial wall.

In conclusion, the results show that these nanoparticles constitute a proof-of-concept of a photothermal system able to activate only in the presence of the pathogen, and to have a bactericidal effect even at low concentrations.

Supplementary Materials: The following supporting information can be downloaded at: https://www.mdpi.com/article/10.3390/nano13040746/s1, Figure S1: DLS measurement; Figure S2: Thermal photo of AuNS@PEG-SH water solution (3.54 μg/mL) irradiated with a laser source (λ = 532 nm, 60 mW). Room temperature 19 °C; Figure S3: Thermal photo of AuNS@PEG-SH-saturated water solution irradiated with a laser source (λ = 532 nm, 60 mW). Room temperature 19 °C; Figure S4: Thermal photo of AuNS@PEG-SH water solution (3.54 μg/mL) irradiated with a laser source (λ = 532 nm, 1500 mW). Room temperature 19 °C; Figure S5: Thermal photo of AuNS@PEG-SH-saturated water solution irradiated with a laser source (λ = 532 nm, 1500 mW). Room temperature 19 °C. Table S1. Results of the original experiments with counting colonies.

Author Contributions: Synthesis of nanoparticles, photophysical and photothermal characterization, validation data curation A.C.; Conceptualization, investigation, validation data curation, writing—editing, supervision M.L.D.; Electron microscopy characterization, validation data curation I.D.P.; Preparation of microorganism suspension, validation and curation of *E. coli* UFC/mL data R.D.R.; Photothermal characterization A.G. All authors have read and agreed to the published version of the manuscript.

Funding: This research received no external funding.

Institutional Review Board Statement: Not applicable.

Informed Consent Statement: Not applicable.

Acknowledgments: The authors are grateful to Mariano Davoli—Dipartimento DiBEST—Centro Microscopia e Microanalisi (CM2) Università della Calabria—Rende (CS) for the acquisition of SEM images, to NLHT- Nanoscience Laboratory for Human Technologies POR Calabria FESR-FSE 14/20 for the photothermal measurement, and to Nicolas Godbert and Iolinda Aiello for DLS measurements.

Conflicts of Interest: The authors declare no conflict of interest.

References

1. Prakash, A.; Ouyang, J.; Lin, J.L.; Yang, Y. Polymer memory device based on conjugated polymer and gold nanoparticles. *J. Appl. Phys.* **2006**, *100*, 054309. [CrossRef]
2. Huang, D.; Liao, F.; Molesa, S.; Redinger, D.; Subramanian, V. Plastic-Compatible Low Resistance Printable Gold Nanoparticle Conductors for Flexible Electronics. *J. Electrochem. Soc.* **2003**, *150*, G412. [CrossRef]
3. Quaresma, P.; Osório, I.; Dória, G.; Carvalho, P.A.; Pereira, A.; Langer, J.; Araújo, J.P.; Pastoriza-Santos, I.; Liz-Marzán, L.M.; Franco, R.; et al. Star-shaped magnetite@gold nanoparticles for protein magnetic separation and SERS detection. *RSC Adv.* **2014**, *4*, 3659–3667. [CrossRef]
4. Liu, F.K. Analysis and applications of nanoparticles in the separation sciences: A case of gold nanoparticles. *J. Chromatogr. A* **2009**, *1216*, 9034–9047. [CrossRef]
5. Das, M.; Shim, K.H.; An, S.S.A.; Yi, D.K. Review on gold nanoparticles and their applications. *Toxicol. Environ. Health Sci.* **2011**, *3*, 193–205. [CrossRef]
6. Madkour, L.H. Applications of gold nanoparticles in medicine and therapy. *Pharm. Pharmacol. Int. J.* **2018**, *6*, 157–174. [CrossRef]
7. Bansal, S.A.; Kumar, V.; Karimi, J.; Singh, A.P.; Kumar, S. Role of gold nanoparticles in advanced biomedical applications. *Nanoscale Adv.* **2020**, *2*, 3764–3787. [CrossRef]
8. Zhang, N.; Xu, A.; Liu, B.; Godbert, N.; Li, H. Lyotropic liquid crystals of tetradecyldimethylaminoxide in water and the in situ formation of gold nanomaterials. *ChemPhysMater* **2022**, in press. [CrossRef]
9. Zeng, Z.; Chen, Y.; Zhu, X.; Yu, L. Polyaniline-supported nano metal-catalyzed coupling reactions: Opportunities and challenges. *Chin. Chem. Lett.* **2023**, *34*, 107728. [CrossRef]
10. Sun, Y.; Jiang, L.; Zhong, L.; Jiang, Y.; Chen, X. Towards active plasmonic response devices. *Nano Res.* **2015**, *8*, 406–417. [CrossRef]
11. Kang, H.; Buchman, J.T.; Rodriguez, R.S.; Ring, H.L.; He, J.; Bantz, K.C.; Haynes, C.L. Stabilization of Silver and Gold Nanoparticles: Preservation and Improvement of Plasmonic Functionalities. *Chem. Rev.* **2019**, *119*, 664–699. [CrossRef]
12. Venkatesh, N. Metallic Nanoparticle: A Review. *Biomed. J. Sci. Tech. Res.* **2018**, *4*, 3765–3775. [CrossRef]
13. Talarico, A.M.; Szerb, E.I.; Mastropietro, T.F.; Aiello, I.; Crispini, A.; Ghedini, M. Tuning solid state luminescent properties in a hydrogen bonding-directed supramolecular assembly of bis-cyclometalated iridium(iii) ethylenediamine complexes. *Dalton Trans.* **2012**, *41*, 4919–4926. [CrossRef]
14. Yue, K.; Nan, J.; Zhang, X.; Tang, J.; Zhang, X. Photothermal Effects of Gold Nanoparticles Induced by Light Emitting Diodes. *Appl. Therm. Eng.* **2016**, *99*, 1093–1100. [CrossRef]
15. Candreva, A.; Di Maio, G.; Parisi, F.; Scarpelli, F.; Crispini, A.; Godbert, N.; Ricciardi, L.; Nucera, A.; Rizzuto, C.; Barberi, R.C.; et al. Luminescent Self-Assembled Monolayer on Gold Nanoparticles: Tuning of Emission According to the Surface Curvature. *Chemosensors* **2022**, *10*, 176. [CrossRef]
16. Candreva, A.; Lewandowski, W.; La Deda, M. Thickness control of the Silica Shell: A way to tune the Plasmonic Properties of isolated and assembled Gold Nanorods. *J. Nanoparticle Res.* **2022**, *24*, 19. [CrossRef]
17. Candreva, A.; Morrone, E.; La Deda, M. Gold Sea Urchin-Shaped Nanoparticles: Synthesis and Characterization of Energy Transducer Candidates. *Plasmonics* **2022**, *18*, 291–298. [CrossRef]
18. Candreva, A.; Parisi, F.; Bartucci, R.; Guzzi, R.; Maio, D. Synthesis and Characterization of Hyper-Branched Nanoparticles with Magnetic and Plasmonic Properties. *Chemistryselect* **2022**, *7*, e202201375. [CrossRef]
19. Candreva, A.; Di Maio, G.; La Deda, M. A quick one-step synthesis of luminescent gold nanospheres. *Soft Matter* **2020**, *16*, 10865–10868. [CrossRef] [PubMed]
20. Wang, Y.; Serrano, A.B.; Sentosun, K.; Bals, S.; Liz-marzán, L.M. Stabilization and Encapsulation of Gold Nanostars Mediated by Dithiols. *Small* **2015**, *11*, 4314–4320. [CrossRef] [PubMed]
21. Hamon, C.; Novikov, S.; Scarabelli, L.; Basabe-Desmonts, L.; Liz-Marzán, L.M. Hierarchical self-assembly of gold nanoparticles into patterned plasmonic nanostructures. *ACS Nano* **2014**, *8*, 10694–10703. [CrossRef] [PubMed]
22. Scarabelli, L.; Coronado-puchau, M.; Giner-casares, J.J.; Langer, J.; Liz-marza, L.M. Monodisperse Gold Nanotriangles: Assembly, and Performance in Surface-Enhanced Raman Scattering. *ACS Nano* **2014**, *8*, 5833–5842. [CrossRef] [PubMed]
23. Scarabelli, L.; Grzelczak, M.; Liz-Marzán, L.M. Tuning gold nanorod synthesis through prereduction with salicylic acid. *Chem. Mater.* **2013**, *25*, 4232–4238. [CrossRef]
24. Serrano-Montes, A.B.; De Aberasturi, D.J.; Langer, J.; Giner-Casares, J.J.; Scarabelli, L.; Herrero, A.; Liz-Marzán, L.M. A General Method for Solvent Exchange of Plasmonic Nanoparticles and Self-Assembly into SERS-Active Monolayers. *Langmuir* **2015**, *31*, 9205–9213. [CrossRef]
25. Wang, P.; Wang, X.; Wang, L.; Hou, X.; Liu, W.; Chen, C. Interaction of gold nanoparticles with proteins and cells. *Sci. Technol. Adv. Mater.* **2015**, *16*, 34610. [CrossRef]

26. Pellas, V.; Hu, D.; Mazouzi, Y.; Mimoun, Y.; Blanchard, J.; Guibert, C.; Salmain, M.; Boujday, S. Gold Nanorods for LSPR Biosensing: Synthesis, Coating by Silica, and Bioanalytical Applications. *Biosensors* **2020**, *10*, 146. [CrossRef]
27. Hammami, I.; Alabdallah, N.M.; Al Jomaa, A.; Kamoun, M. Gold nanoparticles: Synthesis properties and applications. *J. King Saud Univ.—Sci.* **2021**, *33*, 101560. [CrossRef]
28. Shari, M.; Attar, F.; Akbar, A.; Akhtari, K.; Hooshmand, N. Plasmonic gold nanoparticles: Optical manipulation, imaging, drug delivery and therapy. *J. Control. Release* **2019**, *312*, 170–189. [CrossRef]
29. Bhattacharya, S.; Alkharfy, K.M.; Mukhopadhyay, D. Nanomedicine: Pharmacological perspectives. *Nanotechnol. Rev.* **2012**, *1*, 235–253. [CrossRef]
30. Mironava, T.; Hadjiargyrou, M.; Simon, M.; Jurukovski, V.; Rafailovich, M.H. Gold nanoparticles cellular toxicity and recovery: Effect of size, concentration and exposure time. *Nanotoxicology* **2010**, *4*, 120–137. [CrossRef]
31. Li, N.; Zhao, P.; Astruc, D. Anisotropic Gold Nanoparticles: Synthesis, Properties, Applications, and Toxicity. *Angew. Chem. Int. Ed.* **2014**, *53*, 1756–1789. [CrossRef] [PubMed]
32. Henriksen-lacey, M. Cellular Uptake of Gold Nanoparticles Triggered by Host−Guest Interactions. *J. Am. Chem. Soc.* **2018**, *140*, 4469–4472. [CrossRef]
33. Iswarya, V.; Manivannan, J.; De, A.; Paul, S.; Roy, R.; Johnson, J.B.; Kundu, R.; Chandrasekaran, N.; Mukherjee, A. Surface capping and size-dependent toxicity of gold nanoparticles on different trophic levels. *Environ. Sci. Pollut. Res.* **2016**, *23*, 4844–4858. [CrossRef]
34. Chen, Y.-S.; Hung, Y.-C.; Liau, I.; Huang, G.S. Assessment of the In Vivo Toxicity of Gold Nanoparticles. *Nanoscale Res. Lett.* **2009**, *4*, 858–864. [CrossRef] [PubMed]
35. Pan, Y.; Neuss, S.; Leifert, A.; Fischler, M.; Wen, F.; Simon, U.; Schmid, G.; Brandau, W.; Jahnen-Dechent, W. Size-dependent cytotoxicity of gold nanoparticles. *Small* **2007**, *3*, 1941–1949. [CrossRef]
36. Reznickova, A.; Slavikova, N.; Kolska, Z.; Kolarova, K.; Belinova, T.; Kalbacova, M.H.; Cieslar, M.; Svorcik, V. PEGylated gold nanoparticles: Stability, cytotoxicity and antibacterial activity. *Colloids Surf. A* **2019**, *560*, 26–34. [CrossRef]
37. He, X.; Sathishkumar, G.; Gopinath, K.; Zhang, K.; Lu, Z.; Li, C.; Kang, E.; Xu, L. One-step self-assembly of biogenic Au NPs/PEG-based universal coatings for antifouling and photothermal killing of bacterial pathogens. *Chem. Eng. J.* **2021**, *421*, 130005. [CrossRef]
38. Hu, Y.; Wang, R.; Wang, S.; Ding, L.; Li, J.; Luo, Y.; Wang, X.; Shen, M.; Shi, X. Multifunctional Fe_3O_4 @ Au core/shell nanostars: A unique platform for multimode imaging and photothermal therapy of tumors. *Sci. Rep.* **2016**, *6*, 28325. [CrossRef]
39. Sani, A.; Cao, C.; Cui, D. Toxicity of gold nanoparticles (AuNPs): A review. *Biochem. Biophys. Rep.* **2021**, *26*, 100991. [CrossRef]
40. Vecchio, G.; Galeone, A.; Brunetti, V.; Maiorano, G.; Sabella, S.; Cingolani, R.; Pompa, P.P. Concentration-dependent, size-independent toxicity of citrate capped AuNPs in drosophila melanogaster. *PLoS ONE* **2012**, *7*, e29980. [CrossRef]
41. Wang, S.; Lu, W.; Tovmachenko, O.; Rai, U.S.; Yu, H.; Ray, P.C. Challenge in understanding size and shape dependent toxicity of gold nanomaterials in human skin keratinocytes. *Chem. Phys. Lett.* **2008**, *463*, 145–149. [CrossRef] [PubMed]
42. Demir, E. A review on nanotoxicity and nanogenotoxicity of different shapes of nanomaterials. *J. Appl. Toxicol.* **2021**, *41*, 118–147. [CrossRef] [PubMed]
43. Zoroddu, M.; Medici, S.; Ledda, A.; Nurchi, V.; Lachowicz, J.; Peana, M. Toxicity of Nanoparticles. *Curr. Med. Chem.* **2014**, *21*, 3837–3853. [CrossRef] [PubMed]
44. Lapotko, D.O.; Lukianova-Hleb, E.Y.; Oraevsky, A.A. Clusterization of nanoparticles during their interaction with living cells. *Nanomedicine* **2007**, *2*, 241–253. [CrossRef]
45. Barbosa, S.; Agrawal, A.; Rodríguez-Lorenzo, L.; Pastoriza-Santos, I.; Alvarez-Puebla, R.A.; Kornowski, A.; Weller, H.; Liz-Marzán, L.M. Tuning size and sensing properties in colloidal gold nanostars. *Langmuir* **2010**, *26*, 14943–14950. [CrossRef]
46. Borzenkov, M.; Määttänen, A.; Ihalainen, P.; Collini, M.; Cabrini, E.; Dacarro, G.; Pallavicini, P.; Chirico, G. Photothermal effect of gold nanostar patterns inkjet-printed on coated paper substrates with different permeability. *Beilstein J. Nanotechnol.* **2016**, *7*, 1480–1485. [CrossRef]
47. Arguinzoniz, A.G.; Ruggiero, E.; Habtemariam, A.; Hernández-gil, J.; Salassa, L.; Mareque-rivas, J.C. Light Harvesting and Photoemission by Nanoparticles for Photodynamic Therapy. *Part. Part. Syst. Charact.* **2014**, *31*, 46–75. [CrossRef]
48. Sherwani, M.A.; Tufail, S.; Khan, A.A.; Owais, M. Gold Nanoparticle-Photosensitizer Conjugate Based Photodynamic Inactivation of Biofilm Producing Cells: Potential for Treatment of C. albicans Infection in BALB/c Mice. *PLoS ONE* **2015**, *10*, e013168. [CrossRef]
49. Calavia, P.G.; Russell, D.A.; Bruce, G.; Pérez-garcía, L. Photosensitiser-gold nanoparticle conjugates for photodynamic therapy of cancer. *Photochem. Photobiol. Sci.* **2018**, *17*, 1534–1552. [CrossRef]
50. Hwang, S.; Jung, S.; Doh, H.; Kim, S. Gold nanoparticle-mediated photothermal therapy: Current status and future perspective. *Nanomedicine* **2022**, *9*, 2003–2022. [CrossRef]
51. Pallavicini, P.; Donà, A.; Casu, A.; Chirico, G.; Collini, M.; Dacarro, G.; Falqui, A.; Milanese, C.; Sironi, L.; Taglietti, A. Triton X-100 for three-plasmon gold nanostars with two photothermally active NIR (near IR) and SWIR (short-wavelength IR) channels. *Chem. Commun.* **2013**, *49*, 6265–6267. [CrossRef] [PubMed]
52. Annesi, F.; Pane, A.; Losso, M.A.; Guglielmelli, A.; Lucente, F.; Petronella, F.; Placido, T.; Comparelli, R.; Guzzo, M.G.; Curri, M.L.; et al. Thermo-plasmonic killing of *Escherichia coli* TG1 bacteria. *Materials* **2019**, *12*, 1530. [CrossRef] [PubMed]

53. Guglielmelli, A.; Rosa, P.; Contardi, M.; Prato, M.; Mangino, G.; Miglietta, S.; Petrozza, V.; Pani, R.; Calogero, A.; Athanassiou, A.; et al. Biomimetic keratin gold nanoparticle-mediated in vitro photothermal therapy on glioblastoma multiforme. *Nanomedicine* **2021**, *16*, 121–138. [CrossRef] [PubMed]
54. Yeh, Y.C.; Creran, B.; Rotello, V.M. Gold nanoparticles: Preparation, properties, and applications in bionanotechnology. *Nanoscale* **2012**, *4*, 1871–1880. [CrossRef] [PubMed]
55. Umamaheswari, K.; Baskar, R.; Chandru, K.; Rajendiran, N.; Chandirasekar, S. Antibacterial activity of gold nanoparticles and their toxicity assessment. *BMC Infect. Dis.* **2014**, *14*, 2334. [CrossRef]
56. Cui, Y.; Zhao, Y.; Tian, Y.; Zhang, W.; Lü, X.; Jiang, X. Biomaterials The molecular mechanism of action of bactericidal gold nanoparticles on *Escherichia coli* q. *Biomaterials* **2012**, *33*, 2327–2333. [CrossRef] [PubMed]
57. He, Y.; Dong, H.; Li, T.; Wang, C.; Shao, W.; Zhang, Y.; Jiang, L.; Hu, W. Graphene and graphene oxide nanogap electrodes fabricated by atomic force microscopy nanolithography. *Appl. Phys. Lett.* **2010**, *97*, 133301. [CrossRef]
58. Gouyau, J.; Duval, R.E.; Boudier, A.; Lamouroux, E. Investigation of Nanoparticle Metallic Core Antibacterial Activity: Gold and Silver Nanoparticles against *Escherichia coli* and *Staphylococcus aureus*. *Int. J. Mol. Sci.* **2021**, *22*, 1905. [CrossRef]
59. Miller, S.E.; Bell, C.S.; Mejias, R.; Mcclain, M.S.; Cover, T.L.; Giorgio, T.D. Colistin-Functionalized Nanoparticles for the Rapid Capture of *Acinetobacter baumannii*. *J. Biomed. Nanotechnol.* **2016**, *12*, 1806–1819. [CrossRef]
60. Azam, A.; Ahmed, F.; Arshi, N.; Chaman, M.; Naqvi, A.H. One step synthesis and characterization of gold nanoparticles and their antibacterial activities against *E. coli* (ATCC 25922 strain). *Int. J. Theor. Appl. Sci.* **2009**, *1*, 1–4.
61. Liu, M.; Zhang, X.; Chu, S.; Ge, Y.; Huang, T.; Liu, Y.; Yu, L. Selenization of cotton products with NaHSe endowing the antibacterial activities. *Chin. Chem. Lett.* **2022**, *33*, 205–208. [CrossRef]
62. Pissuwan, D.; Cortie, C.H.; Valenzuela, S.M.; Cortie, M.B. Functionalised gold nanoparticles for controlling pathogenic bacteria. *Trends Biotechnol.* **2010**, *28*, 207–213. [CrossRef]
63. Lima, E.; Guerra, R.; Lara, V.; Guzmán, A. Gold nanoparticles as efficient antimicrobial agents for *Escherichia coli* and *Salmonella typhi*. *Chem. Central J.* **2013**, *7*, 11. [CrossRef] [PubMed]
64. Mubdir, D.M.; Al-shukri, M.S.; Ghaleb, R.A. Antimicrobial Activity of Gold Nanoparticles and SWCNT-COOH on Viability of *Pseudomonas aeruginosa*. *Ann. Rom. Soc. Cell Biol.* **2021**, *25*, 5507–5513.
65. Zhang, Y.; Dasari, T.P.S.; Deng, H.; Yu, H. Journal of Environmental Science and Health, Part C: Environmental Carcinogenesis and Ecotoxicology Reviews Antimicrobial Activity of Gold Nanoparticles and Ionic Gold. *J. Environ. Sci. Heath Part C Environ. Carcinog. Ecotoxicol. Rev.* **2015**, *3*, 37–41. [CrossRef]
66. Lai, M.-J.; Huang, Y.-W.; Chen, H.-C.; Tsao, L.-I.; Chang Chien, C.-F.; Singh, B.; Liu, B.R. Effect of Size and Concentration of Copper Nanoparticles on the Antimicrobial Activity in *Escherichia coli* through Multiple Mechanisms. *Nanomaterials* **2022**, *12*, 3715. [CrossRef] [PubMed]
67. Das, B.; Mandal, D.; Dash, S.K.; Chattopadhyay, S.; Tripathy, S.; Dolai, D.P.; Dey, S.K.; Roy, S. Eugenol Provokes ROS-Mediated Membrane Damage-Associated Antibacterial Activity against Clinically Isolated Multidrug-Resistant *Staphylococcus aureus* Strains. *Infect. Dis. Res. Treat.* **2016**, *9*, IDRT.S31741. [CrossRef]
68. Bao, H.; Yu, X.; Xu, C.; Li, X.; Li, Z.; Wei, D.; Liu, Y. New toxicity mechanism of silver nanoparticles: Promoting apoptosis and inhibiting proliferation. *PLoS ONE* **2015**, *10*, 1–10. [CrossRef]
69. Behera, N.; Arakha, M.; Priyadarshinee, M.; Pattanayak, B.S.; Soren, S.; Jha, S.; Mallick, B.C. Oxidative stress generated at nickel oxide nanoparticle interface results in bacterial membrane damage leading to cell death. *RSC Adv.* **2019**, *9*, 24888–24894. [CrossRef]
70. Ahmed, B.; Ameen, F.; Rizvi, A.; Ali, K.; Sonbol, H.; Zaidi, A.; Khan, M.S.; Musarrat, J. Destruction of Cell Topography, Morphology, Membrane, Inhibition of Respiration, Biofilm Formation, and Bioactive Molecule Production by Nanoparticles of Ag, ZnO, CuO, TiO_2, and Al_2O_3 toward Beneficial Soil Bacteria. *ACS Omega* **2020**, *5*, 7861–7876. [CrossRef]
71. Matsuo, N.; Muto, H.; Miyajima, K.; Mafuné, F. Single laser pulse induced aggregation of gold nanoparticles. *Phys. Chem. Chem. Phys.* **2007**, *9*, 6027–6031. [CrossRef] [PubMed]
72. Piella, J.; Bastús, N.G.; Puntes, V. Size-dependent protein-nanoparticle interactions in citrate-stabilized gold nanoparticles: The emergence of the protein corona. *Bioconjug. Chem.* **2017**, *28*, 88–97. [CrossRef] [PubMed]
73. Caligiuri, R.; Di Maio, G.; Godbert, N.; Scarpelli, F.; Candreva, A.; Rimoldi, I.; Facchetti, G.; Lupo, M.G.; Sicilia, E.; Mazzone, G.; et al. Curcumin-based ionic Pt(ii) complexes: Antioxidant and antimicrobial activity. *Dalton Trans.* **2022**, *51*, 16545–16556. [CrossRef]
74. Policastro, D.; Giorno, E.; Scarpelli, F.; Godbert, N.; Ricciardi, L.; Crispini, A.; Candreva, A.; Marchetti, F.; Xhafa, S.; De Rose, R.; et al. New Zinc-Based Active Chitosan Films: Physicochemical Characterization, Antioxidant, and Antimicrobial Properties. *Front. Chem.* **2022**, *10*, 884059. [CrossRef]
75. Mastropietro, T.F.; Meringolo, C.; Poerio, T.; Scarpelli, F.; Godbert, N.; Di Profio, G.; Fontanova, E. Multistimuli Activation of TiO_2/α-alumina membranes for degradation of methylene blue. *Ind. Eng. Chem. Res.* **2017**, *56*, 11049–11057. [CrossRef]
76. Malatesta, M. Transmission electron microscopy as a powerful tool to investigate the interaction of nanoparticles with subcellular structures. *Int. J. Mol. Sci.* **2021**, *22*, 12789. [CrossRef] [PubMed]
77. Schrand, A.M.; Rahman, M.F.; Hussain, S.M.; Schlager, J.J.; Smith, D.A.; Syed, A.F. Metal-based nanoparticles and their toxicity assessment. *Wiley Interdiscip. Rev. Nanomed. Nanobiotechnol.* **2010**, *2*, 544–568. [CrossRef]

78. Cretu, C.; Andelescu, A.A.; Candreva, A.; Crispini, A.; Szerb, E.I.; La Deda, M. Bisubstituted-biquinoline Cu(i) complexes: Synthesis, mesomorphism and photophysical studies in solution and condensed states. *J. Mater. Chem. C* **2018**, *6*, 10073–10082. [CrossRef]
79. La Deda, M.; Di Maio, G.; Candreva, A.; Heinrich, B.; Andelescu, A.A.; Popa, E.; Voirin, E.; Badea, V.; Amati, M.; Costişor, O.; et al. Very intense polarized emission in self-assembled room temperature metallomesogens based on Zn(ii) coordination complexes: An experimental and computational study. *J. Mater. Chem. C* **2022**, *10*, 115–125. [CrossRef]
80. Liguori, P.F.; Ghedini, M.; La Deda, M.; Godbert, N.; Parisi, F.; Guzzi, R.; Ionescu, A.; Aiello, I. Electrochromic behaviour of Ir(III) bis-cyclometalated 1,2-dioxolene tetra-halo complexes: Fully reversible catecholate/semiquinone redox switches. *Dalton Trans.* **2020**, *49*, 2628–2635. [CrossRef]
81. Marsich, E.; Travan, A.; Donati, I.; Di, A.; Benincasa, M.; Crosera, M.; Paoletti, S. Colloids and Surfaces B: Biointerfaces Biological response of hydrogels embedding gold nanoparticles. *Colloids Surf. B Biointerfaces* **2011**, *83*, 331–339. [CrossRef] [PubMed]
82. Harmsen, S.; Huang, R.; Wall, M.A.; Karabeber, H.; Samii, J.M.; Spaliviero, M.; White, J.R.; Monette, S.; O'Connor, R.; Pitter, K.L.; et al. Surface-enhanced resonance Raman scattering nanostars for high-precision cancer imaging. *Sci. Transl. Med.* **2015**, *7*, 271ra7. [CrossRef] [PubMed]
83. Charimba, G.; Hugo, C.J.; Hugo, A. The growth, survival and thermal inactivation of *Escherichia coli* O157:H7 in a traditional South African sausage. *Meat Sci.* **2010**, *85*, 89–95. [CrossRef] [PubMed]
84. FDA Bacterial Pathogen Growth and Inactivation. *Fish Fish. Prod. Hazards Control. Guid.* **2011**, *22*, 417–438.
85. Robinson-Enebeli, S.; Talebi-Moghaddam, S.; Daun, K.J. Time-resolved laser-induced incandescence on metal nanoparticles: Effect of nanoparticle aggregation and sintering. *Appl. Phys. B Lasers Opt.* **2023**, *129*, 25. [CrossRef]
86. Candreva, A.; Parisi, F.; Di, G.; Francesca, M.; Iolinda, S.; Godbert, N.; La, M. Post-Synthesis Heating, a Key Step to Tune the LPR Band of Gold Nanorods Covered with CTAB or Embedded in a Silica Shell. *Gold Bull.* **2022**, *55*, 195–205. [CrossRef]

Disclaimer/Publisher's Note: The statements, opinions and data contained in all publications are solely those of the individual author(s) and contributor(s) and not of MDPI and/or the editor(s). MDPI and/or the editor(s) disclaim responsibility for any injury to people or property resulting from any ideas, methods, instructions or products referred to in the content.

Article

From Nano-Crystals to Periodically Aggregated Assembly in Arylate Polyesters—Continuous Helicoid or Discrete Cross-Hatch Grating?

Cheng-En Yang [1], Selvaraj Nagarajan [1], Widyantari Rahmayanti [1], Chean-Cheng Su [2] and Eamor M. Woo [1,*]

[1] Department of Chemical Engineering, National Cheng Kung University, No. 1, University Road, Tainan 701-01, Taiwan
[2] Department of Chemical and Materials Engineering, National University of Kaohsiung, No. 700, Kaohsiung University Rd., Nan-Tzu Dist., Kaohsiung 811-48, Taiwan
* Correspondence: emwoo@mail.ncku.edu.tw; Tel.: +886-2757575 (ext. 62670)

Abstract: This work used several model arylate polymers with the number of methylene segment n = 3, 9, 10, and 12, which all crystallized to display similar types of periodically banded spherulites at various T_c and kinetic factors. Universal mechanisms of nano- to microscale crystal-by-crystal self-assembly to final periodic aggregates showing alternate birefringence rings were probed via 3D dissection. The fractured interiors of the birefringent-banded poly(decamethylene terephthalate) (PDT) spherulites at T_c = 90 °C revealed multi-shell spheroid bands composed of perpendicularly intersecting lamellae bundles, where each shell (measuring 4 μm) was composed of the interior tangential and radial lamellae, as revealed in the SEM results, and its shell thickness was equal to the optical inter-band spacing (4 μm). The radial-oriented lamellae were at a roughly 90° angle perpendicularly intersecting with the tangential ones; therefore, the top-surface valley band region appeared to be a submerged "U-shape", where the interior radial lamellae were located directly underneath. Furthermore, the universal self-assembly was proved by collective analyses on the three arylate polymers.

Keywords: nano- to micro-patterns; arylate polymers; self-assembly

1. Introduction

Crystal deformation into variety of geometric shapes is common in nature. Almost a century ago, Ferdinand Bernauer [1] certainly was an early-day pioneer on probing variety of crystal deformation by analyzing more than 400 compounds and reached a conclusion that many of these molecular crystals can be made to grow as twist helices or other deformations that could be conveniently observed under polarized-light optical microscopes. Over the past two decades, in a series of systematic investigations and a comprehensive review, Kahr et al. [2–4] probed many representative cases of small-molecule compounds (both organic and inorganic), which are worthy as comparative background information for addressing similar issues in polymer crystals. Long-chain polymeric crystals, other than chain-folding in lamellae, share some interesting common features of crystal deformation as those found in small-molecule compounds. Woo and Lugito [5] further summarized this in a review article demonstrating novel approaches by 3D dissection into interior lamellae with periodic-banded spherulites of many polymers to disclose novel mechanisms with discontinuity in corrugate-board structures, and the review also provided summary evidence for interior periodic assembly matching with optical inter-ring spacing.

Homologous series of arylate polyesters are widely studied. Polymorphic poly(butylene terephthalate) (PBT) reportedly displays crystal transformation [6,7]; yet, it does not form ring-banded aggregates at any T_c. Polyarylates with longer methylene groups than that of

PBT, such as poly(pentamethylene terephthalate) (PPenT), poly(hexamethylene terephthalate) (PHT), and poly(heptamethylene terephthalate) (PHepT), are not as widely studied due to the fewer commercial applications. PPenT can transform from the α form to the β form [8,9], where both the α and β forms are characterized by triclinic chain packing in unit cells, which differ only in the c-axis but remain the same in other lattice parameters. The α-crystal cell is seen in PPenT under zero tension; by contrast, the β-modification dominates if crystallized under high tensions [8,9]. By comparison, PPenT, possessing one more methylene unit than PBT in its repeat unit and, thus, a lower melting point, has been less studied. PPenT has an equilibrium T_m value of 149.4 °C, as reported in the literature [8], much lower than that of PBT due to the even–odd effects in the main chains. Regardless of the similar polymorphism behavior of PBT and PPenT, PBT never crystallizes into periodic ring-banded spherulites; by contrast, PPenT displays distinct ring-banded spherulites [10]. PBT is similar to PPenT in that they both display polymorphic crystal lattices, but the former does not pack into periodic banded aggregate upon crystallization, while the latter does. Polymorphic PHT, similar to polymorphic PBT, does not crystallize into periodic ring-banded aggregates [11]; by contrast, PHepT, with one more methylene segment in the repeating unit than that in PHT, readily crystallizes into ring-banded crystals [12]. The even–odd effect superficially appears to work in the formation of ring-banded polymer spherulites; however, it may be too early to predict. Poly(octamethylene terephthalate) (POT) [13], with an even number of methylene segments in the repeat unit, also easily crystallizes to display periodic bands at a suitable T_c.

A polymorphic polymer can form a ring-banded pattern or a ringless one; conversely, a monomorphic polymer can also form ring-banded aggregates or nonbanded ones. Obviously, the above factual comparison suggests that polymorphism, with more than one crystal lattice form, in polymers is not a determining factor that accounts for the formation of periodic bands or nonbanded ones. The brief survey also indicates that all these arylate polyesters naturally chain-fold upon packing in their long chains into crystalline lamellae of finite thickness (ca. 10–15 nm) and, thus, all inevitably have surface stresses in the lamellae; however, the facts are that some display ring bands but others do not—suggesting that chain-fold stresses might not be a determining factor. From the fact that not all polymers display ring bands, there appears to be a contradiction of the proposition of surface-stress-induced lamellae twist, by applying the classical Aristotle's proof-by-contradiction. By observing from these experimental facts, chemical structural and kinetic factors both can influence the morphology patterns and periodicity in the final crystallized aggregates of aryl polyesters [9–16]. Recently, poly(nonamethylene terephthalate) (PNT) [14–17] was found to display not just one single type of ring band but two dramatically different types of ring bands when crystallized at a specific T_c, suggesting that periodic bands in polymer aggregates cannot be interpreted by the classical continuous helix-twist models. Aryl polyesters may possess polymorphism; however, monomorphism or not in the crystal-unit lattices usually cannot be correlated with the formation of multiple types of spherulites. Monomorphic poly(octamethylene terephthalate) (POT) displays multiple types of spherulites; poly(heptamethylene terephthalate) (PHepT) can possess fractions of α and β crystal lattices and can exhibit many different spherulite patterns with peculiarly varying periodicity [18]. Although commercially available PTT has been widely investigated [19–25], arylate polyesters and their crystal morphologies have been less probed [20–28].

The lamellar assembly of PDT in comparison to a series of other arylate polyester spherulites for more universal proofs was the aim of this work. The governing mechanisms of self-assembly into the intriguing periodically banded patterns were analyzed to understand the universal periodicity commonly seen in arylate polyesters, such as PTT, in comparison to other arylates with a longer methylene segment in repeat units, such as PDT, PNT, POT, and PDoT. The crystal lattice structures of many arylate polyesters might have been classically studied using X-ray [29]; however, the higher-hierarchical lamellar aggregation into periodically crystal assembly has rarely been investigated on these arylate polyesters owing to the complexity of the assembly mechanisms. By following the preliminary and pioneering investigations on periodic assembly of several simple arylate polyesters [26], this work further expounded the mechanisms in fuller and wider detail, and the grating assembly in the nano- to microstructures with suitable cross-bar pitches responsible for photonic iridesence was demonstrated for universal and unprecedented proofs.

2. Materials and Methods

The crystal morphologies of several arylate polyesters in homologous series, with their methylene segments varying systematically from short to long, are discussed here. For comparison purposes, a homologous series of arylate polyesters were analyzed and compared. Poly(decamethylene terephthalate) (PDT or P10T, with 10 methylene segments in each repeating unit) was used. The synthesis procedures for PDT were similar to those used earlier in syntheses of a similar polyester of poly(dodecamethylene terephthalate) (PDoT, m = 12) reported earlier [26,30]. All these polyarylates are not commercially available and had to be synthesized in-house with a two-step polymerization, with procedures the same as those earlier described in a previous work. Briefly, for reference, the procedure is restated here: monomers of 1,10-decanediol and dimethyl terephthalate (DMT) with 0.1% butyl titanate as a catalyst were heated in vessels via ester exchanges. The purification of the products was properly conducted by washing out traces of unreacted monomers or impurities. The weight-averaged molecular weight (M_w) of the synthesized PDT, as determined by gel-permeation chromatography (GPC, Waters), was 21,000 g/mol with PDI = 2.02. Other arylate polyester, poly(dodecamethylene terephthalate) (PDoT, m = 12), was similarly synthesized using respective diols; for brevity, they are not all listed here [26,30]. PDoT (m = 12) has M_n = 9224 g mol^{-1} and the polydispersity index (PDI) = 1.91, as determined by gel-permeation chromatography, which has a low glass transition temperature and a medium melting temperature at -1.3 and 121 °C, respectively. Poly(trimethylene terephthalate) (PTT, m = 3) is commercially available Industrial technology research institute, Hsinchu, Taiwan and has a glass transition (T_g) and melting temperature (T_m) of 45 °C and 228 °C [31], respectively.

The film specimens were prepared by drop-casting. The polymers were dissolved in chloroform at ~4 wt.% by stirring. The homogeneous solution was then cast on the glass slide as thin films and dried by placing at 30 °C in a vacuum oven to remove the residual solvent. The film samples were first heated to a suitable maximum melting temperature (T_{max}) for 2 min on top of a hotplate for erasing the thermal history and then removed rapidly from the hotplate to a temperature-controlled hot stage (temperature precision ± 0.5 °C) being preset at the intended crystallization temperatures till full crystallization. For the interior dissection of the morphology, fracturing on the specimens (thickness 10–20 μm) was performed. The films of the crystallized samples (at a controlled film thickness and T_c) on the glass substrates were fractured by precutting the glass substrates to direct the intended fracture propagation. The fractured pieces of the samples were affixed onto metal stands at proper orientation angles using silver-tape glues at suitable inclination angles with respect to the electron beams in the SEM vacuum chamber, Tokyo, Japan.

Apparatus and Procedures

A polarized-light optical microscope (POM, Nikon Optiphot-2, Tokyo, Japan) equipped with a Nikon Digital Sight (DS-U1) digital camera and a microscopic heating stage (Linkam THMS-600 with T95 temperature programmer, Linkam Scientific Instrument Ltd., Surrey, UK) was used.

Atomic-force microscopy (AFM diCaliber, Veeco Corp., Santa Barbara, CA, USA) investigations on the top surfaces of the cast film samples were made in an intermittent tapping mode with a silicon tip (f_0 = 70 kHz, r = 10 nm) installed. The largest scan range was 150 μm × 150 μm, and the scan was kept at 0.4 Hz for the overview scan and zoomed-in regions.

High-resolution field-emission scanning electron microscopy (Hitachi SU8010, HR-FESEM, Tokyo, Japan) was used for revealing the interior lamellar assembly in the exposed fracture surfaces/interiors, which was to be correlated with the morphology and ring patterns on the top free surfaces. The samples, after fracturing and setting on metal stands, were coated with gold or platinum vapor deposition using vacuum sputtering prior to the SEM characterization.

3. Results and Discussion

3.1. Interior Crystal Assembly in Birefringence-Banded PDT Spherulite

The crystallization temperature (T_c) kinetically influences the mechanisms of nucleation and growth. For PDT, two dramatically different types of ring-banded patterns were present at high versus low T_c [30,32]. For appreciating the difference between these two types of ring bands formed at medium T_c (90–95 °C) versus high T_c (120–125 °C), Figure 1 demonstrates two dramatically different banding patterns of PDT spherulites, with the film thickness being kept at constant 15–20 μm. Figure 1a illustrates the single extinction-band pattern crystallized at relatively high T_c = 115 °C or above, while interestingly, Figure 1b shows an intermediate pattern between these two types at an intermediate T_c = 110 °C, which was composed of "dual-ring bands" in the central core but extinction bands at the outer rims with a unique core–shell pattern. That is, the spherulitic aggregate was actually a composite core–shell morphology of two discrete types of optical birefringence patterns. Figure 1c shows dual-birefringence ring bands crystallized at T_c = 90–95 °C or lower. The mechanisms of the crystal assembly in the dual-birefringence ring-banded PDT at T_c = 90 °C of thicker films were therefore different from that in the epicycloid-extinction PDT at higher T_c = 110–125 °C and in thinner films. As discussed in an earlier work [32], another type of ring band, termed epicycloid-extinction-banded PDT spherulites, at higher T_c = 110–125 °C is sensitively dependent on the film thickness, with the epicycloid-extinction band patterns disappearing completely at film thickness >10 μm, and the inter-band spacing increasing dramatically with respect to the increasing film thickness. Note that the inter-band spacing of the PDT spherulite was 3–5 μm at T_c = 95 °C (Figure 1c), which means that the film thickness should not be lower than 3–5 μm if dual-birefringence bands are to be packed in the PDT specimens. When the film thickness was lower than this critical value (inter-band spacing), then no dual-birefringence bands were observed in the crystallized films; instead, only bands with an optical extinction border (termed "extinction band") were present. This is easy to understand, as thin films constrain lamellae to be normal oriented in cross-hatch patterns.

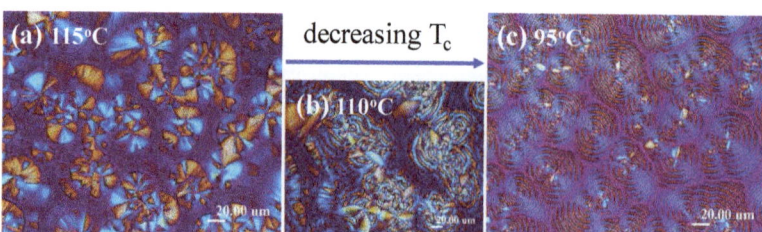

Figure 1. POM micrographs of PDT melt crystallized at decreasing T_c: (**a**) 115 °C; (**b**) 110 °C; (**c**) 90 °C, held till full crystallization by quenching from T_{max} = 165 °C. Reprinted with permission from ref. [32]. Copyright 2021 Elsevier.

Thus, in dramatic contrast to the extinction-banded spherulites (with successive bands bordered with a sharp extinction ring) crystallized at high T_c (120 °C or above), the crystallization of the same PDT films at low to intermediate T_c, such as T_c = 90–95 °C, led to optically dual-birefringent spherulites of distinct alternating blue/orange rings in the spherulites, where the ring bands were not bordered with optical extinction but were featured with alternate birefringence colors in POM with tint plates. This behavior of no dual-birefringence bands in PDT at T_c = 90 °C was opposite to that of PDT at higher T_c = 120 °C, where only extinction bands could be present, as discussed. This fact suggests that the nature of the extinction PDT bands might be significantly different from that in the dual-birefringence PDT bands. Although both extinction bands and dual-birefringent bands were featured with distinct periodicity in the circular rings, the dramatic differences in the optical birefringence properties in these two types of PDT spherulites suggest that the interior crystal-lamellae assembly may significantly differ with respect to their respective optical birefringence patterns. The effect of the film thickness (3–20 μm) did not appear to influence the morphological patterns of the birefringent spherulites with blue/orange bands. Note that for thick PDT films with too high optical retardation (τ), the optical light does not penetrate easily; thus, the blue/orange birefringence could not be contrasted in visible patterns. However, the birefringent ring bands of PDT spherulites at T_c = 90 °C remained similar in pattern and displayed same inter-band spacing (ca. 3–4 μm) for PDT film samples, which did not change much with the increase in the film thickness (from ca. 5–20 μm).

Not only T_c but also confinement by film thickness might influence the lamellar assembly and, thus, birefringence patterns, as seen in POM. A preliminary investigation revealed that a minimum film thickness (3–5 μm) of PDT specimens was required to display dual-birefringence ring bands. Supplementary Materials Figure S1 shows POM graphs of PDT spherulites of four different levels of film thickness: (a) ultra-thin at 300~500 nm, (b) 3~5 μm, (c) 8~10 μm, and (d) 15~20 μm, all at T_c = 90 °C. The very thin PDT film (ca. 300–500 nm) did not exhibit any discernible dual-birefringence ring bands upon crystallization at T_c = 90 °C. The PDT films of all other higher thickness levels (5–20 μm) displayed optically similar dual-birefringent alternating blue/orange bands, with similar inter-band spacing. The effects of the variation of the melt-exposure time (from 1–120 min) at T_{max} = 165 °C on the birefringent PDT spherulites were also investigated. For comparison, Supplementary Materials ESI Figure S2 shows POM graphs for the PDT films (thickness kept constant at 2–3 μm) all melt crystallized at T_c = 95 °C with different times (t_{max}) held at T_{max} for melting to erase the prior thermal histories: (a) 1 min, (b) 15 min, (c) 30 min, (d) 60 min, (e) 90 min, and (f) 120 min. The results show that all PDT birefringent spherulites remained the same or similar with alternate blue/orange bands, with the same inter-band spacing (4 μm). All spherulites remained similar in size with a radius = 30~50 μm. This fact suggests that the melt/thermal exposure at T_{max} = 165 °C had no effect of altering the birefringent bands and that exposure at this temperature caused no degradation. In addition, it should also be commented here that the birefringent spherulites were not influenced by the top-cover confinement on the films during the crystallization at T_c = 95 °C; by contrast, the crystallization of the PDT at T_c = 120 °C with top-cover confinement led to no periodic bands (i.e., ringless), yet extinction bands were visibly present if no top-cover was placed on the PDT films.

Prior to the SEM analysis of the interior lamellae assembly, AFM analysis was performed to reveal the nanopatterns on the top surface of 90 °C crystallized PDT spherulites that displayed dual-birefringent colors with blue/orange bands. Figure 2 shows AFM images of the PDT samples (films of thickness ca. 3–5 μm) crystallized at T_c = 90 °C. The ring bands with the alternate optical birefringence (blue/orange) is shown as an inset in Figure 2A. Note that the pattern of the PDT dual-birefringent PDT spherulites (at T_c = 90 °C) differed completely from those for the extinction-banded PDT spherulites in thin films at T_c = 120 °C, earlier disclosed in a concurrent work on PDT [32]. The zoomed in AFM analysis on the top surface (Figure 2A1,A2) shows a nanograiny feature with alternating low and high bands. The top surface of the 90 °C crystallized birefringent-banded PDT spherulite was composed of apparently grainy polycrystals. Such dual-morphology ring bands are dramatically in contrast to the single-crystalline terrace-like packing on the top surfaces of the 120 °C crystallized PDT extinction-banded spherulite. All these features further reinforce that the nature and mechanisms of the crystals building the birefringent spherulites of the alternating blue/orange bands crystallized at T_c = 95 °C or lower should differ widely from those governing the extinction-banded PDT spherulites crystallized at T_c = 120 °C or higher. However, the detailed lamellae assembly in the interior of the banded spherulites could still not be discerned from the POM patterns or the AFM analysis on the top surface. The circular dot-like grains on the top surfaces of the PDT dual-birefringence bands suggest that they might be the individual terminal ends of the interior lamellae, as they emerge from the inner bulk to top surface. Note also that the darker bands in Figure 2A1,A2 represent the valley, where the grainy dots are oriented in a different direction in contrast to the elongated crystals in the neighboring brighter bands (ridges). This further suggests that the interior lamellae emerge to the top surface in correspondingly different orientations depending on the valley or ridge bands. A later section unveils the interior crystal assembly responsible for such dual-birefringence PDT bands. Interior dissection into the assembly of the interior lamellae crystals in the birefringent PDT spherulites (crystallized at T_c = 90 °C) may shed new light on answering these critical questions. From the AFM phase images, the inter-band spacing = 3–4 μm.

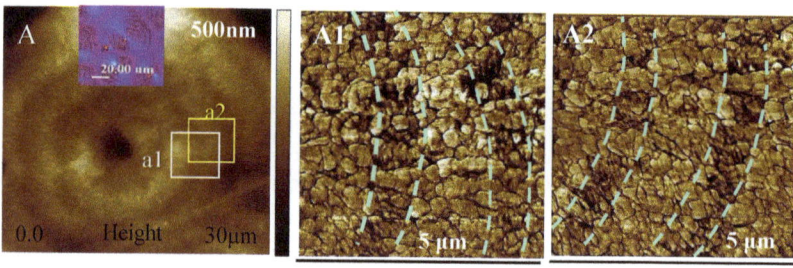

Figure 2. Nanoassembly on the top-surface pattern as viewed in (**A**) AFM height images and (**A1**,**A2**) zoomed-in images of the top surface of the alternate birefringent blue/orange bands in the PDT spherulites at T_c = 90 °C (film thickness = 3–5 μm).

The band patterns on the top surfaces of the blue/orange birefringent PDT spherulites (at T_c = 90 °C) could mislead in the interpretation of the mechanisms of periodic banding if the interior lamellae were not analyzed. Interior dissection by examining the fractured PDT spherulites was performed by SEM characterization. Figure 3A,B show SEM micrographs of the PDT melt crystallized at 90 °C, and Figure 3C,D show schemes illustrating alternate tangential-to-radial lamellae with discontinuity. The PDT films were kept at thickness = ca. 15 μm. All PDT samples were first melted at T_{max} = 165 °C (1 min), then quenched to T_c = 90 °C, and held till full crystallization. Both the top surfaces and interior fractured surfaces of the crystallized PDT specimens were examined. The SEM graphs for the top surfaces revealed similar grainy crystal aggregates on the "ridge region", which is similar to the AFM images discussed earlier. The valley bands on the top surface were of a flat and lower region in comparison to the ridge bands. Apparently, lamellae assembly is not possible by simply examining the top surfaces using either SEM or AFM analysis. As the nucleus center of the spherulite is located near the top surface, the onion-like interior morphology appears like a multi-shell concave-up bowl (i.e., a hemispheroid). The fracture–dissection SEM results in Figure 3B also reveal very critical pieces of evidence showing that the ridge bands on the top surfaces actually correspond to regions where the interior tangential lamellae emerge to the top surfaces; by contrast, the valley bands on the top surfaces correspond to where radial lamellae evolving or bending from the tangential ones. Both the top surface and interior grating-like array clearly reveal that the inter-band spacing was consistently ca. ~3 μm. From the above SEM results for the interiors of the banded PDT (T_c = 90 °C), the correlations between the top periodic bands vs. the interior lamellae assembly can be feasibly constructed. The schemes in Figure 3C show the interior lamellae assembled as a mutually intersecting grating. Apparently, the interior lamellae, as revealed in the SEM result for the interior of the banded PDT, are assembled as a cross-hatch grating structure, whose cross-bar pitch = 3 μm and equal to the optical inter-band spacing in the

POM images. Except for the interfacial layer, the radial lamellae were always roughly a 90° angle perpendicularly intersecting with the tangential ones; therefore, the top-surface valley band region appeared to be a submerged "U-shape", where the interior radial lamellae were located directly underneath. In addition, according to the SEM results, the interior tangential lamellae (or their bundles) were connected to the top-ridge region, while the interior radial lamellae were situated underneath the valley region of the top surface. Upon POM characterization with tint plates, if the interior tangential lamellae (crystals oriented in the perpendicular direction) have an orange color, then the interior radial lamellae (crystals oriented mostly in the horizontal direction) have a blue color. The periodicity repeated to produce optical patterns of orange/blue color rings according to the lamellae's mutual cross-hatching intersections, as shown in Figure 3D.

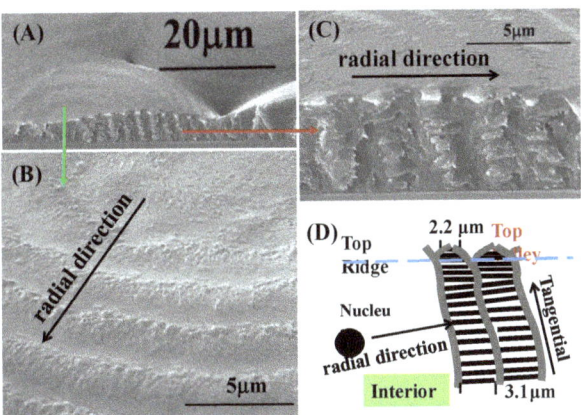

Figure 3. SEM micrographs of the PDT melt crystallized at 90 °C: (**A**) entire fractured specimen; (**B**) top surface; (**C**) fractured surface; (**D**) schemes of shells composed of tangential/radial lamellae alterations assembled as a cross-hatch grating structure (film thickness = ca. 5–10 μm varying from nucleus to periphery).

The dual-birefringence banded PDT spherulites (T_c = 90 °C) can be nucleated on the top surface or interiors of films. Figure 4 shows SEM micrographs of fractured cross-section of top-initiated banded PDT spherulites displaying a distinct layered corrugate-board structure. Of the topology, the interior lamellae with corrugate-board (i.e., multi-shell) assembly reached upward to the top surface to form the periodic banding of inter-band spacing ca. 4 μm that exactly matched with the optical spacing. Each of the interior tangential lamellar bundles corresponded to the "ridge region" of the top band patterns. Finer branching lamellae grew roughly perpendicular to the tangential lamellae; thus, these branching lamellae were aligned in the radial directions. The tangential lamellae in the 3D growth were aligned as multi-shelled hemispheroids of increasing radii, which can be clearly seen in Figure 4A. The interior lamellae of hemispheroid geometry, as revealed in the fractured PDT spherulites, can also be viewed as an "onion-like" structure cut into halves, as illustrated in the inset on the bottom of Figure 4A. The SEM graphs also show that the inter-shell distance (~4–5 μm) equaled exactly the inter-ring spacing (~4–5 μm) in the POM graphs for the PDT at T_c = 90 °C. The perfect match between the morphology and optical birefringence evidence suggests the validity of the proposed assembly mechanism, leading to the final aggregate's banding periodicity. These series of schemes for step-by-step growth in Figure 4B–D illustrate three stages of growth: from initiation of nuclei on or near the top surface to complete 3D growth to form a corrugate-board architecture with a hemispheroid geometry. The 90° angle intersection of two species of lamellae accurately accounts for the periodic optical blue–orange birefringence colors, as illustrated. With the nucleus center on the top surface of the thick PDT films, the alternately concentric shells all took a hemispheroid shape. Regardless of the film thickness or location of the nucleus centers, the analyses yielded consistent results that the interiors of the birefringent-banded PDT spherulites (T_c = 90 °C) were filled with alternate cross-hatch lamellae mutually intersecting at an oblique or nearly perpendicular angle, where the interior tangential lamellae correspond to the ridge region and interior radial lamellae to the valley region on the top surface.

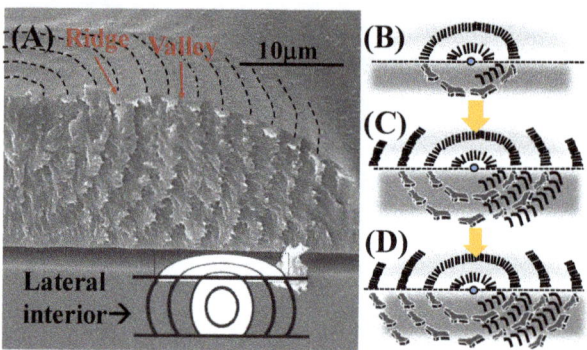

Figure 4. (**A**) SEM graphs of the interiors of the double-ring-banded PDT spherulites packed with hemi-spheroid shells in the tangent circumferential direction interfacing with radial-oriented thinner lamellae; (**B**–**D**) schemes exemplifying the initial growth near the nuclei center eventually to final multiple-packed shells with the interior perpendicularly intersected lamellae.

3.2. Mechanisms of Lamellae Packing into Birefringent Bands

Depending on the PDT film thickness and location of nuclei (top, center, or bottom of the film thickness), the fractured interiors revealed correspondingly different geometries of the alternating shell structures. Figure 5 displays SEM micrographs for fractured PDT films of various thickness. Figure 5A shows the PDT film with a thickness of 7 μm, where five bands were present in one spherulite. As a result of the nucleation on the top surface and the constraint of the narrow film thickness, the 3D-banded spherulite took a shape of a concave-up hemispheroid, and the alternating shells curved up as an "arc" shape. By contrast, for the PDT film ~40 μm thick with the nucleus center near the middle zone of a film, Figure 5B shows five alternate and concentric shells in the banded spherulite that took a spheroid shape. In the SEM evidence for the interior of the birefringent-banded PDT spherulites, it is clear that no interior lamellae underwent continuous helix-twisting, such as DNA's double helices, all originating from a common center. Instead, each of the hemispheroid shells in the banded PDT spherulite was interfaced with discontinuous tangential-to-radial interfaces. Note that some of the tangential-oriented lamellae may branch out or occasionally bend and twist at a ca. 90° angle to merge with the radial lamellae; but such a twist was abrupt at the interfacial regions. If morphological analyses were aimed only on the top surfaces of thin-film specimens, the essential assembly in the 3D-bulk interiors might have been overlooked. It is easy to mistake the occasional branching/twisting on the top surfaces as continuous screw-like helices, when actually there were discontinuous interfaces between the interior crystals that were packed to display successive bands.

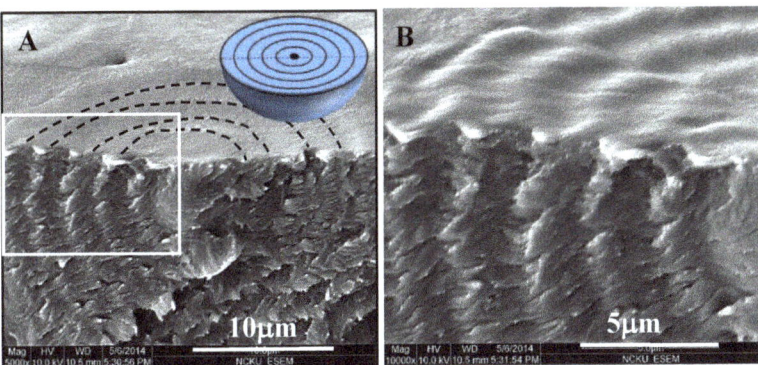

Figure 5. (**A**) SEM micrographs of fractured interior at T_c = 90 °C; (**B**) lamellar assembly in the interior of the banded PDT as a cross-hatch grating structure.

3.3. Top-Surface Morphology Versus Interior Lamellar Assembly

The fracture of samples might randomly cut across various sections of a banded spherulite, which might yield slightly different patterns of assembly. For proof of the universality of the alternating tangential/radial structures in the banded PDT spherulites, alternative fractured interior surfaces were further examined. Figure 6a,b display SEM graphs for the fractured interiors of birefringent-banded PDT spherulites (all at T_c = 90 °C), which clearly revealed a common multi-shelled hemispheroid or spheroid structure (depending on the location of the nucleation sites being near the top surface or middle of the films). In the interior of the birefringent-banded PDT spherulites, the tangential lamellae curved into a bowl shape (i.e., hemispheroid). The interior tangential lamellae were connected to the "ridge region" on the top surface. From the tangential lamellae, some lamellae either flip-twisted at a 90° angle or branch evolved in a perpendicular direction to fill the space between two neighboring tangential shells. The schemes in Figure 6c,d illustrate that the curved tangential shells were attached with a 90° angle bending/twisting or perpendicular branching. The perpendicularly twisted or branched lamellae generally oriented their long axes toward the radial direction of the spherulite. Thus, there were alternative tangential/radial lamellae layers, with each layer being shaped as multi-shelled hemispheroids (i.e., when the curved spheroids are flattened, they become "corrugate-boards"). The radial-oriented lamellae in the banded aggregates were not flat but shaped generally as a concave-up U-shape, forming a valley. From the scheme, the dark wide stripes indicated the "valley region" optical patterns that apparently were situated on top of the interior radial-oriented lamellar plates. The thin, narrow, and solid lines on the top of the scheme indicate the "ridge region" on the top surfaces, which correspond to the protruded spots of the interior tangential-oriented lamellae. The ridge bands on the top surfaces actually correspond to regions where the interior tangential-oriented lamellae emerged to the top surfaces. By contrast, the valley bands on the top surfaces correspond to where the radial lamellae evolved or bent from the tangential-oriented ones.

Apparently, if the analysis was confined to thin films without 3D inner views, investigators might have been misled by the lamellae's superficial assembly on the top surface, while the majority bulk of the assembly in the 3D interiors would have remained hidden. On the top surface of the thin film specimens, investigators might observe some occasional twisting and bending of lamellae from ridge to valley band. However, the interior lamellae (accounting for 8–9 μm of the bulk), accounting for the majority 90% of the entire bulk, would be buried and skipped. The schematic shows how the interior lamellae might emerge and twist while going to the top surface, wherein the interiors were actually composed of discontinuous shells of crystal plates and branches mutually oriented at some oblique angles.

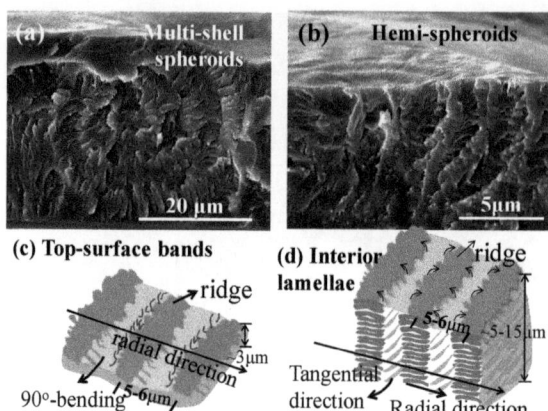

Figure 6. SEM micrographs of the interiors of two different banded PDT spherulites both melt crystallized at 90 °C: (**a**,**b**) nuclei near top surface–multi-shell hemispheroids; (**c**,**d**) schemes for the optical birefringent patterns and top vs. interior assembly (sample film thickness = ca. 20–30 μm).

It has been proved that no continuous helix-twist of single-crystal lamellae is present in the extinction-banded PDT spherulites (crystallized at high T_c = 110–120 °C) [30]. For the birefringent (blue/orange)-banded PDT spherulites crystallized at lower T_c = 90 °C, the interiors were filled with multi-shell concentric spheroids composed of alternating tangential lamellae that periodically 90° angle twisted and/or branched out to form the radial lamellae. During growth, the tangential lamellae first evolved initially from the sheaf-like nucleus center, which subsequently produced branches in perpendicular orientations to fill the expanding space as they grew outward from the nucleus center to the periphery. Toward growth termination, the periphery of the increasingly larger spherulites started to impinge on neighboring spherulites, where the growth terminated. Eventually, a multi-shell hemisphere aggregate formed in the interiors; on the top surface of the films, concentric multi-bands of a fixed inter-band spacing formed. Furthermore, doubly-birefringent banded PDT spherulites were always filled progressively with periodic branching during growth by starting from quasi-single crystals (lamellar sheaf-bundles) at the nucleus center to a final complex hierarchical aggregations of multiple lamellae. Again, in the architecture of the birefringent PDT spherulites (T_c = 90 °C), no continuous helix-twist of the single crystal lamellae was evident, although a sharp 90° twist from the tangential- to radial-oriented lamellae bordering at the discontinuous interfaces was seen in each of the hemi-spheroid shells.

Altogether, the tangential-oriented lamellae and periodically-spawned branches filled the expanding spherulite's space during growth, which increased with the increasing radius as cubic R (i.e., R^3). As the tangentially oriented lamellae emerged to the top surfaces of the spherulites, they protruded upward to become the "ridge band", while the interior radially oriented lamellae, being branches themselves growing at 90° angle from the tangential ones, remained to be submerged but curved up to form a "U-shape" valley band with a flat/smooth texture on the top surface. Furthermore, the tangential-oriented lamellae accounted for the optically blue birefringent ring, while the radial lamellae accounted for the orange rings, as viewed in the POM graphs; vice versa, in the neighboring next quadrant of the POM graphs, the opposite was true.

3.4. Universal Features of 3D Interior Assembly in Arylate Polyesters

For the universal comparison of a series of homologous arylate polyesters with PDT (m = 10), the other polyesters (m = 3, 9, 12) were similarly analyzed and compared [19]. Figure 7 shows the dissected morphology results for the interior assemblies in (A) PDoT (m = 12), (B) PTT (m = 3), and (C) PNT (m = 9), respectively. All specimens of the three arylate polyesters were crystallized as bulks at T_c = 96, 165, and 85 °C, respectively, to develop distinct ring bands (double birefringence). These three (PDoT, PTT and PNT) were fractured in similar ways as PDT (P10T), sputter-coated with gold, and characterized using SEM. Note that all arylate specimens (PDoT, PTT, PNT) for comparative purposes were recharacterized using the same SEM techniques in this work, although similar interior morphologies for them have been earlier reported in the literature [15,33,34]. One can see that these three arylate polyesters all displayed similar cross-hatch architectures (tangential-radial lamellae perpendicularly intersecting at 90° angle), differing only in the inter-layer shell thickness: 4 µm for PDoT, 10 µm for PTT, and 8 µm for PNT Type-1 band, in comparison to the same cross-hatch structure of PDT with an inter-layer shell thickness of ~5 µm in this work. Note here that PNT is more complex in the banding architecture, as it displays not just one but two entirely different types of ring bands (labeled as Type 1 and Type 2, respectively) [15]; for comparison, only Type-1 PNT was used here. All four arylate polyesters displayed similar interior architectures of cross-hatch tangential/radial intersections, proving the universality of the proposed model being fit with all the ring-banded arylate polyesters investigated here. For most arylate polyesters (from PTT to PDT) crystallized at respectively suitable temperature ranges, their banded assemblies all displayed similar periodic birefringence patterns in the interior crystal assemblies, differing mainly in the inter-band spacing and some trivial details of the assemblies.

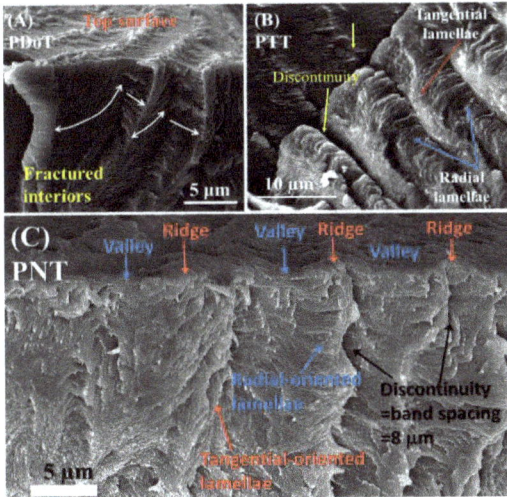

Figure 7. SEM micrographs for comparisons of the assembly analogy of interior lamellae for three arylates: (**A**) PDoT (P12T)-banded spherulite melt crystallized at T_c = 96 °C, with shell thickness: 4 µm (same as optical inter-band spacing); (**B**) PTT (P3T)-banded spherulite melt crystallized at T_c = 165 °C, with shell thickness = 10 µm (=optical inter-band spacing) [31]; (**C**) type-1 ring band of PNT (P9T) crystallized at T_c = 85 °C, with shelled thickness = 8 µm (cross-bar pitch = optical inter-band spacing). (Figure 7B Reprinted with permission from ref. [31]. Copyright 2017 American Chemical Society).

The 3D assembly of PDoT has been analyzed in an earlier work [30]. Polymeric spherulites are 3D aggregates of complex lamellar architectures with periodic branching, sporadic bending, twisting, or scrolling; thus, the interior assembly from inner lamellae to top-surface morphology should not be overlooked in investigating full mechanisms. Figure 8 illustrates three possible assemblies of lamellae in the aggregates of PDT into periodic bands. Figure 8a shows the proposition that the polymeric spherulites are 3D aggregates composed of multi-shells with cross-hatch grating architectures, with complex lamellae of periodic branching and sporadic bending. In this grating architecture, all tangential lamellae are aligned in the same direction and sandwiched in the interfacial layers of two radial-lamellar shells; similarly, the radial-oriented lamellae in the shells are all aligned in the "radial direction". Thus, the tangential-oriented lamellae display a certain birefringence pattern differing from that of the radial-oriented lamellae and collectively displaying alternate double-ring-banded pattern optically in polarized light. Subsequently, another proposition of continuous helix-twist lamellae is checked and proved by contradiction. Figure 8b shows that if all helix-twist lamellae were synchronized in helix pace, then they would have displayed alternate birefringence-to-extinction bands—an expected phenomenon that certainly contradicts with the experimental proofs according to the POM patterns with the dual-birefringence rings (i.e., both valley and ridge bands are packed with crystals but of different orientations). Finally, in a situation where the synchronized pace of these helix-twist screws is not warranted, then the total optical extinction is a result, which again contradicts with the experiment-observed optical patterns. Figure 8c shows a scheme where if all helix-twist lamellae are not in a synchronized pace but offset a fraction of the pitch from one to another in nonsynchronized alignment, then full optical extinction is the result. This of course would oppose with the POM experimental results for PDT at $T_c = 90\ ^\circ C$.

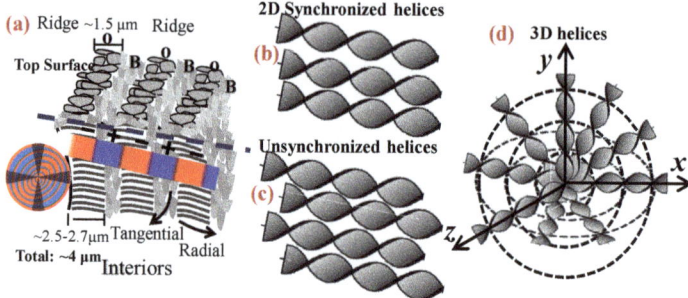

Figure 8. Crystal nanoassemblies of lamellae to higher hierarchical aggregates with the periodicity of PDT: (**a**) 3D cross-hatch with shell layers showing blue/orange birefringence alternate rings; (**b**) 2D helices in perfect synchronized pace showing birefringence/extinction alternate rings; (**c**) 2D helices in disordered pace showing complete optical extinction; (**d**) 3D helices from a common nucleus center. O–orange ring; B–blue rings. Reprinted with permission from ref. [30]. Copyright 2018 Royal Society of Chemistry).

By philosophical articulation for verifying these mutually opposite propositions (shown in Graphs-a–d in Figure 8), the classical Aristotle's proof-by-contradiction for examining the propositions of nanoassembly leading to final hierarchical periodicity was then utilized to testify these propositions by summarizing the results of these three arylate polymers. By deleting the latter two obvious cases of contradiction to the experimental results, naturally, the periodically grating assembly mechanism can be proven to be the only feasible mechanism. In summary, if investigations were conducted only by characterizing the top surfaces of crystallized polymer film specimens, as conventionally done in the long past, then one would see only the patterns of rings and lamellae on top surface but miss the most critical pieces of bulk evidence hidden in the submerged interiors under the top surface. The ridge bands on the top surfaces actually correspond to regions where the interior tangential-oriented lamellae emerge to the top surfaces; by contrast, the valley bands on the top surfaces correspond to zones where radial-oriented lamellae evolving or bending from the tangential-oriented ones. Oppositely, these discussed results collectively show that either 2D or 3D continuous screw-like helices from a nucleus center inevitably result in contradictory cases of nonsynchronized pace in thick films (20 μm), as the helix-twist lamellae, bound in a common center, cannot be physically aligned in perfect pitch pace in 2D or 3D space as they extend outward [30]

In the interior of the dual-birefringence ring-banded PDT spherulites (Figure 9), it is clear that no single-crystalline lamellae continuously helix-twist such as the double-helix conformation of DNA macromolecules from a common nucleation center are seen anywhere in the banded PDT spherulites. Instead, each of the hemispheroid shells in the banded PDT spherulites is bordered with discontinuous tangential/radial interfaces, where the interior lamellae mutually intersect like a cross-hatch grating. Note that some of the tangential-oriented lamellae may either branch out or occasionally twist by ca. 90° angles to merge with the radial lamellae. If investigators had focused their analyses only on the top surfaces of the crystallized polymer films but overlooked the lamellae assembly in the interiors or how the top-surface morphology correlates with interior lamellae, it would be easy to confuse the occasional branching/bending/twisting on the top surfaces with lamellae undergoing 360° continuous helices.

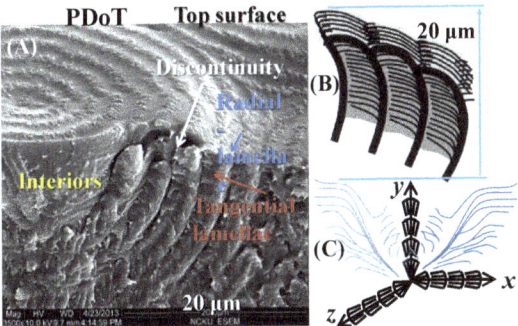

Figure 9. 3D crystal assemblies in PDoT (m = 12) aggregates: (**A**) SEM revealing a multi-shell interior responsible for blue–orange birefringence alternate rings: (**B**) 2D cross-hatch grating assembly; (**C**) cross-hatch (corrugate-board) stacking from a common nucleus center in 3D assembly leading inevitably to onion-shell-like structures in thick films. (Figure 9B,C Reprinted with permission from ref. [30]. Copyright 2018 Royal Society of Chemistry).

The optical bands in polymer spherulites can appear as a double-birefringence ring patterns or rings with extinction. Crystal assembly mechanisms in the optical extinction bands versus double birefringence rings are inherently different, and their assembled architectures are due to the completely different crystal packing mechanisms. Figure 10a shows POM micrograph of double-banded PTT spherulites' (POM for ring patterns as inset) dissected interior (3~5 μm film thickness melt-crystallized at 165 °C), with schematics in Figure 10b revealing the interfaces and crevices between successive bands. The Interfacial discontinuity, as proved by the narrow crevices between the bands, clearly support that lamellae in the double-banded PTT were not continuously helix-twist. The interior lamellae and banded PTT were assembled as periodic gratings, with twist occurring in the interfacial boundary, signaling a discontinuity. It is worth comparing the assembly in the double-banded PTT to the dramatically different morphology of extinction-banded morphology, as represented by PDT crystallized at T_c = 120 °C, as shown in Figure 10c,d. The epicycloid extinction-banded PDT spherulites crystallized at high T_c (>110 °C) were composed of terrace-like single crystals packed along the circumferential direction of the ridge band, whereas the extinction region was due to the periodic growth precipitation [27]. The evidence in this comparative study has collectively reached an advancement of the 3D depiction of PDT dual-birefringence-banded spherulites, which differs significantly from extinction-banded ones.

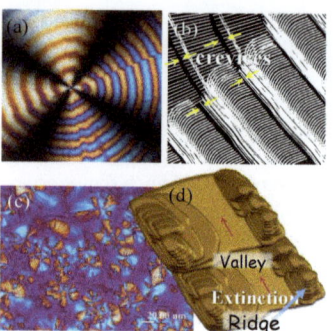

Figure 10. Crystal grating-assembly in double-banded vs. terrace-assembly in extinction-banded polymer spherulites: (**a**) POM micrograph; (**b**) schematic of grating architecture for dissected interior of PTT double-banded spherulites in 3~5 μm film thickness melt-crystallized at T_c = 165 °C, in contrast with terrace-like assembly (**c**) POM and (**d**) schematic for extinction-banded PDT spherulites at T_c = 110–120 °C. (Reprinted with permission from ref. [32]. Copyright 2021 Elsevier).

Figure 11a,b show the general schemes depicting the interior and top-surface vs. interior, respectively, of the universal grating assembly in the periodically assembled crystals. Figure 11c is the SEM micrograph for PDoT revealing both the top-surface bands and interior lamellae directly underneath these periodic bands. Figure 11d shows an enlarged scheme of the fractured interior along the radial direction on the top surface, as well as the interior. The dimensions of the interior radial-oriented lamellae approximately matches with that of the ridge band on the top surface. The exact measures of these two dimensions may differ slightly, which is due to the fact that as the interior tangential lamellae emerge upward to the top surface, they have to bend downward to become radial-oriented lamellae. By contrast, the interior radial lamellae simply branch out horizontally till impinging with the next tangential layer. Thus, the interior radial lamellae tend to be slightly longer (2.6–2.8 μm) than the top-surface radial-oriented lamellae on the ridge band (ca. 2.1 μm). Note that the tangential layer was ca. 1.5–2.0 μm; thus, the total thickness of radial+tangential shells would be roughly 4–5 μm, which is the cross-bar pitch of the corrugate-board structure in PDoT at T_c = 96 °C.

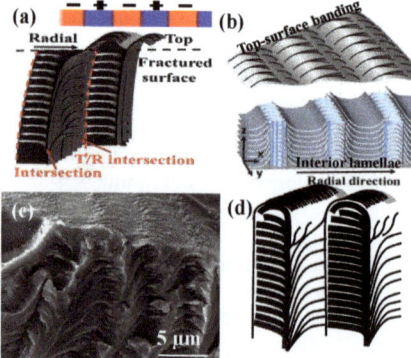

Figure 11. Schemes for SEM-analyzed morphology details in (**a**) lateral view of lamellae, (**b**) top and interior (Reprinted with permission form ref. [34]. Copyright 2021 John Wiley and Sons), (**c**) SEM micrograph, and (**d**) top-surface bands in correlation with interior assembly of ring-banded PDoT spherulites melt crystallized at T_c = 96 °C. (Figure 11b,d Reprinted with permission from ref. [30]. Copyright 2018 Royal Society of Chemistry).

Further dissection–morphology details are justified in constructing the general periodic assembly mechanism for top-surface bands and interior cross-hatch lamellae. The main features in the cyclic growth are that branches are inevitable in growing to fill an ever-expanding space (increasing with respect to cubic radius in 3D or squared radius in 2D) from a tiny nucleus sheaf to final fully grown spherical entity. Figure 12a–c shows the cut blocks of the lamellar stacks in one cycle from ridge to valley, exposing (a) the lateral side of the stacked lamellae, (b) circumferential side of stacked lamellae, and (c) top and lateral views of banded spherulites. It would be erroneous if one sees that the interior lamellae are monotonous single stalks and continuous in growing from nucleus to periphery of spherulites. Branching is a universal feature, as the lamellae self-assemble to fill the aggregated spherulites. Figure 12d shows that the branches are made of crystal-by-crystal assembly both on the top surface and interior. "Brick-by-brick" (single crystal-by-crystal) assembly means that the lamellae are not continuous and there actually is discontinuity in crystal boundaries when final periodic aggregates are formed.

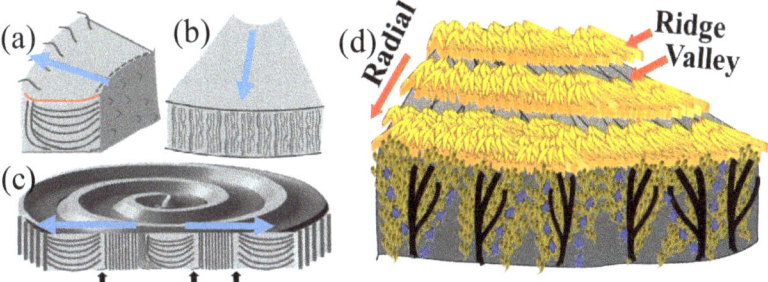

Figure 12. General periodic assembly mechanism for top-surface bands and interior cross-hatch lamellae: (**a**) block of stacked lamellae from valley to ridge cut in radial direction; (**b**) block of stacked lamellae from valley to ridge cut in tangential direction; (**c**) top-surface bands and interior lamellae cut in lateral view; (**d**) branched patterns made of crystal-by-crystal assembly into periodic rings on top and cross-hatch lamellae in interior.

3.5. Periodic Microstructures as Functional Photonic Crystals

Among the series of arylate polyesters, poly(decamethylene terephthalate) (PDT) crystals are taken as a handy example for illustrating the correlation of a periodically assembled microstructure serving as structural crystals for light interference. All PDT spherulites were typically circular in shape and exhibited distinct birefringence rings. The temperature-dependent periodic morphologies of the PDT showed a systematic trend of variation at T_c = 80–120 °C. Iridescence experiments on the PDT crystallized at these Tc were performed in accordance with the procedures and setup for white light interference as documented in the relevant literature [34–36]. Figure 13a–e show POM images (upper row) and iridescence photo shots (lower row), where the iridescence spectra generally took a softer hue color due to the lesser order of the assembly in the crystallized neat PDT films. In the range Tc = 80 °C to 95 °C, the crystallized PDT samples had the dimensions of band-ring spacing, varying from ~2.7 μm to 3.29 μm, and a spherulites radius ranging ~16.4 μm. The lesser order of the rings and smaller size of the spherulites both unfavorably affected the formation of the color, causing it to appear as a softer shade of color. In comparison, the morphology of the PDT could not produce color in the transitional morphology and epicycloid extinction ring-banded morphology. Transitional morphology is characterized by an irregular arrangement of ridges and valleys devoid of the periodic fractal branching of ridges and valleys crystal. By contrast, the fully extinction-banded PDT ridge is entirely packed with terrace-like single-crystal flat plates and the amorphous ingredient accumulates on the valley. These two situations are incapable of causing light interference because the grating architecture's condition is unconfirmed. The inter-bands spacing increased from 80 °C until Tc = 95 °C; with the crystallization temperature at 105 °C or higher, the morphology changes from the regular ring-banded pattern to another transitional morphology, resulting in few nuclei. This indicates that crystallization kinetics strongly governs the formation of ring bands. When the Tc increased above 115 °C, the ring-banded morphology changed to epicycloid extinction ring banded. Only the ordered structure of the ring-banded morphology crystallized between Tc = 80 °C and 95 °C supports effective interference with white light to produce noticeable iridescence spectra.

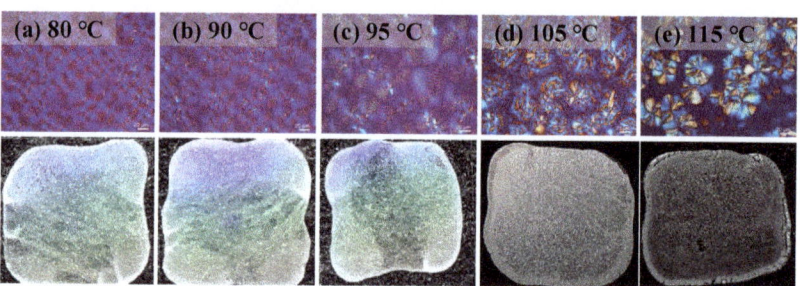

Figure 13. POM images (upper row, scale bar = 20 µm) and photonic iridescence (lower row) of neat PDT crystallized at different crystallized temperature (T_c): (**a**) 80, (**b**) 90, (**c**) 95, (**d**) 105 and (**e**) 115 °C after melting at T_{max} = 165 °C for 1 min, showing the correlation of the periodic bands and qualitative iridescence intensity. (Square box dimension in iridescence = 1.8 cm × 1.8 cm).

4. Conclusions

The birefringent-banded aggregates of PDT in comparison with PTT and PDoT display optically alternating blue/orange rings, which are composed of cross-hatch gratings of multi-shells, with each shell taking a corrugate structure composed of lamellae perpendicularly intersecting each other. The interior dissection results for the birefringent-banded PDT spherulites revealed very critical pieces of evidence for the realistic mechanisms of lamellae starting from nano- to micro-assembly and finally to aggregated crystal entities with alternate birefringent bands, which can be perfectly correlated to the top-surface and interior crystal architectures in dimensions and orientations. The radial lamellae were roughly 90° angle perpendicularly intersecting with the tangential ones; therefore, the top-surface valley band region appeared to be a submerged "U-shape", where the interior radial lamellae were located directly underneath. The total layer thickness (4–5 µm) of the ridge and valley bands, as revealed in the SEM results, equaled the inter-band spacing in the POM results.

Thus, by the results summed up from analyses on several polymers with periodic assemblies, it can be concluded that the periodicity, as revealed in optical patterns and microscopy dissected interiors, are based on discontinuous grating-like lamellar structures rather than continuous screw-like helices. The former is based on crystal-by-crystal self-assembly with periodic branching and discontinuous intersection, which matches with the nature of the formation and assembly from nanocrystals to polycrystalline architectures with periodicity in optical patterns and morphology. By examining the nano- to micro-assembly in hierarchical structures to periodic patterns in crystal aggregates of arylates, this work has provided clear and strong evidence of the absence of continuity or helical geometry. The lamellae in two types of banded PDT spherulites (extinction vs. birefringence types) display diversified mechanisms of banding periodicity that both are different but collectively cannot be described by the classical models of continuous lamellar helix-twisting. The interior lamellae dissection and analysis on distinctly different mechanisms of lamellar assembly in the extinction vs. birefringence-banded PDT spherulites, reinforced by analyses based on other arylate polyesters, provide deeper understanding of mechanisms of periodic crystal assembly phenomena that have long intrigued the science community.

Supplementary Materials: The following supporting information can be downloaded at: https://www.mdpi.com/article/10.3390/nano13061016/s1, Figure S1. Double-birefringence ring bands in PDT spherulites of increasing film thickness: (a) 300–500 nm by spin-casting, (b) 3–5 um, (c) 7–10 um, (d) 15–20 um, all crystallized at T_c = 90 °C; Figure S2. POM micrographs for birefringent-banded spherulites of neat PDT melt-crystallized at T_c = 95 °C with specimens held at T_{max} (165 °C) for different times (t_{max}): (a) 1 min, (b) 15 min, (c) 30 min, (d) 60 min, (e) 90 min, and (f) 120 min. (All film thickness kept at ca. 2–3 μm).

Author Contributions: C.-E.Y. (graduate student at this work) performed the basic experiments, organized the data and figures; S.N. (postdoc) conducted follow-up analysis and contributed in developing some in-depth schemes for interpretations; W.R. (graduate student at this work) performed iridescence measurements; C.-C.S. helped guiding work and proof-read the writing; E.M.W. (Professor) conceived the original research ideas and wrote texts of manuscript into a publishable form. All authors have read and agreed to the published version of the manuscript.

Funding: The authors are grateful for the basic research grants (MOST-110-2811-E-006-509-MY2) funded by the Taiwan Ministry of Science and Technology, reorganized to the National Council of Science and Technology (NSCT).

Data Availability Statement: The data presented in this study are available on request from the corresponding author.

Conflicts of Interest: The authors declare no conflict of interest.

References

1. Bernauer, F. *"Gedrillte" Kristalle; Verbreitung Entstehungsweise Und Beziehungen Zu Optischer Aktivität Und Molekülasymmetrie*; Gebrüder Borntraeger: Berlin, Germany, 1929.
2. Shtukenberg, A.G.; Punin, Y.O.; Gujral, A.; Kahr, B. Growth actuated bending and twisting of single crystals. *Angew. Chem. Int. Ed.* **2014**, *53*, 672–699. [CrossRef]
3. Shtukenberg, A.G.; Freudenthal, J.; Kahr, B. Reversible twisting during helical hippuric acid crystal growth. *J. Am. Chem. Soc.* **2010**, *132*, 9341–9349. [CrossRef] [PubMed]
4. Gunn, E.; Sours, R.; Benedict, J.B.; Kaminsky, W.; Kahr, B. Mesoscale chiroptics of rhythmic precipitates. *J. Am. Chem. Soc.* **2006**, *128*, 14234–14235. [CrossRef] [PubMed]
5. Woo, E.M.; Lugito, G. Origins of periodic bands in polymer spherulites. *Eur. Polym. J.* **2015**, *71*, 27–60. [CrossRef]
6. Yokouchi, M.; Sakakibara, Y.; Chatani, Y.; Tadokoro, H.; Tanaka, T.; Yoda, K. Structures of Two Crystalline Forms of Poly(butylene terephthalate) and Reversible Transition between Them by Mechanical Deformation. *Macromolecules* **1976**, *9*, 266–273. [CrossRef]
7. Chuah, H.H. Orientation and Structure Development in Poly(trimethylene terephthalate) Tensile Drawing. *Macromolecules* **2001**, *34*, 6985–6993. [CrossRef]
8. Hall, I.H.; Pass, M.G.; Rammo, N.N. Structure and properties of oriented fibers of poly(pentamethylene terephthalate)-1. Synthesis of polymer and preparation of two different crystalline phases. *J. Polym. Sci. Polym. Phys. Ed.* **1978**, *16*, 1409–1418. [CrossRef]
9. Hall, I.H.; Rammo, N.N. Structure and properties of oriented fibers of poly(pentamethylene terephthalate)-2. structural analysis of two different crystalline phases by X-ray crystallography. *J. Polym. Sci. Polym. Phys. Ed.* **1978**, *16*, 2189–2214. [CrossRef]
10. Wu, P.L.; Woo, E.M. Crystallization Regime Behavior of Poly(pentamethylene terephthalate). *J. Polym. Sci. Part B Polym. Phys.* **2004**, *42*, 1265–1274. [CrossRef]
11. Woo, E.; Wu, P.L.; Chiang, C.P.; Liu, H.L. Analysis of polymorphism and dual crystalline morphologies in poly(hexamethylene terephthalate). *Macromol. Rapid Commun.* **2004**, *25*, 942–948. [CrossRef]
12. Ni'mah, H.; Woo, E.M. Dendritic Morphology Composed of Stacked Single Crystals in Poly(ethylene succinate) Melt-Crystallized with Poly(p-vinyl phenol). *Cryst. Growth Des.* **2014**, *14*, 576–584. [CrossRef]
13. Chen, Y.F.; Woo, E.M.; Li, S.H. Dual types of spherulites in poly (octamethylene terephthalate) confined in thin-film growth. *Langmuir* **2008**, *24*, 11880–11888. [CrossRef]
14. Tu, C.-H.; Woo, E.M.; Nagarajan, S.; Lugito, G. Sophisticated dual-discontinuity periodic bands of poly(nonamethylene terephthalate). *CrystEngComm* **2021**, *23*, 892–903. [CrossRef]
15. Tu, C.H.; Woo, E.M.; Lugito, G. Structured growth from sheaf-like nuclei to highly asymmetric morphology in poly(nonamethylene terephthalate). *RSC Adv.* **2017**, *7*, 47614–47618. [CrossRef]
16. Woo, E.M.; Nurkhamidah, S. Surface Nanopatterns of Two Types of Banded Spherulites in Poly(nonamethylene terephthalate) Thin Films. *J. Phys. Chem. B* **2012**, *116*, 5071–5079. [CrossRef] [PubMed]
17. Woo, E.M.; Nurkhamidah, S.; Chen, Y.F. Surface and interior views on origins of two types of banded spherulites in poly(nonamethylene terephthalate). *Phys. Chem. Chem. Phys.* **2011**, *13*, 17841–17851. [CrossRef]
18. Yen, K.C.; Woo, E.M.; Tashiro, K. Microscopic fourier transform infrared characterization on two types of spherulite with polymorphic crystals in poly(heptamethylene terephthalate). *Macromol. Rapid Commun.* **2010**, *31*, 1343–1347. [CrossRef]

19. Lugito, G.; Woo, E.M. Three types of banded structures in highly birefringent poly(trimethylene terephthalate) spherulites. *J. Polym. Sci. Part B Polym. Phys.* **2016**, *54*, 1207–1216. [CrossRef]
20. Hong, P.-D.; Chung, W.-T.; Hsu, C.-F. Crystallization kinetics and morphology of poly(trimethylene terephthalate). *Polymer* **2002**, *43*, 3335–3343. [CrossRef]
21. Chuang, W.T.; Hong, P.-D.; Chuah, H.H. Effects of crystallization behavior on morphological change in poly(trimethylene terephthalate) spherulites. *Polymer* **2004**, *45*, 2413–2425. [CrossRef]
22. Yun, J.H.; Kuboyama, K.; Ougizawa, T. High birefringence of poly(trimethylene terephthalate) spherulite. *Polymer* **2006**, *47*, 1715–1721. [CrossRef]
23. Yun, J.H.; Kuboyama, K.; Chiba, T.; Ougizawa, T. Crystallization temperature dependence of interference color and morphology in poly(trimethylene terephthalate) spherulite. *Polymer* **2006**, *47*, 4831–4838. [CrossRef]
24. Chen, H.B.; Chen, L.; Zhang, Y.; Zhang, J.J.; Wang, Y.Z. Morphology and interference color in spherulite of poly(trimethylene terephthalate) copolyester with bulky linking pendent group. *Phys. Chem. Chem. Phys.* **2011**, *13*, 11067–11075. [CrossRef] [PubMed]
25. Rosenthal, M.; Burghammer, M.; Bar, G.; Samulski, E.T.; Ivanov, D.A. Switching chirality of hybrid left-right crystalline helicoids built of achiral polymer chains: When right to left becomes left to right. *Macromolecules* **2014**, *47*, 8295–8304. [CrossRef]
26. Woo, E.M.; Lee, M.-S. Crystallization in arylate polyesters to periodically ringed assembly. *Polym. Cryst.* **2018**, *1*, e10018. [CrossRef]
27. Lugito, G.; Woo, E.M.; Chang, S.-M. Periodic extinction bands composed of all flat-on lamellae in poly(dodecamethylene terephthalate) thin films crystallized at high temperatures. *J. Polym. Sci. Part B Polym. Phys.* **2017**, *55*, 601–611. [CrossRef]
28. Woo, E.M.; Lugito, G.; Yang, C.E.; Chang, S.-M. Atomic-force microscopy analyses on dislocation in extinction bands of poly(Dodecamethylene terephthalate) spherulites solely packed of single-crystal-like lamellae. *Crystals* **2017**, *7*, 274. [CrossRef]
29. Tasaki, M.; Yamamoto, H.; Yoshioka, T.; Hanesaka, M.; Ninh, T.H.; Tashiro, K.; Jeon, H.J.; Choi, K.B.; Jeong, H.S.; Song, H.H.; et al. Microscopically-viewed relationship between the chain conformation and ultimate Young's modulus of a series of arylate polyesters with long methylene segments. *Polymer* **2014**, *55*, 1799–1808. [CrossRef]
30. Woo, E.M.; Lugito, G.; Chang, S.-M. Three-dimensional interior analyses on periodically banded spherulites of poly(dodecamethylene terephthalate). *CrystEngComm* **2018**, *20*, 1935–1944. [CrossRef]
31. Lugito, G.; Woo, E.M. Multishell oblate spheroid growth in poly(trimethylene terephthalate) banded spherulites. *Macromolecules* **2017**, *50*, 5898–5904. [CrossRef]
32. Yang, C.-E.; Woo, E.M.; Nagarajan, S. Epicycloid extinction-band assembly in Poly(decamethylene terephthalate) confined in thin films and crystallized at high temperatures. *Polymer* **2021**, *212*, 123256. [CrossRef]
33. Lugito, G.; Woo, E.M. Novel approaches to study the crystal assembly in banded spherulites of poly(trimethylene terephthalate). *CrystEngComm* **2016**, *18*, 6158–6165. [CrossRef]
34. Liao, Y.; Nagarajan, S.; Woo, E.M.; Chuang, W.; Tsai, Y. Synchrotron X-Ray analysis and morphology evidence for stereo-assemblies of periodic aggregates in poly(3-hydroxybutyrate) with unusual photonic iridescence. *Macromol. Rapid Commun.* **2021**, *42*, 2100281. [CrossRef]
35. Nagarajan, S.; Huang, K.-Y.; Chuang, W.-T.; Lin, J.-M.; Woo, E.M. Thermo-Sensitive Poly(p-dioxanone) Banded Spherulites with Controllable Patterns for Iridescence. *J. Phys. Chem. C* **2023**, *127*, 2628–2638. [CrossRef]
36. Chang, C.-I.; Woo, E.M.; Nagarajan, S. Grating Assembly Dissected in Periodic Bands of Poly(Butylene Adipate) Modulated with Poly (Ethylene Oxide). *Polymers* **2022**, *14*, 4781. [CrossRef] [PubMed]

Disclaimer/Publisher's Note: The statements, opinions and data contained in all publications are solely those of the individual author(s) and contributor(s) and not of MDPI and/or the editor(s). MDPI and/or the editor(s) disclaim responsibility for any injury to people or property resulting from any ideas, methods, instructions or products referred to in the content.

Article

Role of Polyanions and Surfactant Head Group in the Formation of Polymer–Colloid Nanocontainers

Elmira A. Vasilieva *, Darya A. Kuznetsova , Farida G. Valeeva, Denis M. Kuznetsov and Lucia Ya. Zakharova

Arbuzov Institute of Organic and Physical Chemistry, FRC Kazan Scientific Center, Russian Academy of Sciences, Arbuzov Str. 8, 420088 Kazan, Russia
* Correspondence: vasilevaelmira@mail.ru

Abstract: Objectives. This study was aimed at the investigation of the supramolecular systems based on cationic surfactants bearing cyclic head groups (imidazolium and pyrrolidinium) and polyanions (polyacrylic acid (PAA) and human serum albumin (HSA)), and factors governing their structural behavior to create functional nanosystems with controlled properties. Research hypothesis. Mixed surfactant–PE complexes based on oppositely charged species are characterized by multifactor behavior strongly affected by the nature of both components. It was expected that the transition from a single surfactant solution to an admixture with PE might provide synergetic effects on structural characteristics and functional activity. To test this assumption, the concentration thresholds of aggregation, dimensional and charge characteristics, and solubilization capacity of amphiphiles in the presence of PEs have been determined by tensiometry, fluorescence and UV-visible spectroscopy, and dynamic/electrophoretic light scattering. Results. The formation of mixed surfactant–PAA aggregates with a hydrodynamic diameter of up to 190 nm has been shown. Polyanion additives led to a decrease in the critical micelle concentration of surfactants by two orders of magnitude (from 1 mM to 0.01 mM). A gradual increase in the zeta potential of HSA–surfactant systems from negative to positive value indicates that the electrostatic mechanism contributes to the binding of components. Additionally, 3D and conventional fluorescence spectroscopy showed that imidazolium surfactant had little effect on HSA conformation, and component binding occurs due to hydrogen bonding and Van der Waals interactions through the tryptophan amino acid residue of the protein. Surfactant–polyanion nanostructures improve the solubility of lipophilic medicines such as Warfarin, Amphotericin B, and Meloxicam. Perspectives. Surfactant–PE composition demonstrated beneficial solubilization activity and can be recommended for the construction of nanocontainers for hydrophobic drugs, with their efficacy tuned by the variation in surfactant head group and the nature of polyanions.

Keywords: polymer–colloid complex; surfactant; polyelectrolyte; human serum albumin; polyacrylic acid; critical aggregation concentration

Citation: Vasilieva, E.A.; Kuznetsova, D.A.; Valeeva, F.G.; Kuznetsov, D.M.; Zakharova, L.Y. Role of Polyanions and Surfactant Head Group in the Formation of Polymer–Colloid Nanocontainers. *Nanomaterials* **2023**, *13*, 1072. https://doi.org/10.3390/nano13061072

Academic Editor: Frank Boury

Received: 10 February 2023
Revised: 4 March 2023
Accepted: 14 March 2023
Published: 16 March 2023

Copyright: © 2023 by the authors. Licensee MDPI, Basel, Switzerland. This article is an open access article distributed under the terms and conditions of the Creative Commons Attribution (CC BY) license (https://creativecommons.org/licenses/by/4.0/).

1. Introduction

The systems based on oppositely charged surfactants and polyelectrolytes (PEs) are successfully used in the production of cosmetics, household chemicals, and drug delivery [1–3]. This is due to the fact that PEs can regulate surfactant functional properties, such as adsorption at the water–air interface, self-assembly in the bulk solution, rheology, and solubilization effect; the visa verse characteristics of polymer can be purposefully modified by surfactants added [4–6]. Therefore, the formation of mixed polymer–colloid systems and evaluation of factors controlling their properties may be considered an effective and soft tool for the construction of functional materials and tailoring their activity. This encouraged us to focus on such kinds of systems, with special attention paid to cationic surfactants and PE of both synthetic and natural sources.

Interactions between surfactants and PEs in the bulk solution and at the water–air interface have been systematically studied by many research groups, including ours [4,7–10].

Meanwhile, these systems continue to attract the attention of researchers, with the variety of self-assembling nanostructures such as micelles, vesicles, polyelectrolyte capsules, nanolayers, nano- and microparticles involved [11–14]. This research activity emphasizes the urgency of the study of surfactant–PE systems, which is based on the following reasons:

(i) The formation of surfactant–PE complexes is contributed by a variety of noncovalent intermolecular interactions, including hydrophobic effect, hydrogen bonds, Van der Waals, and electrostatic forces. These interactions depend on various factors, such as the structure of components, e.g., chain rigidity, charge density, the molecular weight of PE, the nature of the head group, and hydrophobicity of surfactants [8,15–17], which is responsible for the high specificity and versatility of aggregation characteristics of the systems. Moreover, different values of critical concentrations corresponding to the onset of mixed aggregation may be obtained depending on the methods used [18,19]. Such multifaceted structural behavior is supported by a number of publications focusing on mixed systems upon the variation of the structure of components or solution conditions [20,21]. Importantly, in the case of oppositely charged surfactant–PE mixtures, an extended region of precipitate usually occurs upon charge neutrality, while soluble complexes are formed in the area of an excess surfactant or PE concentration [15]. It was found that in mixed PAA–surfactant systems, the region of heterogeneity is regulated by the hydrocarbon tail length of the amphiphiles [22]. As shown in work [19], the region of turbidity occurring close to zero zeta potential strongly depends on the surfactant–PE ratio and tends to shift toward the higher surfactant concentrations upon the increase in fixed PE concentration. Kogej et al. emphasized that even a slight change in the structure of a polyion significantly affects the binding of surfactants to the PE [20,21];

(ii) It is noteworthy that for mixed polymer–surfactant systems, different models were developed for the description of their structural behavior, e.g., the so-called necklace model [23]. Even though the use of this model in the system based on oppositely charged components is documented [24], this approach is not general. In contrast to surfactant–nonionic polymer systems, the nature of the surfactant head group has a significant effect on the surfactant–PE interaction [18,25]. This can be exemplified by PAA–surfactant systems involving two cationic surfactants with morpholinium and triphenylphosphonium head groups. The former systems demonstrated strong synergetic behavior, with a marked decrease (up to two orders of magnitudes) in the aggregation threshold. On the contrary, only a slight, if any, effect was observed in the latter system. Similarly, the nature of PE may significantly affect the solution behavior of mixed surfactant–PE complexes [26]. Unlike with DABCO-based surfactant and poly(sodium 4-styrenesulfonate) system demonstrating one breakpoint on the surface tension isotherms and an increase in critical concentrations, two breakpoints and pronounced synergetic behavior (decrease in critical aggregation concentration compared to single surfactant) occurred in the case of PAA–surfactant system. Taking into account these considerations, it should be stressed that unlike with surfactant–polymer systems based on uncharged polymers, complicated structural behavior controlled by a variety of factors is typical for the surfactant–PE interactions, especially involving oppositely charged components, with no uniform conception occurring so far. This demonstrates the necessity of additional studies to elucidate the mechanisms responsible for their behavior and develop the optimal model for describing and predicting the interactions and physicochemical properties in such pairs;

(iii) Structural behavior of supramolecular systems contributed by noncovalent interactions tends to demonstrate stimuli responsibility. In this respect, surfactant–PE systems based on unbuffered PAA are of particular interest due to the fact that they demonstrate pH-dependent behavior and can be controlled by changes in solution conditions. Therefore, the information on the self-assembling, morphological, and functional activity of such compositions is of practical importance;

(iv) Since many innovative drugs are lipophilic, great efforts focused on the development of nanosized carriers to increase their solubility and biocompatibility. surfactant–PE nanostructures may be of interest as carriers due to the ease of their formation. According to the literature data, including our studies, mixed systems based on amphiphiles and PE can be used as nanocontainers for the delivery of hydrophobic substances, specifically medicines [3,11–13,27,28];

(v) Polymer–surfactant complexes are biomimetic systems, and their study makes it possible to simulate the interaction of charged amphiphiles with natural biopolymers (nucleic acids, proteins, polysaccharides) and lipids [29,30], factors of enzyme catalysis [31], etc. Moreover, they may be considered the simplest models for studying membrane–drug interactions (e.g., antimicrobial preparations exert their effects by interacting with biological membranes) [32]. Essentially, the self-aggregation study of mixed surfactant–PE compositions is necessary for a comprehensive description of their solution behavior and functional activity. The knowledge gained can provide a better understanding of optimizing the composition and properties of nanocontainers based on surfactants and PEs;

(vi) The systems based on surfactant and natural PE are of great importance, e.g., the interaction of cationic surfactants with globular proteins-albumins-is widely studied due to the bioapplication perspective [33–35]. Albumins are blood transport proteins and have unique properties (commercial availability, water solubility, biodegradability, biogenicity, and ability to accumulate in tumor tissues), which makes them promising candidates for the development of protein-based nanocontainers [36,37]. Complexation involving cationic surfactant–protein systems is of considerable interest since the interaction of proteins with amphiphilic molecules can lead to compaction or unfolding of the proteins, which is an important property for medicine and pharmacology [33];

(vii) Human serum albumin (HSA) is one of the most important plasma proteins. It consists of 585 amino acid residues with three α-helical domains, stabilized by disulfide bonds [36,38]. Importantly, HSA is non-toxic, has a high protein–drug binding ability, and is widely used to transport exogenous and endogenous ligands [39,40]. Currently, HSA is conjugated to numerous drugs such as insulin, paclitaxel, interferon, doxorubicin, etc. [41,42]. As shown in [43,44], the use of mixed surfactant–protein systems leads to a synergistic effect in self-assembly behavior and solubility of lipophilic compounds and, consequently, improves their biocompatibility. Meanwhile, few publications are available on the self-assembly and morphological behavior of the HSA–surfactant systems [45–47], which provide evidence of the deficiency of fundamental information in this field and highlight the novelty of the data expected.

Based on these reasons, PAA and HSA were chosen as synthetic and natural PE for the study of their mixed behavior with cationic surfactants, which aims at elucidating the factors controlling physicochemical characteristics, functionality, and practical potential of such compositions. At the first stage, the complexation between cationic surfactants with a different head group (imidazolium (IA-16) and pyrrolidinium (PS-16)) with (polyacrylic acid and human serum albumin) is studied, with the effect of PAA and HSA on the aggregation and size characteristics of amphiphiles evaluated. The mechanism of the protein–surfactant interaction in an aqueous solution is studied with subsequent calculation of quantitative parameters of component binding: Stern–Volmer (K_{SV}) and binding (K_a) constants, the number of binding sites (n), thermodynamic parameters of the interaction of components (changes in enthalpy ΔH^0, entropy ΔS^0, and Gibbs free energy ΔG^0) are estimated. In the second stage, the possibility of using the polymer–colloid complexes for the solubilization of lipophilic drugs of various types (warfarin, amphotericin B, meloxicam) was assessed. The structural formulas of the compounds used are shown in Figure 1.

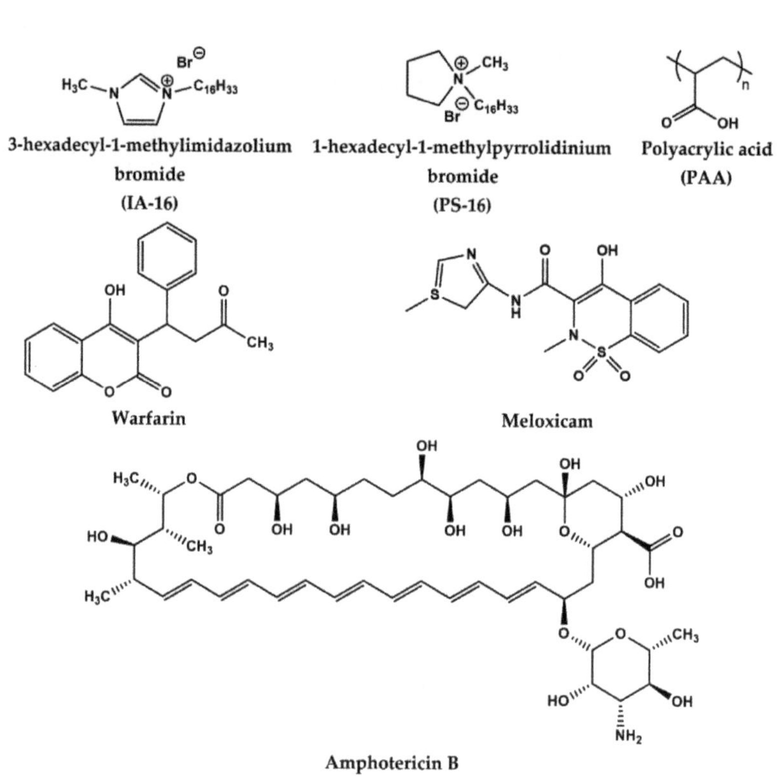

Figure 1. Structural formulas of surfactants, PAA, and drugs.

2. Materials and Methods

2.1. Materials

Polyelectrolytes (PAA, 1800 g/mol, ≥99%), HSA (66 kDa)); fluorescence probe (pyrene, ≥99%), and drugs (warfarin (≥98.0%), meloxicam, amphotericin B) were purchased from Sigma–Aldrich (St. Luis, MO, USA) and used as received. The PS-16 and IA-16 were synthesized similarly to the procedure indicated in refs. [48,49], respectively.

2.2. Sample Preparation

Milli-Q system (Millipore S.A.S. 67120 Molsheim, France) was used for the purification of the solvent (18.2 MΩ·m resistivity) for the preparation of samples. Ultra-purified water was applied as a solvent for the preparation of amphiphile-PAA and amphiphile-HSA mixtures. All solutions (except stock ones) were prepared using a volume dilution method. The concentration of HSA in all binary surfactant–HSA systems was constant and amounted to 5 µM [47]. Meanwhile, in surfactant–PAA systems, the PE concentrations were 1, 3, and 5 mM.

2.3. Surface Tension Measurements

The surface tension of surfactant–HSA and surfactant–PAA mixtures was measured using a Krüss K6 tensiometer (KRUSS GmbH, Hamburg, Germany) via the Du Nouy ring detachment method at 25 °C. Triplicate reproducible surface tension values with a deviation of no more than ±0.1 mN/m were taken into account.

2.4. The Dynamic and Electrophoretic Light Scattering

The dynamic and electrophoretic light scattering (DLS and ELS, Zetasizer Nano, Malvern Instruments Ltd., Worcestershire, UK) methods were used to measure the hydrodynamic diameter (D_h), zeta potential (ζ), and the polydispersity index of aggregates formed based on surfactants and Pes. The concentration of samples was between the critical aggregation concentration-1 (CAC-1) and CAC-2 regions. Measurements were taken in triplicate at 25 °C using a disposable folded capillary cell. D_h and ζ were obtained using Stokes–Einstein and Helmholtz–Smoluchowski equations, respectively [50]. PdI was calculated using cumulative second-order correlation function analysis.

2.5. Fluorescence Spectroscopy

The fluorescence spectra of pyrene (in the concentration of 1 µM) in amphiphile/PAA aqueous solutions were recorded in the range of 350–500 nm using fluorescence spectrophotometer Hitachi F-7100 (Hitachi High-Tech Corporation, Tokyo, Japan) at 25 °C. The excitation wavelength was 335 nm, and the scanning speed was 120 nm/min. The fluorescence intensity of the I peak at 373 nm and the III peak at 384 nm was determined by the spectra obtained.

Registration of fluorescence spectra of amphiphile–HSA mixed systems was performed at 25 °C, 30 °C, 35 °C, 40 °C, and 280 nm excitation wavelength. Emission spectra were recorded in the range of 290–450 nm with 240 nm/min scan speed. All measurements were conducted in a 1 cm cuvette.

Based on the obtained fluorescence spectra, the Stern–Volmer constants for various temperatures were calculated using the following equation [51]:

$$\frac{F_0}{F} = 1 + K_{SV}[Q] = 1 + k_q \cdot \tau_0 [Q], \tag{1}$$

where F_0 and F are the fluorescence intensities in the absence and in the presence of the quencher (surfactant), respectively; K_{SV} is the Stern–Volmer constant; $[Q]$ is the concentration of surfactant; k_q is the bimolecular rate constant of quenching; τ_0 is an average lifetime of fluorophore (HSA) in the excited state in the absence of quencher, which equals 10^{-8} s [51].

The bimolecular rate constant of quenching k_q was evaluated using Equation (2):

$$k_q = \frac{K_{SV}}{\tau_0} \tag{2}$$

Determination of binding constants of components K_a, the number of binding sites n, and thermodynamic characteristics of surfactant–HSA systems was performed on the basis of the primary fluorescent data by Equations (3)–(5) [34]:

$$lg\frac{F_0 - F}{F} = lgK_a + nlg[Q] \tag{3}$$

$$lnK_a = -\frac{\Delta H^0}{RT} + \frac{\Delta S^0}{R} \tag{4}$$

$$\Delta G^0 = \Delta H^0 - T\Delta S^0, \tag{5}$$

where ΔH^0 is the variation of standard enthalpy; ΔS^0 is the variation of standard entropy, R is a universal gas constant (8.314 J·mol^{-1}·K^{-1}); T is absolute temperature; ΔG^0 is the change of standard Gibbs free energy.

Registration of three-dimensional fluorescence spectra was performed in the range of excitation wavelength 200–350 nm with increments of 10 nm and emission wavelengths of 200–450 nm with increments of 10 nm [47]. The scanning speed was 240 nm/min.

2.6. Spectrophotometry

Absorption spectra of the polymer–colloid solutions were recorded using Specord 250 PLUS spectrophotometer (Analytik Jena AG, Jena, Germany). The solubilizing ability of the surfactant-HSA and surfactant-PAA systems was determined by adding an excess of the different drugs (Warfarin, Amphotericin B, and Meloxicam). The absorption spectra of solutes were registered in the range of 200–450 after 24 h at room temperature. Measurements were taken in triplicate at 25 °C using a 0.5 cm quartz cuvette. Primary spectral data were used for extraction of the maximum optical density values at λ = 305 nm (extinction coefficient of Warfarin equals 17,400 $M^{-1} \cdot cm^{-1}$) for Warfarin, at λ = 385 nm (extinction coefficient of Amphotericin B equals 53,935 $M^{-1} \cdot cm^{-1}$) for Amphotericin [47], and at λ = 364 nm (extinction coefficient of Meloxicam equals to 14,610 $M^{-1} \cdot cm^{-1}$) for Meloxicam. The solubilization capacity of complexes (S) was calculated from the concentration dependences of the optical density (D) using the following equation: S = b/ε, where b is the slope of the D/l = f (C) dependence, l is the cuvette thickness, C is the surfactant concentration, and ε is the extinction coefficient.

2.7. Transmission Electron Microscopy (TEM)

TEM images were obtained at the Interdisciplinary Center for Analytical Microscopy of Kazan (Volga Region) Federal University with a Hitachi HT7700 Exalens microscope (Hitachi, Tokyo, Japan). The images were acquired at an accelerating voltage of 100 kV. Samples were dispersed on 300 mesh 3 mm copper grids (Ted Pella) with continuous carbon–formvar support films.

2.8. Data Analysis and Statistics

All measurements were carried out at least three times. Data were performed via OriginPro 8.5 (OriginLab Corporation, Northampton, MA, USA) and expressed as mean values ± standard deviation (SD). The statistical significance of the difference between the values of the solubilization capacity and hydrodynamic diameter was determined using a one-way analysis of variance (ANOVA) test. Differences were considered significant at a p-value < 0.05.

3. Results and Discussion

3.1. Surfactant–PAA Nanocarriers

3.1.1. Phase Behavior of the Surfactant–PAA Systems

Our early investigations demonstrated the complexation of cationic surfactants with pyrrolidinium and imidazolium head groups containing a hydroxyethyl moiety with PAA [52,53]. It was shown that the properties of polymer–colloid systems significantly depend on many factors, including the surfactant head group structure [18] and the presence of functional groups. It was of interest to reveal new structure–property relationships, which may allow elucidating additional factors to control the functional activity of the systems and markedly widen their practical application. Typically, the appearance of turbidity and even the formation of insoluble precipitate occur in a certain concentration range in polymer–colloid solutions. It was shown that the addition of electrolytes could prevent phase separation [54]; however, they significantly reduce the critical micelle concentration (CMC) of cationic surfactants. Therefore, before an investigation of the adsorption of surfactant–PE systems at the water–air interface, it is useful to study the bulk phase behavior of these systems.

Figure 2 shows the spectrophotometry data, which demonstrates the change in the absorbance (turbidity) of the mixed PAA-PS-16 (a) and PAA-IA-16 compositions (b) upon increasing surfactant concentration. Looking ahead, the critical aggregation concentrations (CAC) of surfactant–PAA systems are in the clear solutions region (absorbance up to 0.2). The turbidity increases above the CAC region (~0.05 mM), and close to the surfactant CMC (1 mM), the solutions become clear again. It is noteworthy that in the case of imidazolium surfactants, more cloudy solutions are formed, independently of the polymer concentration (absorbance 1.2), compared to systems with pyrrolidinium surfactant (absorbance 0.4). This may indirectly indicate that stronger electrostatic interactions occur between the components. Meanwhile, phase separation is not observed in either system. The phase behavior of oppositely charged surfactant and PE mixtures is closely related to their charge characteristics. Cloudy solutions are observed in the neutral charge area, with precipitation occurring in some cases, whereas soluble complexes are formed in the presence of a charge (with an excess of surfactant or PE) [15].

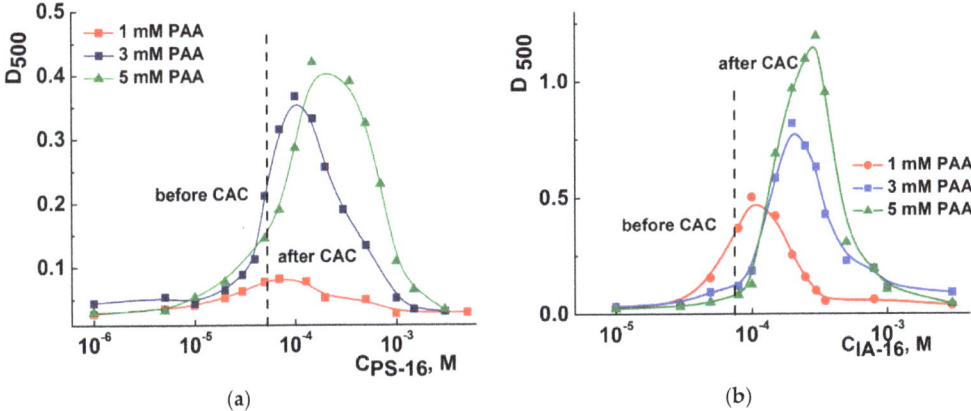

Figure 2. Absorbance of PAA-PS-16 (**a**) and PAA-IA-16 (**b**) systems at 500 nm under different surfactant concentrations and fixed concentrations of PAA; 25 °C.

The charge of the surfactant–PE complexes was estimated by measuring their zeta potential. Figure 3 presents electrophoretic light scattering data reflecting the dependence of the zeta potential of surfactant–PAA systems upon varying the components' ratio. It is known that the presence of a charge prevents the aggregation of particles and determines the stability of colloidal systems [55]. Electrostatic stabilization requires the magnitude of zeta potential of about 30 mV [56]. Gradual increase in the zeta potential from −25 mV to +80 mV, depending on the ratio of the components, was shown for both systems. It should be noted that co-aggregation in the bulk solution always begins in the negatively charged region, i.e., with an excess of the polyanion. Despite the low charge density of PAA (ionization degree is around 8% at pH = 4 [19]), a rather significant zeta potential (Figure 3) indicates that the electrostatic forces are involved in the interaction between the components. At the same time, the hydrophobic effect and hydrogen bonding can also play an important role in the surfactant complexation with PAA. In the case of weak PEs, the presence of a charge depends on the solution pH. In this study, spontaneous solution pH of ca.4 was maintained. According to Manning's theory [57], at a low degree of charge density of the PE chain, the binding of the surfactant to PAA is mainly driven by the hydrophobic effect. Precipitation is assumed to result from hydrogen bonding that occurs between uncharged segments of the PAA chain. At a high charge density of PAA, hydrophobic binding weakens due to the increase in the charged character of PE chains and their binding to the surfactant via electrostatic interactions.

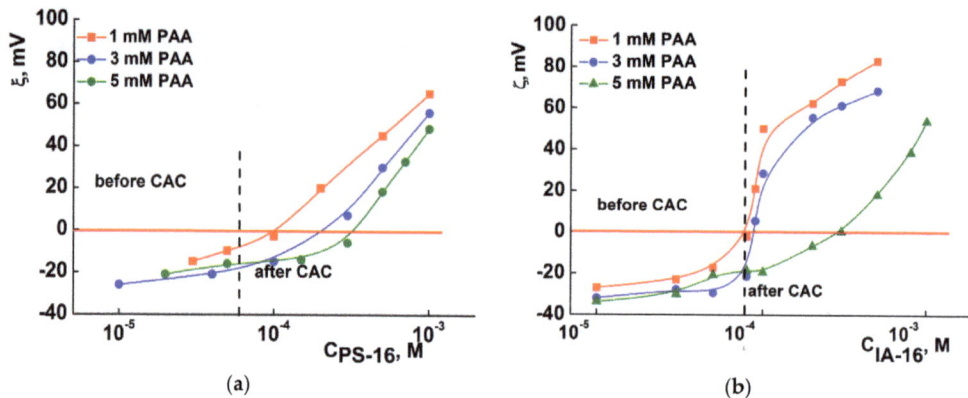

Figure 3. Zeta potential of PS-16-PAA (**a**) and IA-16-PAA (**b**) systems under different surfactant concentrations and fixed concentrations of PAA; 25 °C.

3.1.2. Self-Assembly of the Oppositely Charged Surfactant–PAA Systems

Critical aggregation concentration (CAC) of cationic surfactant–anionic PE mixed systems can be several orders of magnitude lower than the surfactant CMC. This is due to the fact that the addition of the anionic PE significantly reduces the electrostatic repulsion force between the head groups of the cationic surfactant. However, as mentioned above, in the case of weak PE, low CAC values can be caused by not only electrostatic interactions but also by hydrophobic effect and hydrogen bonding. Amphiphiles can bind to PE in both monomeric and micellar forms depending on the concentration. In this study, the critical concentrations were determined by the Du Nouy ring method (tensiometry). Primarily, the tensiometry reflects the surfactant–PE system behavior at the water–air interface and only then by their aggregation properties in the bulk solution. Furthermore, the surface tension curves of the oppositely charged surfactant–PE systems can be more complicated depending on the components ratio and have several breakpoints. In our work, a variety of methods are always involved in determining CAC/CMC.

Figure 4 shows the surface tension isotherms of the PS-16 (a) and IA-16 (b) as a function of the surfactant concentration in the presence of PAA. The PAA-IA-16 concentration curves demonstrate a two-step aggregation, which is typical for oppositely charged surfactant–PE mixtures [58]. The first breakpoint in Figure 4b is ascribed to CAC-1 and characterizes the concentration of the beginning of complexation in the bulk solution. At the second one (CAC-2), a saturation of the PE chain by surfactant aggregates and the formation of pure micelles occurs. In the case of the pyrrolidinium surfactant (Figure 4a), only one breakpoint is clearly visible (CAC) in the surface tension isotherms, and curves do not reach a plateau. Taylor et al. reported that PE additives could initiate the adsorption of surfactant at water–air interfaces in both monomeric and micellar forms, causing a decrease in the surface tension after reaching CAC [58]. As can be seen from Table 1, the CAC-1 of mixed surfactant–PAA systems is 10–15 times lower than the CMC of pure surfactants, independently of their head group. An increase in the concentration of PAA has little effect on the CAC values. It should be noted that the surface tension of mixed systems is much lower compared to pure surfactant up to the aggregation thresholds. This is probably due to the surface activity of the PE, which was shown earlier [19].

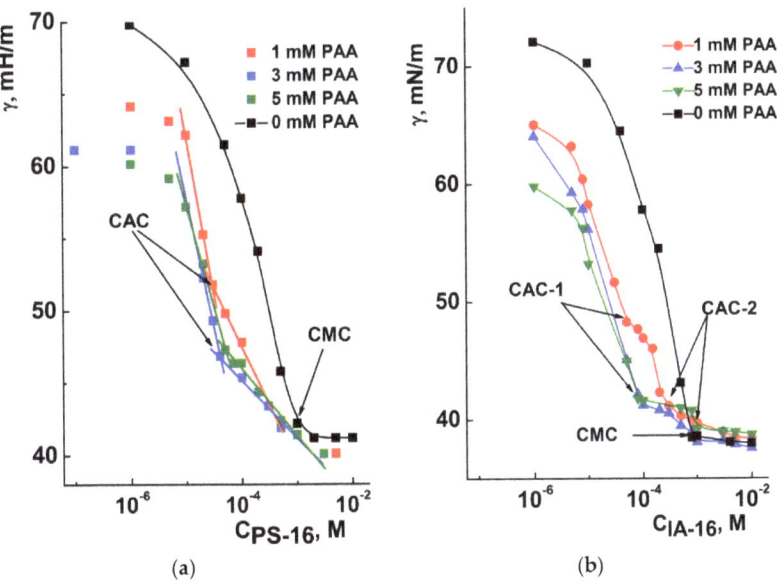

Figure 4. Surface tension isotherms of mixed PS-16-PAA (**a**) and IA-16-PAA (**b**) systems under different surfactant concentrations and fixed PAA additives; 25 °C (CAC and CAC-1 have the same meaning).

Table 1. CMC/CAC values for pure and mixed surfactant–PAA systems under varying PE concentrations obtained by tensiometry and fluorescence spectroscopy.

Surfactant	C_{PAA}, mM	CAC-1/CAC-2 *, mM	
		Tensiometry	Fluorimetry
PS-16	0	1	1
	1	0.060	0.03
	3	0.050	0.02
	5	0.060	0.03
IA-16	0	1	1
	1	0.050/1	0.019
	3	0.080/1	0.015
	5	0.080/1	0.011

* For the case if two breakpoints are observed.

An additional highly sensitive method of fluorescence spectroscopy using pyrene was engaged to confirm the tensiometric measurements. This method is widely used for investigation not only of pure micellar solutions but also of mixed systems based on oppositely charged polymers and surfactants [59]. Furthermore, both methods are well-complemented since tensiometry primarily provides information on the water–air interface, while fluorescence spectroscopy describes bulk aggregation. The pyrene properties depend on the polarity of its microenvironment, and, therefore, the fluorescence spectra provide information about fluorophore location. The polarity index (I_1/I_3) is used as a measure of polarity and is defined as ratios of the first (373 nm) and third (384 nm) peaks in the pyrene spectrum. The decrease in the polarity index indicates the decrease in the polarity of the pyrene microenvironment caused by its solubilization in the hydrophobic core of the micelle. The determination of the CMC of surfactant is based on this property.

Figure 5 demonstrates the I_1/I_3 plots as a function of amphiphiles concentration in the mixed PAA-PS-16 (a) and PAA-IA-16 (b) systems upon various PAA concentrations. In both cases, the addition of PE leads to a decrease in the CMC by two orders of magnitude (from 1 mM to 0.01 mM). The aggregation thresholds of mixtures are slightly depend on the PAA concentration. The CAC values were determined by fluorimetry, which was lower than tensiometry data (Table 1). Discrepancy in results may be due to different sensitivities of the methods and sample preparation. It should be noted that the polarity index noticeably decreases with increasing in PAA concentration. This is consistent with work reporting on the sensitivity of the pyrene microenvironment to the chitosan concentration in the mixed systems [60]. This indicates that the presence of a macromolecule induces the formation of more densely packed micellar aggregates due to a decrease in the electrostatic repulsion of amphiphile head groups. In the case of an imidazolium surfactant (Figure 5b), the concentration dependence of the polarity index has a minimum and begins to rise in the CMC region, followed by a plateau. This is probably due to the morphological transitions and indicates the formation of pure micelles.

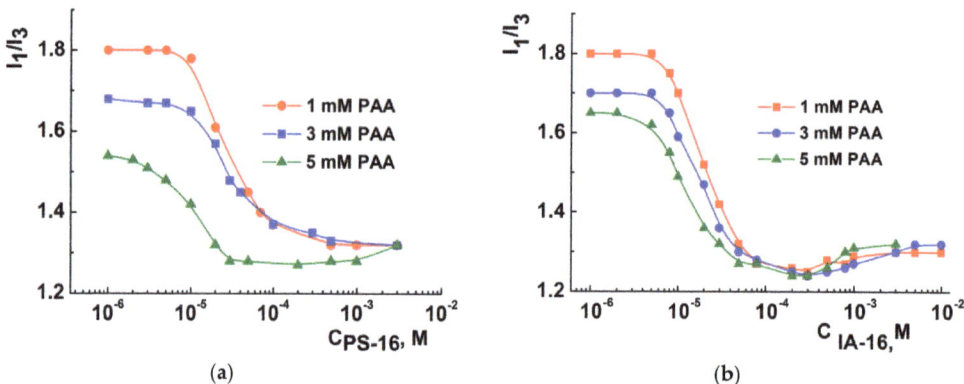

Figure 5. Polarity index (I_1/I_3) as a function of surfactant concentration in the PS-16-PAA (a) and IA-16-PAA (b) mixtures under different PAA concentrations; 25 °C.

3.1.3. Dimensional Characteristics and Morphology of Surfactant–PAA Systems

The hydrodynamic diameter (D_h) of the oppositely charged surfactant–PE systems under the different molar component ratios was determined by dynamic light scattering. D_h of the PS-16 and IA-16 micelles is around 2–5 nm at a concentration of 1 mM (CMC), which corresponds to spherical micelles. Single PAA solution at a concentration range of 1–5 mM contains small aggregates with D_h~10 nm and much larger particles ~250 nm. Probably large aggregates are the combination of several hydrogen-bonded macromolecules, which are predominant at the low degree of the PAA ionization [19]. The D_h, zeta-potential, and polydispersity index (PdI) in the CAC-1/CAC-2 region under excess of PE are summarized in Table 2. Since the CAC-1 values determined by tensiometry and fluorometry are slightly different, average concentrations were chosen to determine the size of the systems. One-way ANOVA testing demonstrated that there were statistical differences between D_h of mixed aggregates in both mixed compositions within a system at a constant concentration of PAA (Table S1).

Aggregates with D_h~90 ± 10 nm were detected in the CAC-1 region in the case of PAA-PS-16 mixtures with 1 mM PAA and 179 ± 3 nm with the increase in PAA content up to 5 mM. This behavior is in line with the increases in the number of PE chains in the solution. The size of mixed aggregates decreases to 37 ± 5 nm with increasing concentration of surfactants to 1 mM. The mechanism of binding PAA to classical cationic surfactant-bearing ammonium head group at a low degree of ionization ($\alpha < 0.3$) is well-described in [57]. In the absence of surfactant, the PAA chain is low-polar and forms a compact structure, while binding of PAA to amphiphile unfolds the PE, thereby contributing to the formation of large complexes through hydrogen bonding. When the surfactant concentration reaches the CAC-2, the PAA–amphiphile complex dissociates, which leads to a decrease in particle size. It should be noted that a monomodal distribution of particle sizes with low PdI not exceeding 0.45 was observed in all cases, which is good enough for oppositely charged systems [61]. A similar trend was observed in the case of imidazolium surfactant: the D_h of complex decreases from 137 ± 5 to 52 ± 19 nm with the increase in amphiphile concentration. The particle size does not change for 2 months at room temperature (25 °C), which indicates the formation of stable mixed nanostructures.

Table 2. The hydrodynamic diameters (D_h), zeta potential (ζ), and polydispersity index (PdI) of PAA-PS-16 and PAA-IA-16 associated under different molar ratio; pH = 4; 25 °C.

No.	C_{PAA}, mM	$C_{surfactant}$, mM	D_h, nm ± SD *	ζ, mV	PdI
			PS-16		
1	1	0.03	90 ± 10	−13 ± 2	0.194 ± 0.102
2	1	0.05	** 186 ± 2	−10 ± 1	0.071 ± 0.019
3	1	0.2	60 ± 3	22 ± 3	0.323 ± 0.009
4	1	0.5	69 ± 2	23 ± 1	0.225 ± 0.012
5	3	0.04	114 ± 12	−22 ± 1	0.298 ± 0.025
6	3	0.5	88 ± 2	32 ± 5	0.109 ± 0.015
7	3	1	37 ± 5	37 ± 5	0.219 ± 0.004
8	5	0.02	137 ± 25	−21 ± 1	0.450 ± 0.037
9	5	0.05	179 ± 3	−17 ± 1	0.430 ± 0.030
10	5	0.5	98 ± 30	21 ± 1	0.104 ± 0.005
11	5	1	59 ± 5	43 ± 2	0.212 ± 0.002
			IA-16		
1	1	0.05	98 ± 1	−17 ± 1	0.073 ± 0.021
2	1	0.5	44 ± 1	−6 ± 1	0.376 ± 0.011
3	1	1	59 ± 15	50 ± 5	0.244 ± 0.028
4	3	0.02	116 ± 1	−36 ± 1	0.333 ± 0.021
5	3	0.1	137 ± 5	−22 ± 2	0.148 ± 0.021
6	3	0.5	94 ± 2	31 ± 1	0.157 ± 0.004
7	3	1	52 ± 19	61 ± 2	0.323 ± 0.041
8	5	0.05	114 ± 11	−21 ± 1	0.483 ± 0.001
9	5	0.08	120 ± 13	−19 ± 1	0.325 ± 0.019
10	5	0.5	135 ± 1	18 ± 1	0.068 ± 0.028
11	5	1	59 ± 5	54 ± 2	0.122 ± 0.003

* Data are represented as mean ± standard deviation (n = 3). ** PS-16-PAA at 0.05–1 mM, respectively, were chosen for visualization.

Morphology of the mixed surfactant–PAA nanostructures was studied with transmission electron microscopy. Nanostructures based on pyrrolidinium surfactant and PAA close to tensiometric CAC (0.05 mM PS-16 and 1 mM PAA) were chosen for the visualization. TEM images are represented in Figure 6. In the presence of PAA, various types of particles with a rather high polydispersity are formed. Mostly spherical particles with a diameter from 80 to 200 nm are observed (Figure 6a,b). It should be noted that, in the case of DLS data, the formation of particles with D_h around 190 nm and low PdI is observed. The presence of highly aggregated particles or separate but smaller sizes is probably due to the removal of water during sample preparation. It is known that pearl-necklace-like aggregates can be formed in surfactant–polymer systems [24].

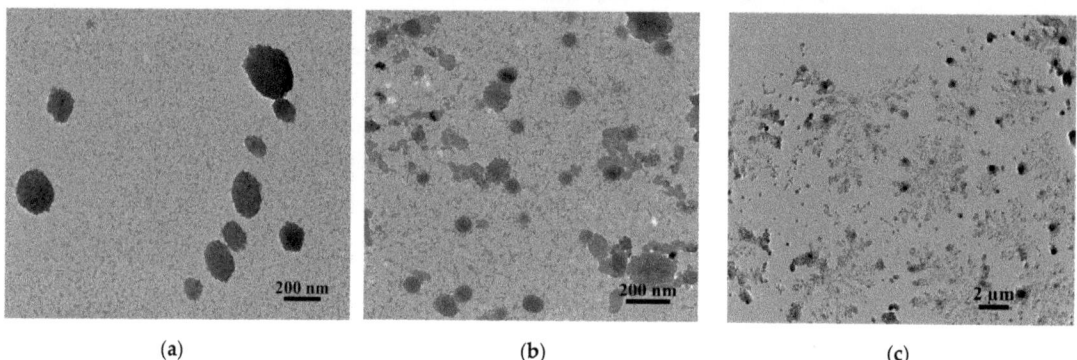

Figure 6. TEM-photo of PS-16-PAA structures at 0.05–1 mM, respectively, and different scales: 200 nm (**a**,**b**); and 2 µm (**c**).

As can be seen from Figure 6a, some of the spherical particles are located in the same row without connection to each other. It was reported that the polymer chains between the beads are exposed to water, which provides low contrast when imaging the sample [62]. It should also be noted that the formation of spherical mixed PS-16-PAA aggregates may indicate the solubilization of the macromolecule in surfactant micelles, which contributes to an increase in the size of mixed systems compared to individual micelles (up to 5 nm for the PS-16 derivative [48]).

Moreover, uniformly distributed large aggregates resembling dendrimers with a diameter of around 3 µm were observed (Figure 6c), similar to those previously shown for chitosan decorated liposomes [55]. However, in this case, the aggregates are 10 times larger than in the liposomal system. It should be noted that in this experiment, the surfactant–PAA aggregation was studied under large excess of PE, i.e., a low concentration of mixed aggregates takes place. Therefore, it is possible that large aggregates correspond to an unbound polymer, which, in turn, are aggregated with each other due to the presence of oppositely charged surfactants and the low degree of PAA ionization.

3.2. Surfactant–HSA Nanocarriers

3.2.1. Self-Assembly of the Oppositely Charged Surfactant–HSA Systems

At the next stage of this work, the complexation of IA-16 and PS-16 with the natural polyanion–human serum albumin (HSA) was studied using a complex of physicochemical methods. The aggregation characteristics of the IA-16-HSA (Figure 7a) and PS-16-HSA (Figure 7b) systems were evaluated via tensiometry. The study of aggregation properties is important since it is known that surfactants can significantly affect the conformational behavior of proteins. This is due to the fact that the binding of proteins to surfactants causes a change in the intramolecular forces that maintain the secondary structure of the protein, causing conformational changes in the macromolecule [63]. However, the addition of proteins to surfactants can significantly change the properties of the adsorption layer and affect the surface tension of the amphiphile solution [44,63].

As can be seen from Figure 7, the surface tension of the mixed surfactant–HSA system is lower than that of pure surfactant, which is probably due to the surface activity of the protein [47,64]. Furthermore, two breakpoints were observed in the concentration dependences of the surface tension of mixed systems (0.08 mM and 1 mM for IA-16-HSA; 0.3 mM and 3 mM for PS-16-HSA). As mentioned above, the first breakpoint at low surfactant concentration is known as CAC-1. It corresponds to the saturation of the water–air interface with amphiphile molecules associated with the protein chain, as well as the beginning of the formation of surfactant–protein complexes [51,63]. The second breakpoint appearance, as in the case of synthetic surfactant–PE complexes, is probably associated with the saturation of HSA macromolecules with amphiphile molecules and corresponds to the formation of free micelles [52]. Interestingly, the formation of mixed IA-16-HSA complexes begins at a lower concentration of amphiphile (0.08 mM), unlike with PS-16-HSA systems (0.3 mM). As has been previously reported [35,63], the binding of the protein to amphiphile at the water–air interface is dominantly observed through electrostatic interactions at low concentrations of cationic surfactant. Hydrophobic interactions begin to predominate with the increase in surfactant concentration. Therefore, the difference in the complexation of the studied amphiphiles with HSA can be related to their different ability for electrostatic binding. Probably, electrostatic interactions are preferable for IA-16 due to the planar structure of the head group [49], compared to cyclic one in PS-16 with low conformational mobility [65].

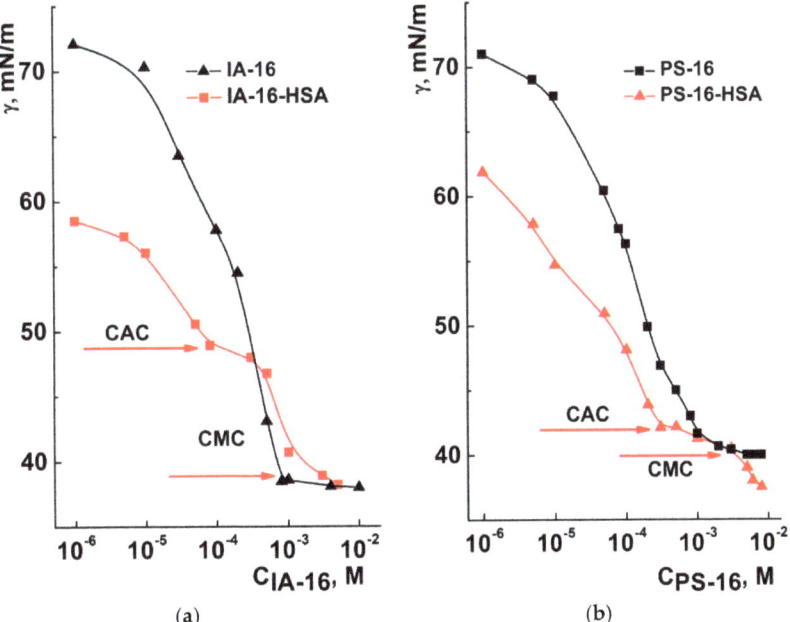

Figure 7. Surface tension isotherms for IA-16-HSA (**a**) and PS-16-HSA (**b**) systems as a function of surfactant concentration under constant HSA concentration (5 µM); 25 °C.

As mentioned above, surfactant binding to HSA can cause various conformational changes in the protein that can affect its size. Therefore, the DLS method was used to determine the D_h of native HSA and mixed IA-16-HSA and PS-16-HSA associates. As can be seen from Figure 8, the D_h of pure HSA is in the range of 6–10 nm, which correlates to previously obtained data [47,66]. Different concentrations of cationic surfactants (IA-16 and PS-16) have a small effect on the size of the polypeptide. Both mixed complexes demonstrated similar sizes in the range of 4–10 nm. This is in agreement with minor changes in protein structure confirming specific binding since a significant increase in the size of the complexes would have indicated the loss of the tertiary and secondary polypeptide structure [67].

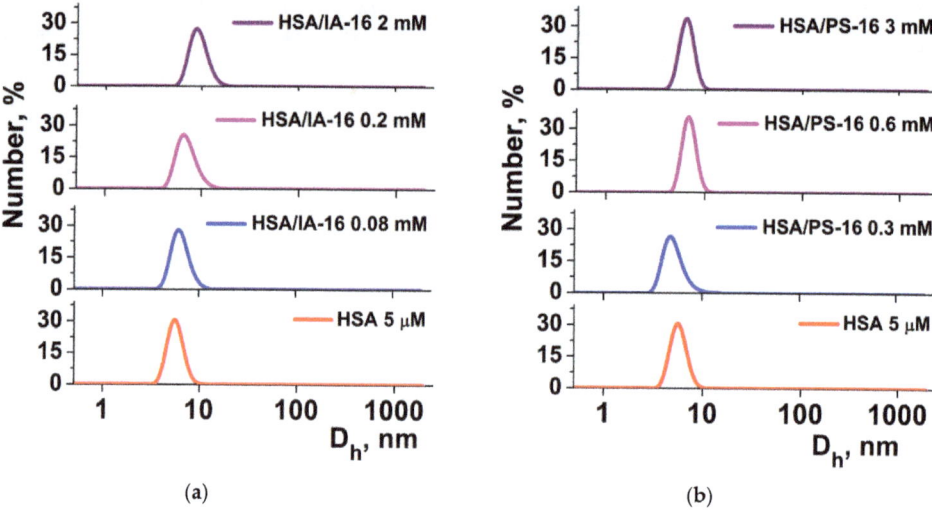

Figure 8. Number-averaged size distribution for mixed IA-16-HSA (**a**) and PS-16-HSA (**b**) systems; 25 °C.

3.2.2. Binding Mechanism Study between Surfactants and HSA

The zeta potential is a useful parameter to assess the surface charge of proteins. Changes in the HSA zeta potential can be the result of surface modifications of the macromolecule, unfolding/denaturation processes, conformational changes in the structure, and also electrostatic interaction in protein–ligand systems [68,69]. Therefore, the zeta potential of pure protein (at physiological pH = 7.4) and mixed IA-16-HSA and PS-16-HSA associates was evaluated by electrophoretic light scattering (Figure 9). The zeta potential of free HSA was ~−10 mV; cationic surfactant additives and an increase in their concentration led to a change of zeta potential from negative to a positive value. Such behavior in mixed systems testifies to the formation of protein–ligand complexes due to the electrostatic interaction [69,70]. Isoelectric points have been observed at 0.04 mM and 0.08 mM amphiphiles for mixed IA-16-HSA and PS-16-HSA systems, respectively. It should be noted that experimental data demonstrated that the electrostatic interactions between imidazolium surfactants and HSA are more pronounced. This is reflected in the value of the isoelectric points, as well as in the total charge compensation of the protein.

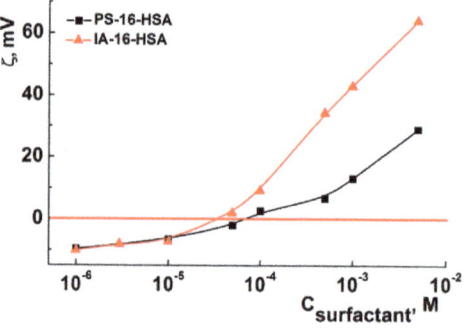

Figure 9. Electrokinetic potential (zeta potential) as a function of IA-16 and PS-16 concentration for mixed IA-16/HSA and PS-16/HSA systems (HSA concentration is 5 μM), 25 °C.

Fluorescence spectroscopy is an important method for studying surfactant–protein binding. Fluorescence quenching is used as a technique for measuring the binding affinity between proteins and ligands and is expressed as a decrease in the fluorescence quantum yield of fluorophore induced via various intermolecular interactions. HSA fluorescence is caused by tryptophan (Trp), tyrosine (Tyr), and phenylalanine (Phe) residues. However, Phe has a very low quantum yield, and Tyr is almost completely quenched if it is ionized or located near the amino-, carboxylic groups, or Trp 214 residue. Therefore, Trp 214 is largely responsible for the intrinsic fluorescence of HSA, which is located deep in the IIA subdomain [69].

Figure 10 reveals that the fluorescence spectrum of HSA has an emission peak at 340 nm (λ_{ex} = 280 nm). The addition of cationic surfactants (IA-16 and PS-16) lead to a hypsochromic shift. Such a blue shift in the fluorescence spectra of proteins corresponds to a decrease in the polarity of the microenvironment after the binding of the components, which indicates that the protein chromophore (Trp and Tyr) goes into a more hydrophobic region [44,64,71,72]. However, the addition of IA-16 quenched the HSA fluorescence, while PS-16 increased the fluorescence intensity. Protein fluorescence quenching is a very common phenomenon, which provides evidence for the binding of components [35,44,64,71–73]. The protein fluorescence enhancement in the presence of surfactants occurs very rarely and is known as the fluorescence sensitization effect. This effect usually appears when biomacromolecules coexist with some ligands [74,75]. It is interesting to note that the same result was previously obtained in the case of the interaction of a pyrrolidinium surfactant containing carbamate fragment with HSA [47].

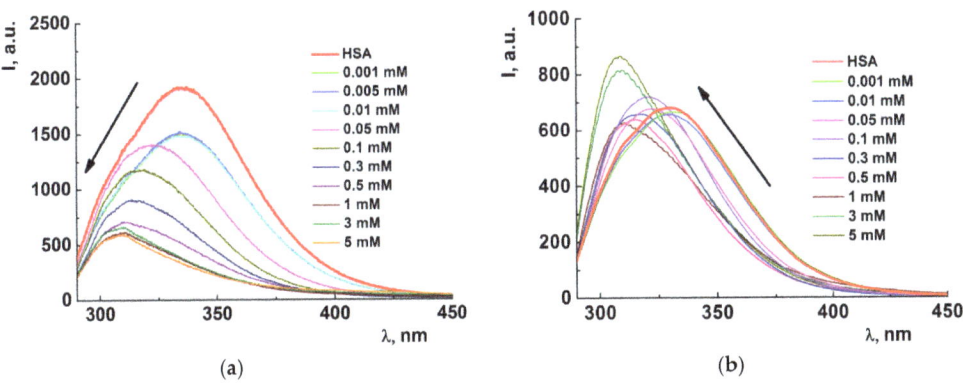

Figure 10. Fluorescence intrinsic spectra of IA-16-HSA (**a**) and PS-16-HSA (**b**) binary systems under various amphiphile concentrations (HSA concentration 5 µM); 25 °C (I-fluorescence intensity).

Fluorescence quenching can occur in two ways: by a collision between the fluorophore and the quencher (dynamic quenching) and by the formation of a complex in the ground state (static quenching). The quenching mechanism can be determined by the value of the bimolecular quenching rate constant, as well as by comparing the values of the Stern–Volmer quenching constant (K_{SV}) obtained at different temperatures. During dynamic quenching, the Stern–Volmer constant is maximum, which is associated with a faster diffusion of molecules at the chosen temperature. In the case of static quenching, the Stern–Volmer constant is lower, which is associated with a decrease in hypersensitivity at a given temperature [76,77].

Since HSA fluorescence quenching occurred only in the presence of IA-16, several quantitative parameters characterizing the efficiency of component binding were obtained only for the IA-16-HSA system. The K_{SV} was calculated graphically using the Stern–Volmer Equation (1) under various temperatures (Table 3). For example, Figure 11 shows the Stern–Volmer dependences for the IA-16-HSA complex at 25 °C. In turn, the Stern–Volmer constant makes it possible to calculate the bimolecular quenching rate constant (K_q), which allows one to draw a conclusion about the predominant quenching mechanism in systems. So, the dynamic quenching mechanism prevails if the value of K_q is less than 2×10^{10} L/mol s and if the static quenching mechanism is more than 2×10^{10} L/mol s [44,51,76,77]. Calculated data on various temperatures are given in Table 3. The Stern–Volmer quenching constant increases with increasing temperature for the IA-16-HSA system, which indicates a dynamic quenching mechanism. However, the calculated values of K_q are at least 20 times higher than the maximum dynamic constant of bimolecular quenching in an aqueous medium (2×10^{10} L/mol s), which indicates a static mechanism of fluorescence quenching. Most likely, the formation of IA-16-HSA complexes is associated with a mixed quenching mechanism, including both the molecular collision and the formation of a non-covalently bound complex due to the adsorption of surfactant molecules on the hydrophobic domains of the protein [34,45,51].

Table 3. Stern–Volmer constants K_{SV}, component binding constants K_a, bimolecular quenching rate constants K_q, and the number of HSA and surfactant binding sites n, changes in enthalpy ΔH^0, entropy ΔS^0, and Gibbs free energy ΔG^0 for the systems under varying temperatures.

T, K	$K_{SV} \cdot 10^3$, L/mol	$K_q \cdot 10^{10}$, L/mol·s	$K_a \cdot 10^3$, L/mol	n	ΔH^0, kJ/mol	ΔS^0, J/mol·K	ΔG^0, kJ/mol
298	4.169	42.54	0.151	0.6			−12.40
303	4.417	44.17	0.126	0.6	−21.32	−29.93	−12.25
308	4.443	44.43	0.110	0.6			−12.10
313	4.547	45.47	0.100	0.6			−11.95

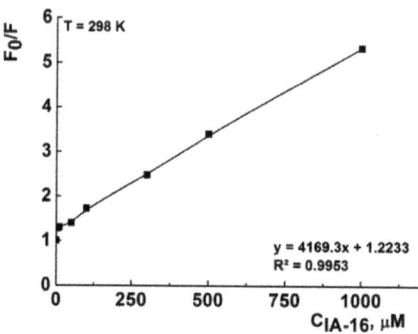

Figure 11. Stern–Volmer plots for the quenching of HSA fluorescence by IA-16 at 298 K. Symbols show the ratio of the fluorescence intensities in the absence and in the presence of the quencher (surfactant) on the surfactant concentration.

A modified version of the Stern–Volmer equation (Equation (1)) can be applied to the calculation of the binding constants of surfactants with HSA (K_a) and the number of binding sites (n). Table 3 shows that the binding constant decreases with increasing temperature, indicating lower stability of the complex in the ground state at a higher temperature due to the pronounced mobility of the molecules [34,77]. In addition, it should be noted that the K_a values obtained for IA-16-HSA suggest a lower binding affinity compared to other strong protein–ligand complexes [34,78]. The n values for IA-16-HSA are at the level of 0.6, which testifies to the presence of one binding site in HSA for surfactants upon their interaction. Further, using the values of the coupling constants of the components K_a, the thermodynamic parameters of the system were calculated: the change in enthalpy ΔH^0, entropy ΔS^0, and Gibbs free energy ΔG^0 (Equations (3)–(5), Table 3). It is known that, depending on the nature of the change in the thermodynamic functions ΔH^0 and

ΔS^0, various intermolecular interactions can dominate in the formation of surfactant–HSA complexes [34,51]:

- $\Delta H^0 < 0$ and $\Delta S^0 < 0$—hydrogen bonding and Van der Waals interactions predominate;
- $\Delta H^0 > 0$ and $\Delta S^0 > 0$—hydrophobic interactions predominate;
- $\Delta H^0 < 0$ and $\Delta S^0 > 0$—electrostatic interactions predominate.

Summing up all this, hydrogen bonding and Van der Waals interactions predominate in the IA-16-HSA systems, which is in accordance with the quite low binding constants of surfactants to HSA (K_a). Negative Gibbs energy demonstrates the thermodynamic advantage of the formation of IA-16-HSA complexes and the spontaneity of the process.

At the next stage, the IA-16 and HSA interactions were studied using three-dimensional fluorescence spectroscopy. A 3D fluorescence is an important method for studying conformational changes in proteins, and it also allows to evaluate the microenvironment around aromatic residues in macromolecules. In this method, the fluorescence intensity is measured for all possible combinations of excitation and emission wavelengths and summarized on a three-dimensional graph [79].

The 3D fluorescence spectra of pure HSA and in the presence of IA-16 are shown in Figure 12. There are three characteristic peaks in the HSA spectra, namely, peak A, peak I, and peak II. Peak A corresponds to the Rayleigh scattering peak ($\lambda_{ex} = \lambda_{em}$). Peak I at $\lambda_{ex} = 280$ nm and $\lambda_{em} = 345$ nm demonstrates the characteristic HSA fluorescence associated with Tyr and Trp residues, respectively ($\pi \rightarrow \pi^*$ transition of aromatic amino acids). Peak II indicates the spectral behavior of the peptide bond skeleton in HSA obtained by the transition of the n–π^*electron of the carbonyl group ($\lambda_{ex} = 230$ nm, $\lambda_{em} = 345$ nm) [47,80]. It can be seen that an increase in the surfactant concentration in the system leads to fluorescence quenching of peak I, as well as to its slight hypsochromic shift (by ~35 nm). This indicates the transition of Tyr and Trp residues into a less polar environment and, consequently, an increase in hydrophobicity around amino acid fragments [79]. The results of 3D fluorescence spectra are in good agreement with the results of conventional fluorescence, which evidence that the interaction of IA-16 and HSA occurs with little effect on protein conformation and the formation of a non-covalently bound complex due to the adsorption of surfactant molecules on the hydrophobic domains of the protein.

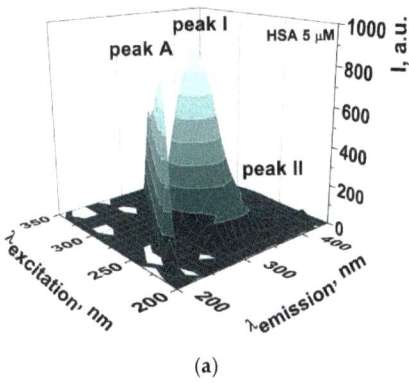
(a)

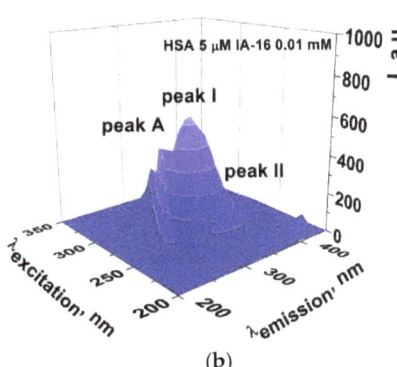
(b)

Figure 12. *Cont.*

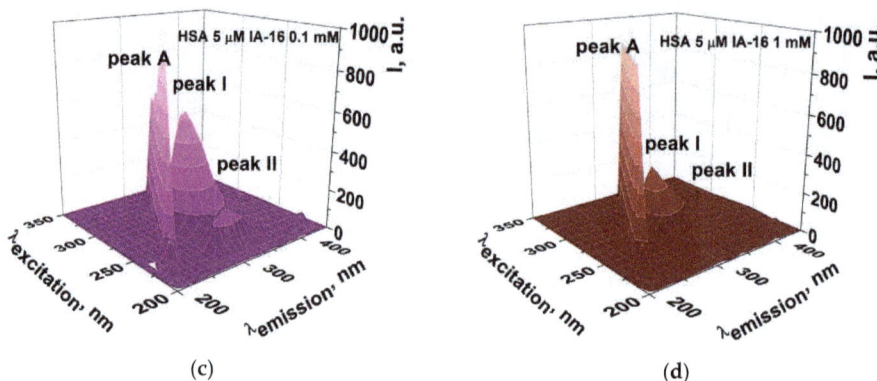

(c)   (d)

Figure 12. 3D fluorescence spectra of pure HSA (**a**) and in the presence of 0.01 mM (**b**), 0.1 mM (**c**), 1 mM (**d**) IA-16.

3.3. Drug Solubilization Study in Surfactant–PE Nanocontainers

Poor water solubility is one of the important limiting factors for drug administration and successful treatment. The most common ways to increase the solubility of drugs are the use of hydrotropes, surfactant-based nanocarriers, and changing environmental conditions, such as temperature, pH, etc. [81]. Micelles attract wide attention, providing multiple increases in solubility of different practically relevant substrates (drugs, food, chemical dyes, cosmetics, pesticides, etc.), which is one of the key properties of amphiphiles from the practical viewpoint. Polymer–colloid complexes are an alternative system for dissolving insoluble substances [13]. However, the solubilization properties of surfactant are more studied than surfactant–PE mixtures. It is most likely that drug solubilization occurs therein at a lower surfactant concentration compared to pure one due to the formation of micellar aggregates at a lower surfactant concentration in the presence of oppositely charged PE.

In accordance with the literature, including our works, the solubilization activity of cationic surfactants depends on the nature of their head group and the type of drug. Moreover, in many cases, the solubilization capacity of cationic amphiphiles with a cyclic head group is higher compared to open head group analogs [81,82]. Therefore, the solubilization ability of mixed surfactant–PE systems toward poorly water-soluble drugs of various therapeutic effects (non-steroidal anti-inflammatory agent (NSAID)-meloxicam (Figure 13), anticoagulant-warfarin (Figure 14), and antifungal antibiotic-amphotericin B (Figure 15)) was investigated.

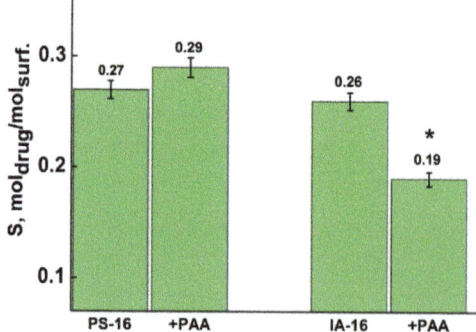

Figure 13. Solubilization capacity of the pure surfactant and mixed surfactant–PAA systems toward Meloxicam. Data are represented as mean ± standard deviation (n = 3); statistically significant differences between PS-16-PAA and IA-16-PAA systems were represented as *—$p = 0.04$.

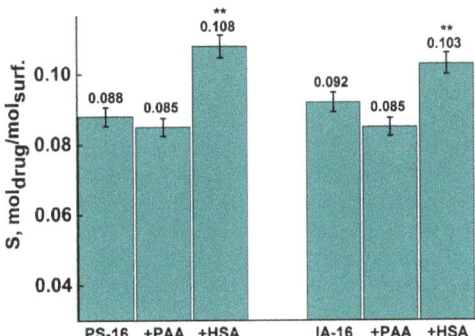

Figure 14. Solubilization capacity of the pure surfactant and mixed surfactant–PE systems toward Warfarin. Data are represented as mean ± standard deviation (n = 3); statistically significant differences in S value between pure surfactant and mixed systems were represented as **—$p < 0.01$.

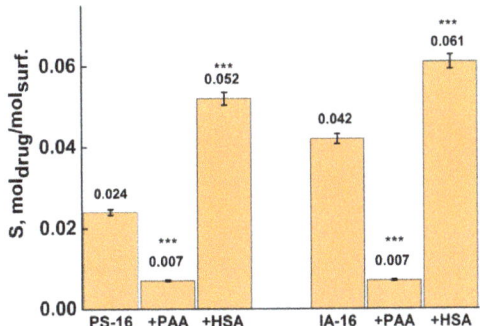

Figure 15. Solubilization capacity of the pure surfactant and mixed surfactant–PE systems toward Amphotericin B. Data are represented as mean ± standard deviation (n = 3); statistically significant differences in S value between pure surfactant and mixed systems were represented as ***—$p < 0.001$.

In accordance with [83], oppositely charged surfactant–PE mixed solutions were incubated with an excess of drugs for 24 h. S is used for quantitative estimation, which is the ratio of the mole of solubilized drug per mole of surfactant. Parameter S of the mixed surfactant–PE systems in relation to the presented drugs was evaluated by UV-vis spectrophotometry. Absorption spectra were recorded (Figures S1, S2, S4–S6 and S8–S10), followed by the construction of concentration dependences of the drug absorbance (Figures S3, S7 and S11). Meloxicam is an NSAID that has analgesic, anti-inflammatory, and antipyretic effects. It is a pH-dependent drug, the solubility of which is determined by the pK_a values and the existence ranges of the charged forms of the compound. Therefore, studies were carried out in acetate buffer at pH 4.4; under these conditions, the drug is insoluble in water but soluble in micellar solutions.

As can be seen from Figure S1, meloxicam is highly soluble in micellar solutions of individual surfactants independent of the structure of the surfactant head group. The peak at 364 nm in the spectra of meloxicam showed maximum absorption and was taken for study (Figures S1 and S2). However, the solubilization capacity (S) of the studied pyrrolidinium surfactant (0.27 mole of meloxicam/mole of PS-16) is much worse compared to PS-16 containing a hydroxyethyl fragment (0.8 mole of meloxicam/mole of PS-16) [83]. It should be noted that solubilization of the drug begins before tensiometric CMC (1 mM) at a concentration of 0.3 mM (Figure S3). It was previously reported that the acetate buffer reduces twice the aggregation thresholds of PS-16 with the hydroxyethyl fragment. The addition of PAA to micellar solutions impairs the solubility of meloxicam. The solubility is three–five times higher in a pure solution compared to mixed systems at comparable surfactant concentrations (Figure S3). It is likely that the polyanion creates a steric barrier and prevents the penetration of the drug into micellar aggregates, despite the fact that the dimensions of polymer–colloidal aggregates are much larger than micelles. At the same time, for the mixed systems containing imidazolium surfactant, the S values are lower by 30% compared to pyrrolidinium surfactant (Figure 13). To estimate the statistical significance, a one-way analysis of variance (ANOVA), where $p = 0.04$.

Next, spectrophotometric studies of the solubility of warfarin in micellar solutions of pyrrolidinium (Figure S4b) and imidazolium surfactants (Figure S4a) in the presence of PAA (Figure S5) and HSA (Figure S6) were carried out. Warfarin is an indirect anticoagulant, a coumarin derivative, which is used for the treatment and prevention of thrombosis and hematogenous embolism. The peak at 307 nm in the spectra of warfarin showed maximum absorbance and was taken to study the solubilization capacity (Figures S4–S6). The studied polyanions have the opposite effect on the solubilization capacity of amphiphiles (Figure 14). PAA additives slightly reduce the S value of the micellar systems, whereas HSA leads to an increase in S by 10–20% with statistically significant differences between the pure surfactant and mixed systems ($p < 0.01$). It is likely that HSA provides favorable hydrophobic binding sites for the drug, as previously was shown for other drugs [84,85]. In addition, binding site 1 of albumins showed a strong affinity for heterocyclic compounds such as warfarin, indomethacin, etc. [86].

Amphotericin B is an effective polyene broad-spectrum antibiotic against fungal diseases with poor solubility in water. Currently, several liposomal forms of amphotericin B have been developed; however, the high cost is the disadvantage of this approach [87]. The application of micellar nanocontainers can significantly reduce the cost and toxicity of the antibiotic. There are three maxima in the absorption spectra of the drug; the peak at 389 nm was chosen to study the solubilization capacity (Figures S8–S10). The most pronounced PE effect on the solubilization ability of amphiphiles is observed in the case of amphotericin B (Figure 15). Modification of nanocontainers with a protein increases the solubilization capacity more than twice, while the synthetic polyanion reduces the drug solubility by six times compared to pure surfactants. This is probably due to the strong affinity of amphotericin B for serum proteins. In work [86], a variety of spectroscopic methods showed a significant decrease in the HSA and BSA intrinsic fluorescence with an increase in the amphotericin concentration, which confirms the effectiveness of the interaction between components. Synthetic polyelectrolytes, specifically PAA, can initiate the formation of mixed surfactant–PE aggregates, similar to a pearl-necklace structure [24], wherein micelles look as strung on a PE chain. In this case, the hydrophobic PE chains probably compete with the hydrophobic drug for the solubilization in micelles.

4. Conclusions

Multifunctional self-assembling systems based on cationic surfactants, synthetic polyanion (polyacrylic acid), and protein (human serum albumin) were formed under varying the structure of the head group of amphiphiles (pyrrolidinium and imidazolium). The properties of the mixed systems have been systematically studied using a variety of modern complementary physicochemical methods. The polyanion additives increased the aggregation activity of the studied amphiphiles in terms of critical aggregation concentrations by two orders of magnitude (from 1 mM to 0.01 mM). The results also revealed a correlation between aggregation thresholds of mixed surfactant–HSA systems and the structure of the surfactant head group; in particular, lower CAC was demonstrated in the case of IA-16. Synthetic macromolecule PAA gives the possibility of obtaining nanoscale mixed micellar aggregates with controlled size and shape depending on the surfactant concentration, while no significant changes in size behavior were observed in the case of HSA. For the mixed system with IA-16, classical quenching of HSA fluorescence was shown, while for the system with PS-16, on the contrary, an unusual increase in fluorescence was observed, which indicates a different binding mechanism of surfactant to the protein. According to the structure–properties relationships revealed, polymer–colloid nanocontainers with controlled parameters and different loading efficiencies of hydrophobic drugs have been developed. It was shown that the solubilization capacity of surfactant–PE mixtures depends on the structure of the surfactant head group and the nature of polyanions. It can be increased more than twice, which was demonstrated on several drugs (meloxicam, warfarin, and amphotericin B). The obtained experimental data extend the understanding of the mechanism of action of polyanions and the biomedical application of amphiphilic compounds.

Supplementary Materials: The following supporting information can be downloaded at: https://www.mdpi.com/article/10.3390/nano13061072/s1, Figure S1: Absorbance spectra of Meloxicam in the IA-16 (a) and PS-16 (b) systems under varying surfactant concentration; acetic buffer pH = 4.4; 25 °C.; Figure S2: Absorbance spectra of Meloxicam in the oppositely charged IA-16-PAA (a) and PS-16-PAA (b) systems under varying surfactant concentration; acetic buffer pH = 4.4; 1 mM PAA; 25 °C; Figure S3: Meloxicam absorbance at 305 nm wavelength upon amphiphile concentration plot for pure surfactant and surfactant–PAA (b) systems; l = 0.5 cm; 25 °C; Figure S4: Absorbance spectra of Warfarin in the IA-16 (a) and PS-16 (b) aqueous solutions under varying surfactant concentration; 25 °C; Figure S5: Absorbance spectra of Warfarin in the oppositely charged IA-16-PAA (a) and PS-16-PAA (b) systems under varying surfactant concentration; 1 mM PAA; l = 0.5 cm; 25 °C; Figure S6: Absorbance spectra of Warfarin in the oppositely charged PS-16-HSA (a) and IA-16-HSA (b) systems under varying surfactant concentration; 0.5 mM HSA; l = 0.5 cm; 25 °C; Figure S7: Warfarin absorbance at 305 nm wavelength versus amphiphile concentration plot for the pure surfactant, surfactant–HSA systems (a) and surfactant–PAA (b) systems; l = 0.5 cm; 25 °C; Figure S8: Absorbance spectra of Amphotericin B in the IA-16 (a) and PS-16 (b) aqueous solutions under varying surfactant concentration; 25 °C; Figure S9: Absorbance spectra of Amphotericin B in the oppositely charged IA-16-PAA (a) and PS-16-PAA (b) systems under varying surfactant concentration; 1 mM PAA; l = 0.5 cm; 25 °C; Figure S10: Absorbance spectra of Amphotericin B in the oppositely charged PS-16-HSA (a) and IA-16-HSA (b) systems under varying surfactant concentration; 0.5 mM HSA; l = 0.5 cm; 25 °C; Figure S11: Amphotericin B absorbance at 385 nm wavelength upon amphiphile concentration plot for pure surfactant, surfactant–HSA (a) and surfactant–PAA (b) systems; l = 0.5 cm; 25 °C; Table S1: The hydrodynamic diameters (D_h), zeta potential (ζ) and polydispersity index (PdI) of PAA-PS-16 and PAA-IA-16 associates under different molar ratio; pH = 4; 25 °C; $p < 0.05$ means statistically significant difference values, $p > 0.05$ no significance.

Author Contributions: E.A.V.: investigation, data curation, visualization, formal analysis, writing—original draft, writing—review and editing; D.A.K.: investigation, writing—review and editing; F.G.V.: investigation; D.M.K.: investigation; L.Y.Z.: conceptualization, methodology, supervision, project administration. All authors have read and agreed to the published version of this manuscript.

Funding: This research was carried out with financial support from the government assignment for FRC Kazan Scientific Center of RAS.

Data Availability Statement: Not applicable.

Acknowledgments: TEM images were obtained at the Interdisciplinary Center for Analytical Microscopy, Kazan (Volga Region) Federal University, Russia.

Conflicts of Interest: The authors declare that they have no known competing financial interests or personal relationships that could have appeared to influence the work reported in this paper.

References

1. Bourganis, V.; Karamanidou, T.; Kammona, O.; Kiparissides, C. Polyelectrolyte complexes as prospective carriers for the oral delivery of protein therapeutics. *Eur. J. Pharm. Biopharm.* **2017**, *111*, 44–60. [CrossRef] [PubMed]
2. Lankalapalli, S.; Kolapalli, V.R.M. Polyelectrolyte complexes: A review of their applicability in drug delivery technology. *Indian J. Pharm. Sci.* **2009**, *71*, 481. [CrossRef] [PubMed]
3. Oh, K.T.; Bronich, T.K.; Bromberg, L.; Hatton, T.A.; Kabanov, A.V. Block ionomer complexes as prospective nanocontainers for drug delivery. *J. Control. Release* **2006**, *115*, 9–17. [CrossRef] [PubMed]

4. Penfold, J.; Thomas, R.K. Counterion condensation, the gibbs equation, and surfactant binding: An integrated description of the behavior of polyelectrolytes and their mixtures with surfactants at the air-water interface. *J. Phys. Chem. B* **2020**, *124*, 6074–6094. [CrossRef] [PubMed]
5. Jiang, R.; Liu, C.; Tan, L.T.; Lin, C. Formation of carboxymethylchitosan/gemini surfactant adsorption layers at the air/water interface: Effects of association in the bulk. *J. Dispers. Sci. Technol.* **2020**, *41*, 11–23. [CrossRef]
6. Petkov, J.T.; Penfold, J.; Thomas, R.K. Surfactant self-assembly structures and multilayer formation at the solid-solution interface induces by electrolyte, polymers and proteins. *Curr. Opin. Colloid Interface Sci.* **2022**, *57*, 101541. [CrossRef]
7. Davidson, M.L.; Walker, L.M. Interfacial properties of polyelectrolyte-surfactant aggregates at air/water interfaces. *Langmuir* **2018**, *34*, 12906–12913. [CrossRef]
8. Khan, N.; Brettmann, B. Intermolecular interactions in polyelectrolyte and surfactant complexes in solution. *Polymers* **2018**, *11*, 51. [CrossRef]
9. Vasilieva, E.A.; Vasileva, L.A.; Valeeva, F.G.; Karimova, T.R.; Zakharov, S.V.; Lukashenko, S.S.; Kuryashov, D.A.; Gaynanova, G.A.; Bashkirtseva, N.Y.; Zakharova, L.Y. Aggregation of a pyrrolidinium surfactant in the presence of polymers and hydrotropes. *Surf. Innov.* **2020**, *8*, 190–199. [CrossRef]
10. Bodnár, K.; Fegyver, E.; Nagy, M.; Mészáros, R. Impact of polyelectrolyte chemistry on the thermodynamic stability of oppositely charged macromolecule/surfactant mixtures. *Langmuir* **2016**, *32*, 1259–1268. [CrossRef]
11. Mirtič, J.; Paudel, A.; Laggner, P.; Hudoklin, S.; Kreft, M.E.; Kristl, J. Polyelectrolyte-surfactant-complex nanoparticles as a delivery platform for poorly soluble drugs: A case study of ibuprofen loaded cetylpyridinium-alginate system. *Int. J. Pharm.* **2020**, *580*, 119199. [CrossRef]
12. Zakharova, L.Y.; Ibragimova, A.R.; Vasilieva, E.A.; Mirgorodskaya, A.B.; Yackevich, E.I.; Nizameev, I.R.; Kadirov, M.K.; Zuev, Y.F.; Konovalov, A.I. Polyelectrolyte capsules with tunable shell behavior fabricated by the simple layer-by-layer technique for the control of the release and reactivity of small guests. *J. Phys. Chem. C* **2012**, *116*, 18865–18872. [CrossRef]
13. Gradzielski, M. Polyelectrolyte-surfactant complexes as a formulation tool for drug delivery. *Langmuir* **2022**, *38*, 13330–13343. [CrossRef] [PubMed]
14. Braun, L.; Uhlig, M.; Löhmann, O.; Campbell, R.A.; Schneck, E.; von Klitzing, R. Insights into extended structures and their driving force: Influence of salt on polyelectrolyte/surfactant mixtures at the air/water interface. *ACS Appl. Mater. Interfaces* **2022**, *14*, 27347–27359. [CrossRef] [PubMed]
15. Hill, C.; Abdullahi, W.; Dalgliesh, R.; Crossman, M.; Griffiths, P.C. Charge modification as a mechanism for tunable properties in polymer–surfactant complexes. *Polymers* **2021**, *13*, 2800. [CrossRef]
16. Raffa, P. Interactions between an associative amphiphilic block polyelectrolyte and surfactants in water: Effect of charge type on solution properties and aggregation. *Polymers* **2021**, *13*, 1729. [CrossRef]
17. Owiwe, M.T.; Ayyad, A.H.; Takrori, F.M. Surface tension of the oppositely charged sodium poly(styrene sulfonate) /benzyldimethylhexadecylammonium chloride and sodium poly(styrene sulfonate)/polyallylamine hydrochloride mixtures. *Colloid Polym. Sci.* **2020**, *298*, 1197–1204. [CrossRef]
18. Vasilieva, E.A.; Samarkina, D.A.; Gaynanova, G.A.; Lukashenko, S.S.; Gabdrakhmanov, D.R.; Zakharov, V.M.; Vasilieva, L.A.; Zakharova, L.Y. Self-assembly of the mixed systems based on cationic surfactants and different types of polyanions: The influence of structural and concentration factors. *J. Mol. Liq.* **2018**, *272*, 892–901. [CrossRef]
19. Vasilieva, E.; Ibragimova, A.; Lukashenko, S.; Konovalov, A.; Zakharova, L. Mixed self-assembly of polyacrylic acid and oppositely charged gemini surfactants differing in the structure of head group. *Fluid Phase Equilibria* **2014**, *376*, 172–180. [CrossRef]
20. Kogej, K. Study of the effect of polyion charge density on structural properties of complexes between poly(acrylic acid) and alkylpyridinium surfactants. *J. Phys. Chem. B* **2003**, *107*, 8003–8010. [CrossRef]
21. Kogej, K.; Škerjanc, J. Fluorescence and conductivity studies of polyelectrolyte-induced aggregation of alkyltrimethylammonium bromides. *Langmuir* **1999**, *15*, 4251–4258. [CrossRef]
22. Fernández-Peña, L.; Abelenda-Nuñez, I.; Hernández-Rivas, M.; Ortega, F.; Rubio, R.G.; Guzmán, E. Impact of the bulk aggregation on the adsorption of oppositely charged polyelectrolyte-surfactant mixtures onto solid surfaces. *Adv. Colloid Interface Sci.* **2020**, *282*, 102203. [CrossRef] [PubMed]
23. Nagarajan, R. Thermodtnamics of surfactant-polymer interactions in dilute aqueous solutions. *Chem. Phys. Lett.* **1980**, *76*, 282–286. [CrossRef]
24. Chiappisi, L.; Hoffmann, I.; Gradzielski, M. Complexes of oppositely charged polyelectrolytes and surfactants—Recent developments in the field of biologically derived polyelectrolytes. *Soft Matter* **2013**, *9*, 3896. [CrossRef]
25. Asadov, Z.H.; Nasibova, S.M.; Ahmadova, G.A.; Zubkov, F.I.; Rahimov, R.A. Head-group effect of surfactants of cationic type in interaction with propoxylated sodium salt of polyacrylic acid in aqueous solution. *Colloids Surf. Physicochem. Eng. Asp.* **2017**, *527*, 95–100. [CrossRef]
26. Vasilieva, E.A.; Zakharov, S.V.; Kuryashov, D.A.; Valeeva, F.G.; Ibragimova, A.R.; Bashkirtseva, N.Y.; Zakharova, L.Y. Supramolecular systems based on cationic surfactants: An influence of hydrotropic salts and oppositely charged polyelectrolytes. *Russ. Chem. Bull.* **2015**, *64*, 1901–1905. [CrossRef]
27. Ristroph, K.D.; Prud'homme, R.K. Hydrophobic ion pairing: Encapsulating small molecules, peptides, and proteins into nanocarriers. *Nanoscale Adv.* **2019**, *1*, 4207–4237. [CrossRef]

28. Ferreira, G.A.; Loh, W. Liquid crystalline nanoparticles formed by oppositely charged surfactant-polyelectrolyte complexes. *Curr. Opin. Colloid Interface Sci.* **2017**, *32*, 11–22. [CrossRef]
29. Zakharova, L.Y.; Gabdrakhmanov, D.R.; Ibragimova, A.R.; Vasilieva, E.A.; Nizameev, I.R.; Kadirov, M.K.; Ermakova, E.A.; Gogoleva, N.E.; Faizullin, D.A.; Pokrovsky, A.G.; et al. Structural, biocomplexation and gene delivery properties of hydroxyethylated gemini surfactants with varied spacer length. *Colloids Surf. B Biointerfaces* **2016**, *140*, 269–277. [CrossRef]
30. Kuznetsova, D.A.; Kuznetsov, D.M.; Amerhanova, S.K.; Buzmakova, E.V.; Lyubina, A.P.; Syakaev, V.V.; Nizameev, I.R.; Kadirov, M.K.; Voloshina, A.D.; Zakharova, L.Y. Cationic imidazolium amphiphiles bearing a methoxyphenyl fragment: Synthesis, self-assembly behavior, and antimicrobial activity. *Langmuir* **2022**, *38*, 4921–4934. [CrossRef]
31. Zakharova, L.Y.; Valeeva, F.G.; Kudryavtsev, D.B.; Ibragimova, A.R.; Kudryavtseva, L.A.; Timosheva, A.P.; Kataev, V.E. Catalytic effect of multicomponent supramolecular systems in phosphoryl-group transfer reactions. *Kinet. Catal.* **2003**, *44*, 547–551. [CrossRef]
32. Usman, M.; Siddiq, M. Surface and micellar properties of chloroquine diphosphate and its interactions with surfactants and human serum albumin. *J. Chem. Thermodyn.* **2013**, *58*, 359–366. [CrossRef]
33. Fatma, I.; Sharma, V.; Thakur, R.C.; Kumar, A. Current trends in protein-surfactant interactions: A review. *J. Mol. Liq.* **2021**, *341*, 117344. [CrossRef]
34. Kuznetsova, D.A.; Kuznetsov, D.M.; Zakharov, V.M.; Zakharova, L.Y. Interaction of bovine serum albumin with cationic imidazolium surfactants containing a methoxyphenyl fragment. *Russ. J. Gen. Chem.* **2022**, *92*, 1262–1270. [CrossRef]
35. Lal, H.; Akram, M. Physico-chemical characterization of bovine serum albumin-cationic gemini surfactant interaction. *J. Mol. Liq.* **2022**, *361*, 119626. [CrossRef]
36. Lamichhane, S.; Lee, S. Albumin nanoscience: Homing nanotechnology enabling targeted drug delivery and therapy. *Arch. Pharm. Res.* **2020**, *43*, 118–133. [CrossRef]
37. Cantelli, A.; Malferrari, M.; Mattioli, E.J.; Marconi, A.; Mirra, G.; Soldà, A.; Marforio, T.D.; Zerbetto, F.; Rapino, S.; Di Giosia, M.; et al. Enhanced uptake and phototoxicity of C60@albumin hybrids by folate bioconjugation. *Nanomaterials* **2022**, *12*, 3501. [CrossRef]
38. de Guzman, A.C.V.; Razzak, M.A.; Cho, J.H.; Kim, J.Y.; Choi, S.S. Curcumin-loaded human serum albumin nanoparticles prevent parkinson's disease-like symptoms in c. elegans. *Nanomaterials* **2022**, *12*, 758. [CrossRef]
39. Liu, Y.; Cao, Z.; Wang, J.; Zong, W.; Liu, R. The interaction mechanism between anionic or cationic surfactant with HSA by using spectroscopy, calorimetry and molecular docking methods. *J. Mol. Liq.* **2016**, *224*, 1008–1015. [CrossRef]
40. Karami, E.; Behdani, M.; Kazemi-Lomedasht, F. Albumin nanoparticles as nanocarriers for drug delivery: Focusing on antibody and nanobody delivery and albumin-based drugs. *J. Drug Deliv. Sci. Technol.* **2020**, *55*, 101471. [CrossRef]
41. Steiner, D.; Merz, F.W.; Sonderegger, I.; Gulotti-Georgieva, M.; Villemagne, D.; Phillips, D.J.; Forrer, P.; Stumpp, M.T.; Zitt, C.; Binz, H.K. Half-life extension using serum albumin-binding DARPin®domains. *Protein Eng. Des. Sel.* **2017**, *30*, 583–591. [CrossRef] [PubMed]
42. Elsadek, B.; Kratz, F. Impact of albumin on drug delivery—New applications on the horizon. *J. Control. Release* **2012**, *157*, 4–28. [CrossRef]
43. Paul, M.; Itoo, A.M.; Ghosh, B.; Biswas, S. Current trends in the use of human serum albumin for drug delivery in cancer. *Expert Opin. Drug Deliv.* **2022**, *19*, 1449–1470. [CrossRef] [PubMed]
44. Samarkina, D.A.; Gabdrakhmanov, D.R.; Lukashenko, S.S.; Nizameev, I.R.; Kadirov, M.K.; Zakharova, L.Y. Homologous series of amphiphiles bearing imidazolium head group: Complexation with bovine serum albumin. *J. Mol. Liq.* **2019**, *275*, 232–240. [CrossRef]
45. ud din Parray, M.; Mir, M.U.H.; Dohare, N.; Maurya, N.; Khan, A.B.; Borse, M.S.; Patel, R. Effect of cationic gemini surfactant and its monomeric counterpart on the conformational stability and esterase activity of human serum albumin. *J. Mol. Liq.* **2018**, *260*, 65–77. [CrossRef]
46. Fatma, I.; Sharma, V.; Ahmad Malik, N.; Assad, H.; Cantero-López, P.; Sánchez, J.; López-Rendón, R.; Yañez, O.; Chand Thakur, R.; Kumar, A. Influence of HSA on micellization of NLSS and BC: An experimental-theoretical approach of its binding characteristics. *J. Mol. Liq.* **2022**, *367*, 120532. [CrossRef]
47. Kuznetsova, D.A.; Kuznetsov, D.M.; Vasileva, L.A.; Toropchina, A.V.; Belova, D.K.; Amerhanova, S.K.; Lyubina, A.P.; Voloshina, A.D.; Zakharova, L.Y. Pyrrolidinium surfactants with a biodegradable carbamate fragment: Self-assembling and biomedical application. *J. Mol. Liq.* **2021**, *340*, 117229. [CrossRef]
48. Vasilieva, E.A.; Lukashenko, S.S.; Voloshina, A.D.; Strobykina, A.S.; Vasileva, L.A.; Zakharova, L.Y. The synthesis and properties of homologous series of surfactants containing the pyrrolidinium head group with hydroxyethyl moiety. *Russ. Chem. Bull.* **2018**, *67*, 1280–1286. [CrossRef]
49. Samarkina, D.A.; Gabdrakhmanov, D.R.; Lukashenko, S.S.; Khamatgalimov, A.R.; Kovalenko, V.I.; Zakharova, L.Y. Cationic amphiphiles bearing imidazole fragment: From aggregation properties to potential in biotechnologies. *Colloids Surf. Physicochem. Eng. Asp.* **2017**, *529*, 990–997. [CrossRef]
50. Kuznetsova, D.A.; Kuznetsov, D.M.; Vasileva, L.A.; Amerhanova, S.K.; Valeeva, D.N.; Salakhieva, D.V.; Nikolaeva, V.A.; Nizameev, I.R.; Islamov, D.R.; Usachev, K.S.; et al. Complexation of oligo- and polynucleotides with methoxyphenyl-functionalized imidazolium surfactants. *Pharmaceutics* **2022**, *14*, 2685. [CrossRef]

51. Kuznetsova, D.A.; Gabdrakhmanov, D.R.; Lukashenko, S.S.; Faizullin, D.A.; Zuev, Y.F.; Nizameev, I.R.; Kadirov, M.K.; Kuznetsov, D.M.; Zakharova, L.Y. Interaction of bovine serum albumin with cationic imidazolium-containing amphiphiles bearing urethane fragment: Effect of hydrophobic tail length. *J. Mol. Liq.* **2020**, *307*, 113001. [CrossRef]
52. Kuznetsova, D.A.; Gabdrakhmanov, D.R.; Kuznetsov, D.M.; Lukashenko, S.S.; Zakharov, V.M.; Sapunova, A.S.; Amerhanova, S.K.; Lyubina, A.P.; Voloshina, A.D.; Salakhieva, D.V.; et al. Polymer-colloid complexes based on cationic imidazolium amphiphile, polyacrylic acid and DNA decamer. *Molecules* **2021**, *26*, 2363. [CrossRef] [PubMed]
53. Vasilieva, E.A.; Lukashenko, S.S.; Vasileva, L.A.; Pavlov, R.V.; Gaynanova, G.A.; Zakharova, L.Y. Aggregation behavior of the surfactant bearing pyrrolidinium head group in the presence of polyacrylic acid. *Russ. Chem. Bull.* **2019**, *68*, 341–346. [CrossRef]
54. Ouverney Ferreira, M.; Câmara de Assis, H.F.; Percebom, A.M. Cocamidopropyl betaine can behave as a cationic surfactant and electrostatically associate with polyacids of high molecular weight. *Colloids Surf. Physicochem. Eng. Asp.* **2022**, *654*, 130123. [CrossRef]
55. Vasilieva, E.A.; Kuznetsova, D.A.; Valeeva, F.G.; Kuznetsov, D.M.; Zakharov, A.V.; Amerhanova, S.K.; Voloshina, A.D.; Zueva, I.V.; Petrov, K.A.; Zakharova, L.Y. Therapy of organophosphate poisoning via intranasal administration of 2-PAM-loaded chitosomes. *Pharmaceutics* **2022**, *14*, 2846. [CrossRef]
56. Colino, C.I.; Velez Gomez, D.; Alonso Horcajo, E.; Gutierrez-Millan, C. A comparative study of liposomes and chitosomes for topical quercetin antioxidant therapy. *J. Drug Deliv. Sci. Technol.* **2022**, *68*, 103094. [CrossRef]
57. Wang, C.; Tam, K.C. Interaction between polyelectrolyte and oppositely charged surfactant: Effect of charge density. *J. Phys. Chem. B* **2004**, *108*, 8976–8982. [CrossRef]
58. Taylor, D.J.F.; Thomas, R.K.; Penfold, J. Polymer/surfactant interactions at the air/water interface. *Adv. Colloid Interface Sci.* **2007**, *132*, 69–110. [CrossRef]
59. Aguiar, J.; Carpena, P.; Molina-Bolívar, J.A.; Carnero Ruiz, C. On the determination of the critical micelle concentration by the pyrene 1:3 ratio method. *J. Colloid Interface Sci.* **2003**, *258*, 116–122. [CrossRef]
60. Tai, K.; Rappolt, M.; Mao, L.; Gao, Y.; Li, X.; Yuan, F. The stabilization and release performances of curcumin-loaded liposomes coated by high and low molecular weight chitosan. *Food Hydrocoll.* **2020**, *99*, 105355. [CrossRef]
61. Luan, Y.; Ramos, L. Role of the preparation procedure in the formation of spherical and monodisperse surfactant/polyelectrolyte complexes. *Chem. Eur. J.* **2007**, *13*, 6108–6114. [CrossRef] [PubMed]
62. Kwiatkowski, A.L.; Molchanov, V.S.; Kuklin, A.I.; Orekhov, A.S.; Arkharova, N.A.; Philippova, O.E. Structural transformations of charged spherical surfactant micelles upon solubilization of water-insoluble polymer chains in salt-free aqueous solutions. *J. Mol. Liq.* **2022**, *347*, 118326. [CrossRef]
63. Geng, F.; Zheng, L.; Yu, L.; Li, G.; Tung, C. Interaction of bovine serum albumin and long-chain imidazolium ionic liquid measured by fluorescence spectra and surface tension. *Process Biochem.* **2010**, *45*, 306–311. [CrossRef]
64. Sinha, S.; Tikariha, D.; Lakra, J.; Yadav, T.; Kumari, S.; Saha, S.K.; Ghosh, K.K. Interaction of bovine serum albumin with cationic monomeric and dimeric surfactants: A comparative study. *J. Mol. Liq.* **2016**, *218*, 421–428. [CrossRef]
65. Karukstis, K.K.; McDonough, J.R. Characterization of the aggregates of N-Alkyl-N-methylpyrrolidinium bromide surfactants in aqueous solution. *Langmuir* **2005**, *21*, 5716–5721. [CrossRef] [PubMed]
66. Liu, T.; Liu, M.; Guo, Q.; Liu, Y.; Zhao, Y.; Wu, Y.; Sun, B.; Wang, Q.; Liu, J.; Han, J. Investigation of binary and ternary systems of human serum albumin with oxyresveratrol/piceatannol and/or mitoxantrone by multipectroscopy, molecular docking and cytotoxicity evaluation. *J. Mol. Liq.* **2020**, *311*, 113364. [CrossRef]
67. Cheema, M.A.; Taboada, P.; Barbosa, S.; Juárez, J.; Gutiérrez-Pichel, M.; Siddiq, M.; Mosquera, V. Human serum albumin unfolding pathway upon drug binding: A thermodynamic and spectroscopic description. *J. Chem. Thermodyn.* **2009**, *41*, 439–447. [CrossRef]
68. Chaves, O.A.; Mathew, B.; Cesarin-Sobrinho, D.; Lakshminarayanan, B.; Joy, M.; Mathew, G.E.; Suresh, J.; Netto-Ferreira, J.C. spectroscopic, zeta potential and molecular docking analysis on the interaction between human serum albumin and halogenated thienyl chalcones. *J. Mol. Liq.* **2017**, *242*, 1018–1026. [CrossRef]
69. Atarodi Shahri, P.; Sharifi Rad, A.; Beigoli, S.; Saberi, M.R.; Chamani, J. Human serum albumin-amlodipine binding studied by multi-spectroscopic, zeta-potential, and molecular modeling techniques. *J. Iran. Chem. Soc.* **2018**, *15*, 223–243. [CrossRef]
70. Bakaeean, B.; Kabiri, M.; Iranfar, H.; Saberi, M.R.; Chamani, J. Binding effect of common ions to human serum albumin in the presence of norfloxacin: Investigation with spectroscopic and zeta potential approaches. *J. Solut. Chem.* **2012**, *41*, 1777–1801. [CrossRef]
71. Samarkina, D.A.; Gabdrakhmanov, D.R.; Lukashenko, S.S.; Khamatgalimov, A.R.; Zakharova, L.Y. Aggregation capacity and complexation properties of a system based on an imidazole-containing amphiphile and bovine serum albumin. *Russ. J. Gen. Chem.* **2017**, *87*, 2826–2831. [CrossRef]
72. Iranfar, H.; Rajabi, O.; Salari, R.; Chamani, J. Probing the interaction of human serum albumin with ciprofloxacin in the presence of silver nanoparticles of three sizes: Multispectroscopic and ζ potential investigation. *J. Phys. Chem. B* **2012**, *116*, 1951–1964. [CrossRef] [PubMed]
73. Kuznetsova, D.A.; Gabdrakhmanov, D.R.; Lukashenko, S.S.; Voloshina, A.D.; Sapunova, A.S.; Kashapov, R.R.; Zakharova, L.Y. Self-assembled systems based on novel hydroxyethylated imidazolium-containing amphiphiles: Interaction with DNA decamer, protein and lipid. *Chem. Phys. Lipids* **2019**, *223*, 104791. [CrossRef]
74. Naik, D.V.; Paul, W.L.; Threatte, R.M.; Schulman, S.G. Fluorometric determination of drug-protein association constants. binding of 8-anilino-1-naphthalenesulfonate by bovine serum albumin. *Anal. Chem.* **1975**, *47*, 267–270. [CrossRef]

75. Zhang, S.; Chen, X.; Ding, S.; Lei, Q.; Fang, W. Unfolding of human serum albumin by gemini and single-chain surfactants: A comparative study. *Colloids Surf. Physicochem. Eng. Asp.* **2016**, *495*, 30–38. [CrossRef]
76. Branco, M.A.; Pinheiro, L.; Faustino, C. Amino acid-based cationic gemini surfactant-protein interactions. *Colloids Surf. Physicochem. Eng. Asp.* **2015**, *480*, 105–112. [CrossRef]
77. Zhou, T.; Ao, M.; Xu, G.; Liu, T.; Zhang, J. Interactions of bovine serum albumin with cationic imidazolium and quaternary ammonium gemini surfactants: Effects of surfactant architecture. *J. Colloid Interface Sci.* **2013**, *389*, 175–181. [CrossRef]
78. Mandeville, J.-S.; Froehlich, E.; Tajmir-Riahi, H.A. Study of curcumin and genistein interactions with human serum albumin. *J. Pharm. Biomed. Anal.* **2009**, *49*, 468–474. [CrossRef]
79. Bortolotti, A.; Wong, Y.H.; Korsholm, S.S.; Bahring, N.H.B.; Bobone, S.; Tayyab, S.; van de Weert, M.; Stella, L. On the purported "backbone fluorescence" in protein three-dimensional fluorescence spectra. *RSC Adv.* **2016**, *6*, 112870–112876. [CrossRef]
80. Zhang, M.; Chai, Y.; Han, B. Mechanistic and conformational studies on the interaction between myriocin and human serum albumin by fluorescence spectroscopy and molecular docking. *J. Solut. Chem.* **2019**, *48*, 835–848. [CrossRef]
81. Zakharova, L.Y.; Vasilieva, E.A.; Mirgorodskaya, A.B.; Zakharov, S.V.; Pavlov, R.V.; Kashapova, N.E.; Gaynanova, G.A. Hydrotropes: Solubilization of nonpolar compounds and modification of surfactant solutions. *J. Mol. Liq.* **2023**, *370*, 120923. [CrossRef]
82. Zakharova, L.Y.; Kashapov, R.R.; Pashirova, T.N.; Mirgorodskaya, A.B.; Sinyashin, O.G. Self-assembly strategy for the design of soft nanocontainers with controlled properties. *Mendeleev Commun.* **2016**, *26*, 457–468. [CrossRef]
83. Vasileva, L.A.; Kuznetsova, D.A.; Valeeva, F.G.; Vasilieva, E.A.; Lukashenko, S.S.; Gaynanova, G.A.; Zakharova, L.Y. Micellar nanocontainers based on cationic surfactants with a pyrrolidinium head group for increasing drug bioavailability. *Russ. Chem. Bull.* **2021**, *70*, 1341–1348. [CrossRef]
84. Farooqi, M.J.; Penick, M.A.; Negrete, G.R.; Brancaleon, L. Human serum albumin as vehicle for the solubilization of perylene diimides in aqueous solutions. *Int. J. Biol. Macromol.* **2017**, *94*, 246–257. [CrossRef] [PubMed]
85. Kinoshita, R.; Ishima, Y.; Chuang, V.T.G.; Watanabe, H.; Shimizu, T.; Ando, H.; Okuhira, K.; Otagiri, M.; Ishida, T.; Maruyama, T. The therapeutic effect of human serum albumin dimer-doxorubicin complex against human pancreatic tumors. *Pharmaceutics* **2021**, *13*, 1209. [CrossRef]
86. Temboot, P.; Usman, F.; Ul-Haq, Z.; Khalil, R.; Srichana, T. Biomolecular interactions of amphotericin B nanomicelles with serum albumins: A combined biophysical and molecular docking approach. *Spectrochim. Acta. A. Mol. Biomol. Spectrosc.* **2018**, *205*, 442–456. [CrossRef]
87. Gurudevan, S.; Francis, A.P.; Jayakrishnan, A. Amphotericin B-albumin conjugates: Synthesis, toxicity and anti-fungal activity. *Eur. J. Pharm. Sci.* **2018**, *115*, 167–174. [CrossRef]

Disclaimer/Publisher's Note: The statements, opinions and data contained in all publications are solely those of the individual author(s) and contributor(s) and not of MDPI and/or the editor(s). MDPI and/or the editor(s) disclaim responsibility for any injury to people or property resulting from any ideas, methods, instructions or products referred to in the content.

Article

Synthesis of PDMS-µ-PCL Miktoarm Star Copolymers by Combinations () of Styrenics-Assisted Atom Transfer Radical Coupling and Ring-Opening Polymerization and Study of the Self-Assembled Nanostructures

Yi-Shen Huang [1,†], Dula Daksa Ejeta [1,†], Kun-Yi (Andrew) Lin [2], Shiao-Wei Kuo [3], Tongsai Jamnongkan [4,*] and Chih-Feng Huang [1,*]

1. Department of Chemical Engineering, i-Center for Advanced Science and Technology (iCAST), National Chung Hsing University, Taichung 40227, Taiwan; yishen617@gmail.com (Y.-S.H.); duladaksa@gmail.com (D.D.E.)
2. Department of Environmental Engineering, i-Center for Advanced Science and Technology (iCAST), National Chung Hsing University, Taichung 40227, Taiwan; linky@nchu.edu.tw
3. Department of Materials and Optoelectronic Science, Center for Nanoscience and Nanotechnology, National Sun Yat-Sen University, Kaohsiung 80424, Taiwan; kuosw@faculty.nsysu.edu.tw
4. Department of Fundamental Science and Physical Education, Faculty of Science at Sriracha, Kasetsart University, Chonburi 20230, Thailand
* Correspondence: jamnongkan.t@ku.ac.th (T.J.); huangcf@dragon.nchu.edu.tw (C.-F.H.)
† These authors contributed equally to this work.

Citation: Huang, Y.-S.; Ejeta, D.D.; Lin, K.-Y.; Kuo, S.-W.; Jamnongkan, T.; Huang, C.-F. Synthesis of PDMS-µ-PCL Miktoarm Star Copolymers by Combinations () of Styrenics-Assisted Atom Transfer Radical Coupling and Ring-Opening Polymerization and Study of the Self-Assembled Nanostructures. *Nanomaterials* 2023, *13*, 2355. https://doi.org/10.3390/nano13162355

Academic Editor: Pavel Padnya

Received: 2 August 2023
Revised: 14 August 2023
Accepted: 15 August 2023
Published: 17 August 2023

Copyright: © 2023 by the authors. Licensee MDPI, Basel, Switzerland. This article is an open access article distributed under the terms and conditions of the Creative Commons Attribution (CC BY) license (https://creativecommons.org/licenses/by/4.0/).

Abstract: Due to their diverse and unique physical properties, miktoarm star copolymers (µ-SCPs) have garnered significant attention. In our study, we employed α-monobomoisobutyryl-terminated polydimethylsiloxane (PDMS-Br) to carry out styrenics-assisted atom transfer radical coupling (SA ATRC) in the presence of 4-vinylbenzyl alcohol (VBA) at 0 °C. By achieving high coupling efficiency ($\chi_c = 0.95$), we obtained mid-chain functionalized PDMS-VBA$_m$-PDMS polymers with benzylic alcohols. Interestingly, matrix-assisted laser desorption/ionization time of flight mass spectrometry (MALDI-TOF MS) analysis revealed the insertion of only two VBA coupling agents (m = 2). Subsequently, the PDMS-VBA$_2$-PDMS products underwent mid-chain extensions using ε-caprolactone (ε-CL) through ring-opening polymerization (ROP) with an efficient organo-catalyst at 40 °C, resulting in the synthesis of novel (PDMS)$_2$-µ-(PCL)$_2$ µ-SCPs. Eventually, novel (PDMS)$_2$-µ-(PCL)$_2$ µ-SCPs were obtained. The obtained PDMS-µ-PCL µ-SCPs were further subjected to examination of their solid-state self-assembly through small-angle X-ray scattering (SAXS) experiments. Notably, various nanostructures, including lamellae and hexagonally packed cylinders, were observed with a periodic size of approximately 15 nm. As a result, we successfully developed a simple and effective reaction combination () strategy (i.e., SA ATRC--ROP) for the synthesis of well-defined PDMS-µ-PCL µ-SCPs. This approach may open up new possibilities for fabricating nanostructures from siloxane-based materials.

Keywords: polydimethylsiloxane; poly(ε-caprolactone); miktoarm star copolymers; styrenics-assisted atom transfer radical coupling; combination

1. Introduction

Star copolymers (SCPs) are non-linear macromolecules with more than two linear polymer chains (arms) with a central core [1,2]. Among SCPs, homogeneous arms (e.g., homopolymer, random, or block copolymer segments) and heterogeneous arms (i.e., miktoarm SCPs (µ-SCPs)) are the two major classes of SCPs [3,4]. The former consists of arms with identical chemical structures, chain ends, and comparable compositions and molecular weights (MWs). The latter encompasses dissimilar arms with varying MWs, compositions, and chain-end functionalities linked at a central junction [5]. Their superior features of high functional group density and low solution viscosity, arising from a spatially defined and congested 3D globular structure, generally do not exhibit in typical linear copolymers [6].

Moreover, μ-SCPs have recently attracted significant attention due to their diversity, exhibiting unique physical properties. Originating from the architecture effects [5], the variability of compositions and functionality within μ-SCPs also grants access to a captivating range of novel properties, including distinct microdomain morphologies, the capacity to generate stable unimolecular micelles, and the formation of exceptional supramolecular assemblies. These unprecedented properties of μ-SCPs enable them in various potential applications, including thermoplastic elastomers, nanostructured thin films, solid-state electrolytes, gene and drug delivery systems, and stimuli-responsive materials [5,7–9]. Proper polymer designs with well-defined MWs, low molecular weight polydispersity (PDI), and controllable architectures thus play an essential role in establishing physical property relationships. This is vital for polymer chemists to achieve one of the most important goals of designing polymers with predetermined structures and manipulating materials with desired properties [10,11].

Based on the designing strategies, μ-SCPs can be synthesized via the core-first approach, coupled with the grafting-from or arm-first methods associated with the grafting onto/through approach. Through combinations (symbolized as "") of two or more controlled/"living" polymerization (CLP) techniques, we can achieve control of the arm compositions, MWs, PDIs, and functionalities [6]. The CLP techniques include living anionic/cationic polymerization (LAP/LCP), reversible-deactivation radical polymerizations (RDRPs), ring-opening (metathesis) polymerization (ROP), catalyst transfer polymerization (CTP), chain-growth condensation polymerization (CGCP), etc. Using the core-first approach, μ-SCPs can be directly synthesized by initiating arm growth from a multifunctional initiator with heterogeneous sites through sequential initiation systems. Using the "core-first" approach, we might face the arduous synthesis of a multifunctional core and the later orthogonal initiating functionalities [12]. In addition, the synthesis of μ-SCPs with high MWs could be bargained by the occurrence of unsought terminations (e.g., intra-arms or inter-arms coupling reactions) during polymerization, leading to undesired molecular structures [13]. The arm-first approaches are more diverse. Several approaches can be utilized: (i) A conventional method is that heterogeneous polymer chains with a reactive end can be coupled with efficient organic reactions to conduct a grafting-onto strategy among the different types of polymer chain ends and the multifunctional core. (ii) A late-developed method is that macromonomer (MM) precursors can be first synthesized by CLPs or modified by a chain end of condensation-type polymers to possess a vinyl group. Subsequently, an appropriate amount of chain-end cross-linking via a CLP initiator can lead to μ-SCPs. (iii) Vice versa, macroinitiator (MI) precursors can be synthesized by CLPs or modified by the chain ends of condensation-type polymers to possess an initiating site. Subsequently, chain-end cross-linking through the CLP mechanism in the presence of divinyl compounds can also result in μ-SCPs [4,5]. (iv) A recently developed and innovative "in-out" strategy method was demonstrated. In this method, polymer arms were created, and a core was convergently formed through in-situ additions of divinyl crosslinking agents. Within the core of the SCP, dormant initiating sites were preserved. Subsequently, this star copolymer was utilized as a multifunctional MI to initiate CLP of another monomer, resulting in the production of μ-SCPs.

Common drawbacks of the conventional arm-first approach are lack of chain mobility and steric hindrance in high MWs, leading to slow and inefficient reactions. Thus, it is usually employed for the syntheses of μ-SCPs with low MWs or fewer arms [13–15]. Conversely, the two recently developed arm-first methods offer a highly convenient and efficient means of preparing μ-SCPs. This is achieved by separately synthesizing functional arms with distinct compositions and crosslinking them together, leading to high molecular weight and a substantial number of arms (>100) [13,16]. It allows us to control the number of arms, which is dictated by parameters including amount of crosslinker, small amounts of monomers, timing of crosslinker addition, MW of arms, ratio of MMs/initiator, ratio of MIs/crosslinker, functionality, and composition of the polymeric arms [15,17,18].

In the previous literature, LAP has been extensively and successfully employed to synthesize various types of μ-SCPs since its pioneer position of CLPs [19–29]. It is still a challenge to extend the arm numbers beyond six in all cases. In recent decades, RDRP (e.g., atom transfer radical polymerization (ATRP), reversible addition–fragmentation chain-transfer (RAFT) polymerization, and nitroxide-mediated polymerization (NMP)) and ROP methods started a revolution owing to their orthogonal reactivity, gentle reaction conditions, and expansive monomer range in comparison to LAP [13,14,30–37]. Some reports on synthesizing polyester-based μ-SCPs by combining other chemistry and contemporary polymerization techniques and ROP are available. However, only a few works have reported using the arm-first route with the facile based on efficient radical coupling reactions, such as styrenics-assisted atom transfer radical coupling (SA ATRC) [5,16,38] and photoirradiation organotellurium-mediated radical coupling ($h\nu$TERC) [39,40] to synthesize polyester-based μ-SCPs.

Several kinds of literature illustrate that SA ATRC is a unique method of introducing polymers with mid-chain functionality at a low temperature with high coupling efficiency (χ_C) [13,31]. For example, Huang and co-workers progressively developed various A_nB_m-type μ-SCPs by using the arm-first approach via combinations of three techniques (e.g., ROP--SA ATRC--ATRP or ROP--SA ATRC--CGCP) [14,36]. In another case, the CGCP was first applied for the synthesis of well-defined poly(N-octyl benzamide) (PBA) and converted the chain end to an effective initiating site of 2-bromoisobutyryl (i.e., PBA-Br: M_n = 3000 and PDI = 1.1). A high coupling efficiency (>0.95) and doubling of the MW with low PDI (M_n = 5770 and PDI = 1.13) were obtained through the SA ATRC of PBA-Br with 4-vinylbenzyl alcohol (VBA) coupling agent. Through the combined strategy of CGCP--SA ATRC--ROP (i.e., final ROP mid-chain extension of PBA-VBA$_m$-PBA MI with ε-caprolactone (ε-CL)), a novel well-defined (PBA)$_2$-μ-(PCL)$_4$ μ-SCP ($M_{n,NMR}$ = ca. 13,400 and PDI = 1.17) was successfully obtained [41].

Herein, we develop a simple and effective SA ATRC--ROP strategy for synthesizing well-defined PDMS-μ-PCL (PDMS: polydimethylsiloxane) μ-SCPs. As shown in Scheme 1, the chain-end modification and SA ATRC of the PDMS-based precursors were conducted and examined by gel permeation chromatography (GPC) and matrix-assisted laser desorption/ionization time of flight mass spectrometry (MALDI-TOF MS). We employed a styrenic monomer, VBA, as a coupling agent to synthesize PDMS-VBA$_m$-PDMS MI. An organo-catalyst catalyzed the ROP mid-chain extension of PDMS-VBA$_m$-PDMS MI with ε-CL to afford well-defined μ-SCPs. Eventually, the self-assembly behaviors and the microstructures in the obtained solid-state PDMS-μ-PCL μ-SCPs were investigated by small-angle X-ray scattering (SAXS).

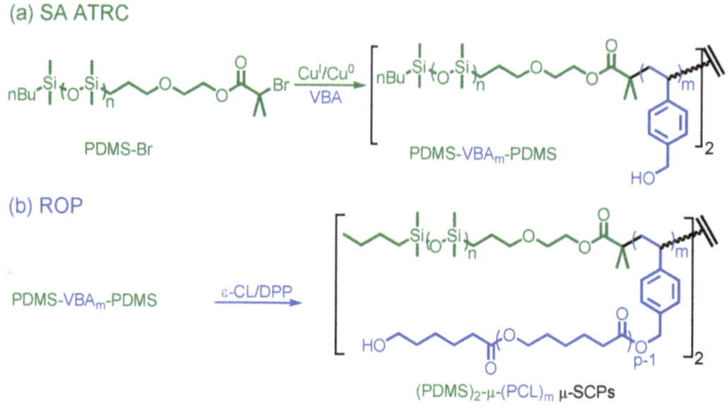

Scheme 1. Reaction routes for the synthesis of μ-SCPs: (**a**) SA ATRC at 0 °C and (**b**) ROP at 40 °C.

2. Experimental Section

2.1. Materials

Triethylamine (TEA, 99.5%), 2-bromoisobutyryl bromide (BiB, 97%), copper bromide (CuBr, 99%), diphenyl phosphate (DPP, 97%), copper wire (Cu wire (diameter = 1.0 mm), 99.9%), hydrochloric acid (37%), formic acid (88%), magnesium sulfate anhydrous ($MgSO_4$, 99%), dichloromethane (DCM, 99.9%), dimethyl sulfoxide (DMSO, 99%), diethyl ether (99%), methanol (MeOH, 99%), and tetrahydrofuran (THF, 99%) were procured from Sigma-Aldrich (St. Louis, MO, USA) and were used without purification. Monohydroxyl-terminated polydimethylsiloxane (PDMS-OH: M_n = 1250, PDI = 1.17) was purchased from Gelest. ε-Caprolactone (ε-CL, 99%) and toluene (99%) were dried by distillation with sodium hydride and sodium prior to use. According to the previous literature, we prepared 4-vinylbenzyl alcohol (VBA) [42–45] and tris(2-(dimethylamino)ethyl)amine (Me_6TREN) [46–48] (Scheme S1a,b, respectively (see the ESI)). Their characterization of ^1H NMR spectra is shown in Figure S1 (see the ESI).

2.2. Characterization

The progress of monomer conversion was tracked using a Hewlett Packard 5890 series II gas chromatograph (GC) featuring an FID detector and a CNW CD-5 column (30 m), with toluene employed as an internal standard. Gel permeation chromatography (GPC) was carried out at 40 °C in THF (flow rate: 1 mL/min) using a Waters 515 pump, a Waters 410 differential refractometer, a Waters 486 absorbance detector, and two PSS SDV columns (Linear S and 100 Å pore size) to determine M_n, M_w and M_w/M_n (i.e., PDI). Monodisperse polystyrene (PSt) standards were used for calibrations. Recycling preparative gel permeation chromatography (rGPC) was carried out at ambient temperature in THF (flow rate: 10 mL/min) using a LaboACE LC-5060 equipped with a P-LA60 pump, a RI-700 LA differential refractometer, and JAIGEL-2.5HR columns (exclusion limit 20,000). Polymer solutions (10 mg/mL) were purified to obtain low PDI samples. FT-IR spectra were captured using a Nicolet Avatar 320 FT-IR spectrometer operating at a resolution of 4 cm^{-1}, with 24 scans. The samples were dissolved in THF and applied onto a KBr plate. Proton nuclear magnetic resonance (^1H NMR) spectra were acquired using a Varian 400 NMR and calibrated against an internal standard of $CDCl_3$ (δ = 7.26 ppm). Matrix-assisted laser desorption/ionization time-of-flight mass spectrometry (MALDI-TOF MS) was recorded on a Bruker Autoflex III spectrometer plus in the reflectron ion mode. Trans-2-[3-(4-tert-butylphenyl)-2-methyl-2-propenylidene] malononitrile (DCTB) and sodium chloride (NaCl) were, respectively, used as the matrix and ionizing agent for the MALDI-TOF mass measurements. The dried samples were placed in an N_2-purged chamber to maintain environmental dryness. Small-angle X-ray scattering (SAXS) analysis was conducted at the Endstation BL23A1 National Synchrotron Radiation Research Center (NSRRC) in Hsinchu, Taiwan. The X-ray source operated at 15 kV, and the sample-to-detector distance was 3 m. The scattering profile, obtained using a Pilatus-1MF detector, was plotted in terms of scattering intensity (I) as a function of the scattering vector magnitude [q = (4π/λ) sin (θ/2)]. The d-spacing was determined from the first-order scattering peak (q*) using the formula d = 2π/q*.

2.3. Synthesis of PDMS-Br Macroinitiator (MI)

An acid scavenger TEA (1 mL, 9.9 mmol), PDMS-OH (5.5 g, 5 mmol), and DCM (50 mL) were mixed in a reaction flask. BiB (2.3 mL, 9.9 mmol) was added dropwise into the solution. The reaction mixture was stirred at room temperature for 24 h. The reaction mixture was diluted with 1 M HCl solution (100 mL) and extracted with DCM 100 mL 3 times. The collected organic phase was washed with deionized water, dried with $MgSO_4$, and concentrated to obtain PDMS-Br precursor ($M_{n,NMR}$ = 1280, $M_{n,GPC}$ = 1280, PDI = 1.14, yield 80%). ^1H NMR (400 MHz, $CDCl_3$, δ = ppm): 4.32 (t, -CH_2OCO-, 2H), 3.67 (t, -CH_2CH_2OCO-, 2H), 3.45 (t, -CH_2CH_2CH$_2$O-, 2H), 1.94 (s, -OCOC(CH_3)$_2$Br, 6H), 1.60 (m,

-CH$_2$CH$_2$CH$_2$O-, 4H), 1.25–1.37 (m, CH$_3$CH$_2$CH$_2$CH$_2$-, 4H), 0.88 (t, CH$_3$CH$_2$CH$_2$CH$_2$-, 3H), 0.53 (m, -Si(CH$_3$)$_2$CH$_2$-, 4H), 0.1 (br, -(CH$_3$)$_2$ of PDMS, 6H).

2.4. Synthesis of PDMS-VBA$_m$-PDMS through Styrenics-Assisted Atom Transfer Radical Coupling (SA ATRC)

In an example of the SA ATRC reaction, CuBr (0.43 g, 3 mmol), toluene (100 mL), and a stirring bar wrapped with fresh copper wire (5 cm) were added to a septum-sealed Schlenk flask. The mixture was subsequently deoxygenated by nitrogen purging treatment for 60 min. A mixture of VBA (3.8 mL, 30 mmol), PDMS-Br (3.8 g, 3 mmol with $M_{n,0}$ = 1280, PDI = 1.14), and Me$_6$TREN (2 mL, 7.5 mmol) was then injected into the flask after treating in the nitrogen atmosphere and the reaction was cooled to 0 °C. Samples were taken periodically under a nitrogen blanket and passed through a short column of neutral alumina to remove dissolved copper salts before the analysis of GPC. The reaction was exposed to an ambient atmosphere and the crude product was diluted with THF and passed through a neutral alumina column. The solution was concentrated under a vacuum to remove the residual solvent, monomer, and ligand. The collected polymer was vacuum-dried to afford PDMS-VBA$_m$-PDMS (M_n = 2440, PDI = 1.31, yield 25%; coupling efficiency (x_c) = 0.95 where $x_c = 2 \times (1 - M_{n,0}/M_n)$).

2.5. Chain Extension of PDMS-VBA$_m$-PDMS with ε-CL through ROP

An oven-dried Schlenk flask and magnetic stirrer bar with DPP (50 mg, 0.2 mmol) and PDMS-VBA$_m$-PDMS (0.5 g, 0.2 mmol) and toluene (10 mL) were charged. The flask was sealed and deoxygenated using freeze–pump–thaw for three cycles and refilled with nitrogen. ε-CL (1.1 mL, 9.8 mmol) was injected into the solution ([ε-CL]$_0$ = 0.8 M), and the polymerization was allowed to commence at 40 °C. An initial sample was taken via a syringe to trace the chain extension through ROP at ambient temperature. The reaction was stopped by adding an acidic Amberlyst® A21 and diluted with THF. The ion-exchange resin was removed by filtration and the clear solution was precipitated in MeOH. Using similar procedures, two crude PDMS-μ-PCL were collected and dried under vacuum (i.e., $M_{n,GPC}$ = 4300, PDI = 1.72, yield 75%; $M_{n,GPC}$ = 7990, PDI = 1.52, yield 73%). These samples were further purified by rGPC to remove the impurities and well-defined PDMS-μ-PCL samples can be acquired (i.e., μ-SCP1: $M_{n,GPC}$ = 7140, PDI = 1.24 and μ-SCP2: $M_{n,GPC}$ = 11,230, PDI = 1.20).

3. Results and Discussion

Figure 1A demonstrates the GPC traces of PDMS-OH and PDMS-Br. Before conducting effective SA ATRC [49–52], the PDMS-Br precursor was synthesized through the acylation of the PDMS-OH with 2-bromoisobutyryl bromide (Scheme S1c (see the ESI)) [47]. The GPC traces of the corresponding PDMS-OH and PDMS-Br products (i.e., curves a1 and a2 in Figure 1A) showed only a minor shift and remained similar in molecular weight (MW) characteristics. After acylation of PDMS-OH, the MW slightly shifted from 1250 to 1280 with low PDI values (PDI < 1.17). Characterized by IR spectroscopy, Figure S2a (see the ESI) displays the bending of the Si-CH$_3$ bond, Si-O, Si-CH$_3$ bond, and the stretching of the C-H bond (790, 1010, 1260, and 2960 cm^{-1}, respectively). As shown in Figure S2b (see the ESI), the stretching of C=O (1740 cm^{-1}) was additionally observed, revealing the successful acylation of the PDMS-OH. As shown in Figure 1B–D, various ratios of VBA/PDMS-Br were then conducted to examine the efficacy of SA ATRCs (VBA/PDMS-Br/CuBr/Me$_6$TREN = x/1/1/2.5 at 0 °C with 0.05 cm Cu wire/mL toluene (x = 0, 1, 4, and 10)). Without VBA (Figure 1B), the M_n increased to approximately 1830 with a low PDI value (1.33) after SA ATRC of PDMS-Br. The coupling efficiency (x_c) can be estimated to be ca. 0.6 (where $x_c = 2 \times (1 - M_{n,0}/M_n)$, $M_{n,0}$: the initial number average MW, M_n: the number average MW after coupling). For SA ATRCs with excess amounts of VBA (Figure 1C,D), the M_ns increased to approximately double the $M_{n,0}$ (ca. 2450) with low PDI values (<1.30). High x_c values of approximately 0.95 were acquired in both cases. With a high ratio of VBA/PDMS-Br = 10, the result stayed at the coupling reaction instead of the chain extension of the PDMS-Br with VBA. This is ascribed to the low-temperature condition for SA ATRC.

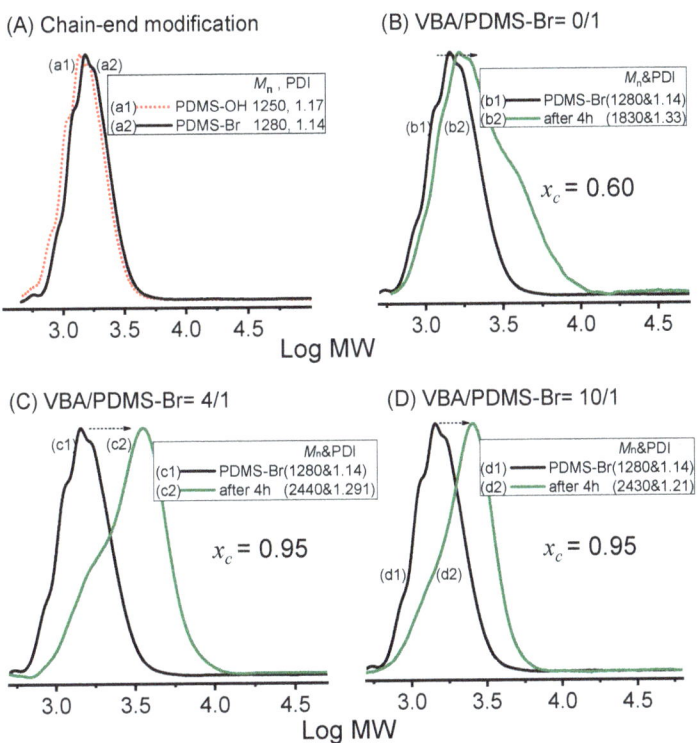

Figure 1. (**A**) GPC traces for the PDMS chain-end modification. Monitoring the coupling efficiency (x_c) by GPC traces with various VBA/PDMS-Br ratios: (**B**) 0/1, (**C**) 4/1, and (**D**) 10/1 (VBA/PDMS-Br/CuBr/Me$_6$TREN = x/1/1/2.5 at 0 °C with 0.05 cm copper wire/mL toluene; [PDMS-Br]$_0$ = 0.03 M in toluene; $x_c = 2 \times (1 - M_{n,0}/M_n)$).

Figure 2 displays ^{1}H NMR spectroscopy analysis of the PDMS-based polymers and their assigned relevant characteristic peaks. Compared the spectra of 2A and 2B, quantitative chain end modification from PDMS-OH to PDMS-Br can be confirmed (i.e., disappearance of peaks a (~3.65 ppm (t, -CH$_2$CH$_2$OH, 2H)) and b (~3.45 ppm (t, -CH$_2$CH$_2$OH, 2H)) and appearance of peaks a' (~4.32 ppm (t, -CH$_2$OCO-, 2H)), b' (~3.67 ppm (t, -CH$_2$CH$_2$OCO-, 2H)), and j (~1.94 ppm (s, -OCOC(CH$_3$)$_2$Br, 6H))). After SA ATRC of PDMS-Br with VBA (Figure 2C), the disappearance of peak j and appearance of peaks k and l were observed (~4.6 and 6.8–7.2 ppm, respectively). The ^{1}H NMR spectra implied the occurrence of the convergent reaction that inserted only a few VBA units.

To further understand the structural details after SA ATRC, the PDMS-Br precursor and a well-coupled PDMS-VBA$_m$-PDMS polymer were analyzed by MALDI-TOF MS using DCTB as the matrix and NaCl as the ionizing agent. Figure 3a displays the MALDI-TOF spectrum of PDMS-Br (M_n = 1280, PDI = 1.14). Interestingly, a series of major peaks (i.e., star marks) with a range of approximately 222 Da, corresponding to the hexamethyl trisiloxane (HMTS) molar mass, was observed. This should intrinsically originate from the ROP of the 6-member ring of HMTS for PDMS preparations. Taking an example of the peak at m/z = 1131.5, the PDMS-Br with rational functionalities can be deduced (i.e., m/z (cal.) = MW$_{head}$ (115.27) + n × MW$_{DMS\,unit}$ (10 × 74.15) + MW$_{end}$ (252.13) + MW$_{Na^+}$ (22.99) = 1131.89), indicating alternative evidence of the chain end modification through robust MALDI-TOF MS analysis. Figure 3b shows the MALDI-TOF spectrum of PDMS-VBA$_m$-PDMS (M_n = 2430 g/mol, PDI = 1.21), which also displayed a series of major peaks corresponding to the HMTS repeating unit (ca. 222 Da). Taking an example of the peak at m/z = 2348.3, structural details of the coupling product can be estimated (i.e., m/z (cal.) = MW$_{chain-ends}$ (2 × 115.27) + n × MW$_{DMS\ unit}$ (20 × 74.15) + m × MW$_{VBA}$ (2 × 134.18) + MW$_{mid-linkages}$ (2 × 172.22) + MW$_{Na^+}$ (22.99) = 2349.3). These results importantly and surprisingly revealed that high x_c of SA ATRC can be attained even at 0 °C and only two VBA coupling agents (m = 2) were inserted.

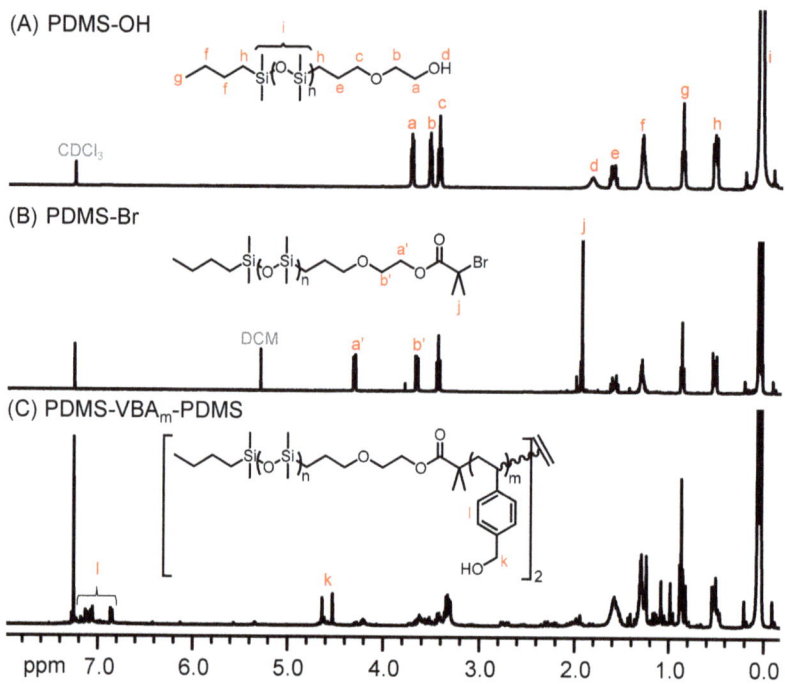

Figure 2. ^{1}H NMR spectra (400 MHz, CDCl$_3$) of (**A**) PDMS-OH, (**B**) PDMS-Br, and (**C**) PDMS-VBA$_m$-PDMS.

The scenario can be rationalized based on kinetic details to discuss such interesting results. For simplicity, the rate constants of activation (k_{act}), deactivation (k_{dea}), and termination (k_t) of PDMS-Br with the bromoisobutyryl group are classified to those of the similar structures of ethyl 2-bromoisobutyrate (EBiB) and methyl methacrylate (MMA) (i.e., $k_{act,EBiB}$, $k_{dea,EBiB}$, and $k_{t,MMA}$, respectively). For the role of VBA, the rate constants were amended from 1-phenylethylbromide (PEBr) and styrene (St) (i.e., $k_{act,PEBr}$, $k_{dea,PEBr}$, k_p, and $k_{t,St}$). Table 1 summarizes the relevant reaction rate constants. In the absence of VBA, disproportionation termination between PDMS• methacrylic macroradicals was dominated [53,54], leading to the formation of PDMS-= and PDMS-H products. Once VBA was added, disproportionation termination of the PDMS• methacrylic macroradical was significantly suppressed and guided the overall reaction to conduct radical–radical coupling based on the following kinetic balances: (i) A high $k_{dea,EBiB}$ (= ca. 5.1 × 10^5 M^{-1} s^{-1}) [55–57] that can conduct efficient deactivation reaction and thus keep the PDMS-Br chain-end integrity; (ii) A rapid cross-reaction between PDMS• and VBA (k_{cr} = ca. 2.5 × 10^3 M^{-1} s^{-1}) [58] to significantly transform the nature of the methacrylic macroradical to styrenic macroradical; (iii) The propagation rate constant of VBA is maintained at a significantly low value of approximately 40 M^{-1} s^{-1} at 0 °C. This critical control guarantees minimal chain extension during the reaction, thereby minimizing the consecutive addition of VBA to the newly-formed PDMS-VBA• macroradical (i.e., resulting in nearly a monoadduct). The prompt coupling reaction can be effectively dominated ($k_{t,St}$ = ca. 10^8 M^{-1} s^{-1}) due to the nature of the styrenic radical termination [54,59,60]. Rationally, the $k_{act,PEBr}$ and $k_{dea,PEBr}$ are relatively low, which would be insufficient to establish the activation–deactivation reactions of PDMS-VBA-Br. Thus, PDMS-VBA$_2$-PDMS was mainly formed.

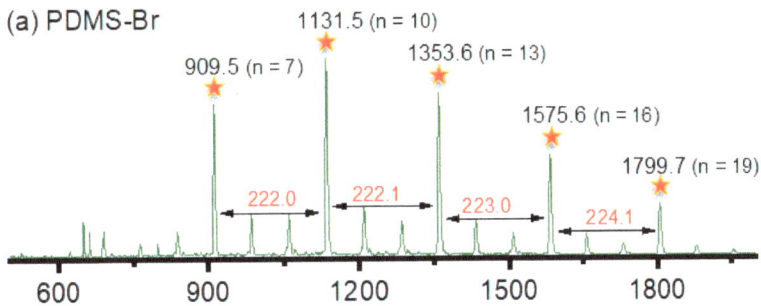

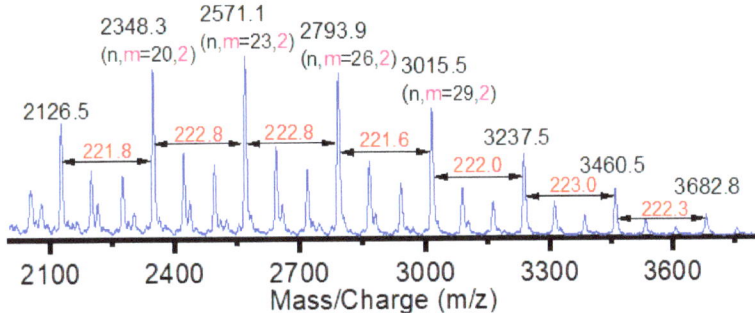

Figure 3. MALDI-TOF MS of (**a**) PDMS-Br and (**b**) PDMS-VBA$_m$-PDMS polymers (n: DMS units, m: VBA units).

Table 1. Reaction rate constants at 0 °C based on the relevant initiators (EBiB and PEBr) and monomers (MMA and St).

Rate Constant	Value (M^{-1} s^{-1})	Reference
$k_{act,EBiB}$	13	[55–57]
$k_{dea,EBiB}$	5.1 × 10^5	[55–57]
$k_{t,MMA}$	6.0 × 10^8	[61]
k_{cr}	2.5 × 10^3	[58]
k_p	42	[55–57]
$k_{t,St}$	4.0 × 10^8	[59]
$k_{act,PEBr}$	0.8	[55–57]
$k_{dea,PEBr}$	1.8 × 10^4	[55–57]

k_{act}: ATRP activation rate constant; k_{dea}: ATRP deactivation rate constant. These rate constants were determined based on the conditions outlined in the literature (specifically, values for EBiB and PEBr with CuBr/Me$_6$TREN in MeCN). k_{cr}: Rate constant for cross-reaction from methacrylic radical to St (value calculated from the activation energy and frequency factor). k_p: Propagation rate constant for styrenics; k_t: Termination rate constant.

After achieving high efficiency of SA ATRC with VBA as a coupling agent, mid-chain functionalized polymers with benzylic alcohols were obtained. The PDMS-VBA$_2$-PDMS MI with two hydroxyl groups allows us to perform chain extension with ε-CL through living-ROP (ε-CL/PDMS-VBA$_2$-PDMS/DPP = 50/1/1 at 40 °C; [ε-CL]$_0$ = 0.8 M). As shown in Figure 4A, the kinetics revealed a linear first-order plot with an apparent rate constant (k_{app}) of approximately 1.03×10^{-4} s^{-1}. Figure 4B showed moderate increases in PDIs and Figure 4C showed gradual increases in MWs and the M_ns relating to conversion. As shown in Figure 4D, the GPC traces showed significant growth in MW. However, a shoulder was observed after 2 h, which might be attributed to the unignorable self-polymerization of ε-CL via the cationic catalyst. To obtain the well-defined PDMS-μ-PCL μ-SCPs with low PDI, we employed a recycling GPC (rGPC) to separate the undesired portions. As shown in Figure S3, the trace during the first cycle (elution time: 7.7–13 min) revealed poor separation due to insufficient elution time. The trace displayed sufficient separation time in the second cycle (elution time: 17.7–27 min). The desired PDMS-μ-PCL was collected within an elution time of 18–23 min. As shown in Figure 5, curves a and c with shoulders possessed relatively large PDIs (>1.5). After the rGPC purifications, curves b and d displayed low PDIs (<1.3) with mono-modal traces, accompanied by increased MWs, indicating the undesired products had been effectively purified.

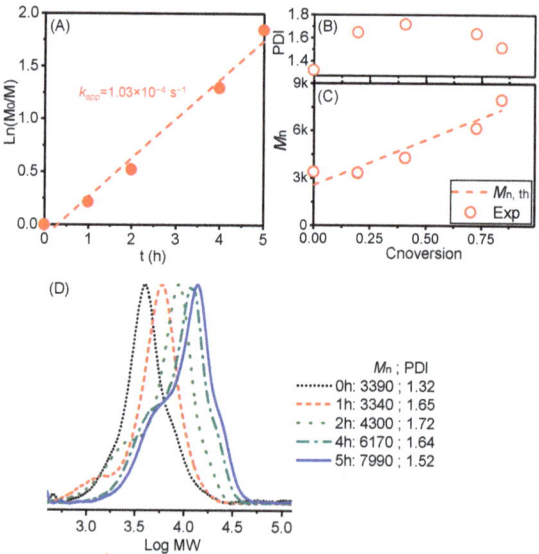

Figure 4. Chain extension of PDMS-VBA$_2$-PDMS with ε-CL: (**A**) Ln vs. t, (**B**,**C**) PDI/M_n vs. Conversion, and (**D**) GPC traces for the ROP chain extension (ε-CL/PDMS-VBA$_2$-PDMS/DPP = 50/1/1 at 40 °C; [ε-CL]$_0$ = 0.8 M).

The obtained PDMS-μ-PCL μ-SCP was also subjected to structural elucidation using ^{1}H NMR spectroscopy. Figure 6A displays the spectrum of PDMS-VBA$_2$-PDMS MI, which contains benzylic alcohol (-Ph-CH_2-OH) at peak b (4.4–4.7 ppm). In Figure 6B, the spectrum of PDMS-μ-PCL μ-SCP reveals the presence of characteristic signals from the PDMS backbone. Additionally, peaks b', j, l, and q from benzyl ester and PCL backbone (4.9–5.2, 4.06, 2.30, 1.2–1.7 ppm, respectively) were observed, while peak b corresponding to the benzylic alcohol disappeared. Two samples were thus acquired and the volume fractions of PDMS in μ-SCP1 and μ-SCP2 were estimated as f^v_{PDMS} = 0.46 and 0.32, respectively. The successful acquisition of μ-SCP1 ($M_{n,GPC}$ = 7140, PDI = 1.24) and μ-SCP2 ($M_{n,GPC}$ = 11,230, PDI = 1.20) was achieved using the rGPC instrument, and both samples exhibited well-defined structures (see again curves b and d in Figure 5).

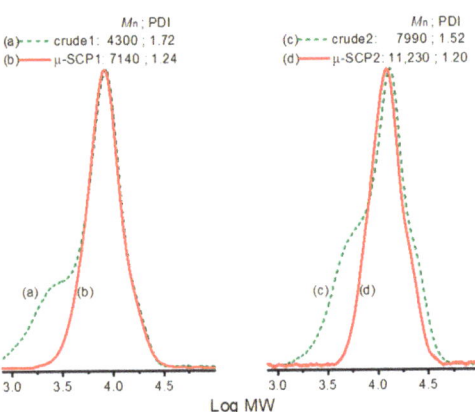

Figure 5. GPC traces for the (a, c) crudes and the corresponding rGPC-purified products of (b) μ-SCP1 and (d) μ-SCP2.

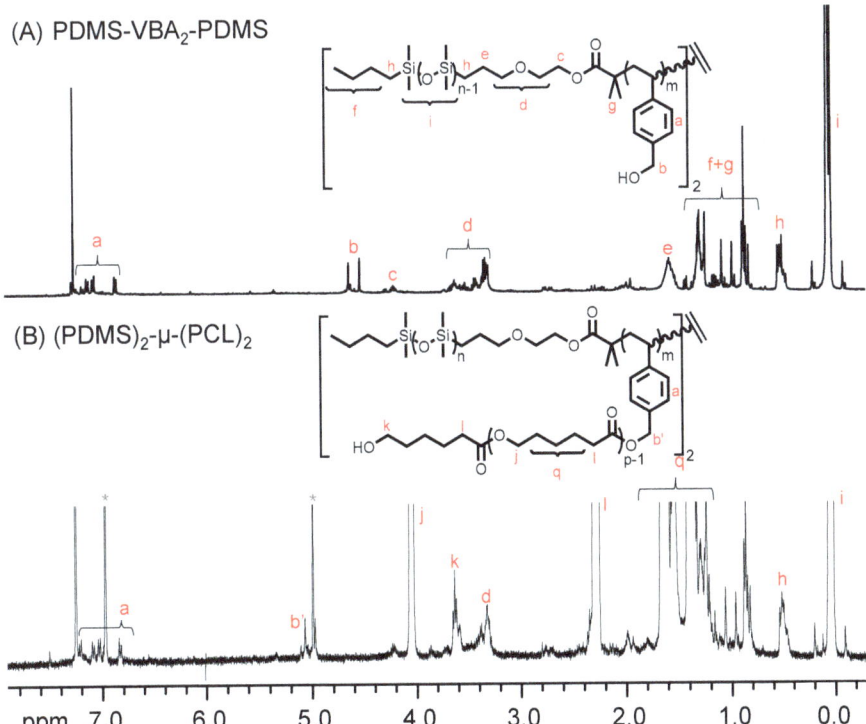

Figure 6. ^{1}H NMR spectra (400 MHz, CDCl$_3$) of (**A**) PDMS-VBA$_2$-PDMS and (**B**) (PDMS)$_2$-μ-(PCL)$_2$ (*: inhibitor peaks originating from THF solvent).

The obtained μ-SCP1 and μ-SCP2 were further investigated for their solid-state self-assembly. Bulk film samples were individually prepared by solution-casting method (5 wt% μ-SCPs/THF solution) on Teflon plates. The specimens were dried under ambient temperature for 24 h and then thermal annealed at 60 °C under vacuum for 12 h. The obtained bulk film samples were measured using SAXS under ambient conditions to analyze their microstructure. As shown in Figure 7A, a lamellar (LAM) structure (i.e., q/q* peaks of 1:2) were obtained in the annealed sample of μ-SCP1 [(PDMS$_{13}$)$_2$-μ-(PCL$_{20}$)$_2$, $f^v$$_{PDMS}$ = 0.46]. The d-spacing of the nanostructures was approximately 15.3 nm. As shown in Figure 7B, a hexagonally packed cylinder (HEX) structure (i.e., q/q* peaks of 1:√3) having a periodic size of approximately 16.5 nm was detected in the case of the μ-SCP2 sample [(PDMS$_{13}$)$_2$-μ-(PCL$_{38}$)$_2$, $f^v$$_{PDMS}$ = 0.32]. It is worth noting that the "bottom-up" approaches to forming significantly ordered nanostructures by block copolymer (BCP) have a common strategy to use overall high molecular weight copolymers (i.e., high degrees of polymerization (N)) to conduct self-assembly behaviors. It is generally because most polymer–polymer interaction parameters (χ) are small. However, Kim et al. [62]. introduced an alternative strategy using a star block copolymer (SBCP) with 18-armed PMMA-b-PSt (i.e., (PMMA-b-PSt)$_{18}$). They introduced a self-neutralization approach to achieve structured nanoarchitectures by increasing the "entropy penalty". By innovatively controlling the polymer architecture in such a low χ system (ca. 0.04) [63], they beautifully fabricated oriented lamellar and cylindrical nanostructures ranging from approximately 50–150 nm on silicon wafers. Similarly, Ho et al. [64] combined two strategies to produce aligned structures (i.e., low entropy and large χ). By controlling the polymer architecture of (PSt$_x$-b-PDMS$_y$)$_n$ SBCPs (arm numbers (n) = 1–4 and repeating units (x + y) < 20), they ensured a high value of χ (ca. 0.27) between the blocks. As a result, these SBCPs spontaneously aligned into the cylinder and lamellar structures of sub-20 nm feature size on silicon wafers. A similar two-strategies method was interestingly demonstrated by a unique system of maltoheptaose-based SBCPs [65,66], which showed sub-10 nm feature size. Our PDMS-μ-PCL system has the attributes of low N, large χ, and low entropy. Accordingly, such peculiarities can easily lead to performing self-assembly on a sub-nano scale. By minimizing the molecular spatial stability through MM2 molecular modeling method, contour lengths (L) of the μ-SCPs can be estimated. As shown in Figure S4 (see the ESI), L$_{μ-SCP1}$ = 22.7 and L$_{μ-SCP2}$ = 36.9 nm were acquired based on a set of end groups, PDMS$_n$ segments, PCL$_p$ segments, linkages, and VBA units. Figure 8 displays our proposed microstructure to indicate the roles and features of PDMS (i.e., only in the form of amorphous) and PCL (i.e., with the forms of amorphous and crystalline) chains. Consequently, the measured periodic dimensions of the μ-SCPs using SAXS are smaller than the estimated contour lengths. This phenomenon can be logically attributed to the strong crystallization integrity of the PCL chains, leading to a densely packed PCL-rich area, whether in a LAM or HEX structure. Our ongoing efforts involve examining the intrinsic properties and self-assembly tendencies of these μ-SCPs. More comprehensive insights and specific findings will be unveiled in an upcoming publication.

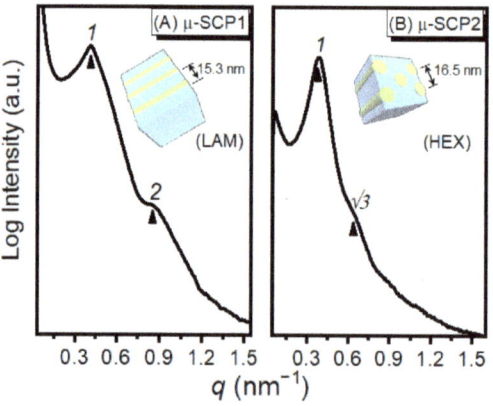

Figure 7. SAXS profiles of thermal annealed PDMS-μ-PCL of (**A**) μ-SCP1 ($M_{n,GPC}$ = 7140, $f^v$$_{PDMS}$ = 0.46) and (**B**) μ-SCP2 ($M_{n,GPC}$ = 11,230, $f^v$$_{PDMS}$ = 0.32) bulk film samples.

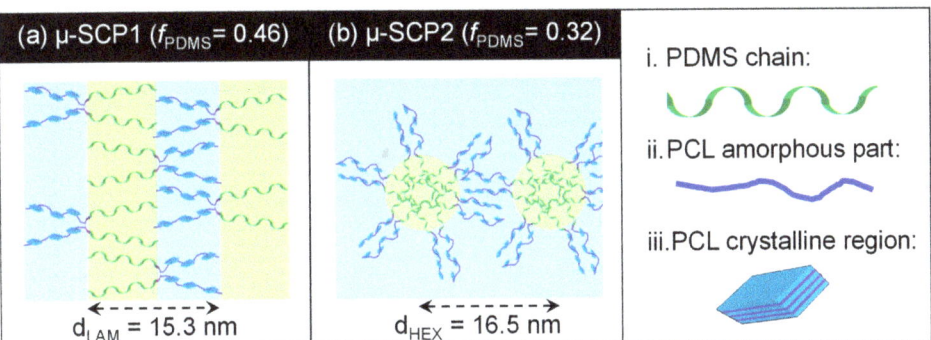

Figure 8. Proposed microstructures of the annealed μ-SCP bulk film samples referred to the SAXS analysis and estimated contour length.

4. Conclusions

We employed a styrenic coupling agent, VBA, in the SA ATRC process to obtain PDMS-VBA$_m$-PDMS products. By analyzing the outcomes through GPC, we achieved a high coupling efficiency (χ_c = 0.95) with a monomodal MW distribution. Intriguingly, MALDI-TOF MS analysis of the PDMS-VBA$_m$-PDMS product confirmed the insertion of only two VBA units (m = 2). To extend chains, we subjected PDMS-VBA$_2$-PDMS to mid-chain extensions with ε-CL through ROP, employing the organo-catalyst DPP. To remove any undesirable side products, such as linear PCL homopolymer, we employed a rGPC process. This allowed us to acquire well-defined μ-SCPs with low PDIs (<1.30). Further examination of the obtained PDMS-μ-PCLs through SAXS measurements revealed microphase separations, with LAM and HEX structures observed in bulk film samples of μ-SCP1 ($M_{n,GPC}$ = 7140, f^v_{PDMS} = 0.46) and μ-SCP2 ($M_{n,GPC}$ = 11,230, f^v_{PDMS} = 0.32), respectively. The achievement of sub-15 nm microstructures can be attributed to the attributes of low PDI (<1.3), low N of arms (PDMS$_{13}$ and PCL$_{20}$/PCL$_{38}$), large χ, and low entropy in the μ-SCPs. In brief, our approach demonstrates high coupling efficiency and provides a means to introduce mid-chain functionality for conducting post-reactions. This convergent strategy can be extended to synthesize architecturally distinct and well-defined novel macromolecules. The characterization of structural details and diversities is currently underway. We will present the relevant data shortly.

Supplementary Materials: The following supporting information can be downloaded at: https://www.mdpi.com/article/10.3390/nano13162355/s1, Scheme S1. Reaction steps for the syntheses of (a) VBA, (b) Me$_6$TREN, and (c) PDMS-Br (BiB: 2-bromoisobutyryl bromide). (d) Cyclic monomer of hexamethyl trisiloxane. Figure S1. ^{1}H NMR spectra (400 MHz, CDCl$_3$) of (a) VBA and (b) Me$_6$TREN. Figure S2. FT-IR spectra of (a) PDMS-OH and (b) PDMS-Br. Figure S3. A consecutive rGPC trace for purifying a PDMS-μ-PCL crude product [red region: collections of well-defined μ-SCPs (i.e., elution time = ca. 18–23 min)]. Figure S4. Spatial contour lengths of the relevant functionalities and monomer (estimated through MM2 molecular modeling) for $L_{μ-SCP1}$ = 22.7 and $L_{μ-SCP2}$ = 36.9 nm [estimation of the extended distance (L) using one set of PDMS end, PDMS$_n$ segment, PCL$_p$ segment, linkage, and VBA unit].

Author Contributions: Y.-S.H., T.J. and C.-F.H. designed and conceived the experiments; Y.-S.H. conducted the experiments and measurements; Y.-S.H. and D.D.E. analyzed the data and wrote the relevant contents; T.J., S.-W.K., K.-Y.L. and C.-F.H. summarized and edited all the contents. All authors have read and agreed to the published version of the manuscript.

Funding: This research was funded by the National Science and Technology Council (NSTC110-2221-E-005-001-MY3 and NSTC111-2811-E-005-016) and TCUS Exchange Project. This research was also supported by the two-institution co-research project provided by Kasetsart University and National Chung Hsing University.

Data Availability Statement: Figure 7 is the data from the NSRRC.

Acknowledgments: The SAXS experiments conducted in the National Synchrotron Radiation Research Center ((NSRRC), Taiwan) are deeply acknowledged.

Conflicts of Interest: The authors declare no conflict of interest.

References

1. Aloorkar, N.; Kulkarni, A.; Patil, R.; Ingale, D. Star polymers: An overview. *Int. J. Pharm. Sci. Nanotechnol* **2012**, *5*, 1675–1684.
2. Feng, H.; Lu, X.; Wang, W.; Kang, N.-G.; Mays, J.W. Block copolymers: Synthesis, self-assembly, and applications. *Polymers* **2017**, *9*, 494. [CrossRef] [PubMed]
3. Nitta, N.; Kihara, S.I.; Haino, T. Synthesis of Supramolecular A_8B_n Miktoarm Star Copolymers by Host-Guest Complexation. *Angew. Chem. Int. Ed.* **2023**, *62*, e202219001. [CrossRef]
4. Gao, H.; Matyjaszewski, K. Synthesis of Low-Polydispersity Miktoarm Star Copolymers via a Simple "Arm-First" Method: Macromonomers as Arm Precursors. *Macromolecules* **2008**, *41*, 4250–4257. [CrossRef]
5. Liu, M.; Blankenship, J.R.; Levi, A.E.; Fu, Q.; Hudson, Z.M.; Bates, C.M. Miktoarm Star Polymers: Synthesis and Applications. *Chem. Mater.* **2022**, *34*, 6188–6209. [CrossRef]
6. Khanna, K.; Varshney, S.; Kakkar, A. Miktoarm star polymers: Advances in synthesis, self-assembly, and applications. *Polym. Chem.* **2010**, *1*, 1171–1185. [CrossRef]
7. Bates, M.W.; Barbon, S.M.; Levi, A.E.; Lewis, R.M., III; Beech, H.K.; Vonk, K.M.; Zhang, C.; Fredrickson, G.H.; Hawker, C.J.; Bates, C.M. Synthesis and Self-Assembly of AB_n Miktoarm Star Polymers. *ACS Macro Lett.* **2020**, *9*, 396. [CrossRef]
8. Lotocki, V.; Kakkar, A. Miktoarm Star Polymers: Branched Architectures in Drug Delivery. *Pharmaceutics* **2020**, *12*, 827. [CrossRef] [PubMed]
9. Aghajanzadeh, M.; Zamani, M.; Rostamizadeh, K.; Sharafi, A.; Danafar, H. The role of miktoarm star copolymers in drug delivery systems. *J. Macromol. Sci. Part A* **2018**, *55*, 559–571. [CrossRef]
10. Hadjichristidis, N. Synthesis of miktoarm star (μ-star) polymers. *J. Polym. Sci. Part A Polym. Chem.* **1999**, *37*, 857. [CrossRef]
11. Kalow, J.A.; Swager, T.M. Synthesis of Miktoarm Branched Conjugated Copolymers by ROMPing In and Out. *ACS Macro Lett.* **2015**, *4*, 1229–1233. [CrossRef]
12. Tunca, U.; Ozyurek, Z.; Erdogan, T.; Hizal, G. Novel miktofunctional initiator for the preparation of an ABC-type miktoarm star polymer via a combination of controlled polymerization techniques. *J. Polym. Sci. Part A Polym. Chem.* **2004**, *42*, 4228–4236. [CrossRef]
13. Oliveira, A.S.R.; Mendonça, P.V.; Simões, S.; Serra, A.C.; Coelho, J.F.J. Amphiphilic well-defined degradable star block copolymers by combination of ring-opening polymerization and atom transfer radical polymerization: Synthesis and application as drug delivery carriers. *J. Polym. Sci.* **2021**, *59*, 211. [CrossRef]
14. Huang, C.-F.; Aimi, J.; Lai, K.-Y. Synthesis of Novel μ-Star Copolymers with Poly(N-Octyl Benzamide) and Poly(ε-Caprolactone) Miktoarms through Chain-Growth Condensation Polymerization, Styrenics-Assisted Atom Transfer Radical Coupling, and Ring-Opening Polymerization. *Macromol. Rapid Commun.* **2017**, *38*, 1600607. [CrossRef]
15. Gao, H.; Matyjaszewski, K. Arm-First Method As a Simple and General Method for Synthesis of Miktoarm Star Copolymers. *J. Am. Chem. Soc.* **2007**, *129*, 11828–11834. [CrossRef] [PubMed]
16. Ren, J.M.; McKenzie, T.G.; Fu, Q.; Wong, E.H.H.; Xu, J.; An, Z.; Shanmugam, S.; Davis, T.P.; Boyer, C.; Qiao, G.G. Star Polymers. *Chem. Rev.* **2016**, *116*, 6743. [CrossRef] [PubMed]
17. Li, H.; Zhao, H.; Yao, L.; Zhang, L.; Cheng, Z.; Zhu, X. Photocontrolled bromine–iodine transformation reversible-deactivation radical polymerization: Facile synthesis of star copolymers and unimolecular micelles. *Polym. Chem.* **2021**, *12*, 2335–2345. [CrossRef]
18. Ding, H.; Park, S.; Zhong, M.; Pan, X.; Pietrasik, J.; Bettinger, C.J.; Matyjaszewski, K. Facile Arm-First Synthesis of Star Block Copolymers via ARGET ATRP with ppm Amounts of Catalyst. *Macromolecules* **2016**, *49*, 6752–6760. [CrossRef]
19. Kakkar, A. *Miktoarm Star Polymers: From Basics of Branched Architecture to Synthesis, Self-Assembly and Applications*; Royal Society of Chemistry: London, UK, 2017.
20. Ito, S.; Goseki, R.; Ishizone, T.; Hirao, A. Successive synthesis of well-defined multiarmed miktoarm star polymers by iterative methodology using living anionic polymerization. *Eur. Polym. J.* **2013**, *49*, 2545–2566. [CrossRef]
21. Higashihara, T.; Hayashi, M.; Hirao, A. Synthesis of well-defined star-branched polymers by stepwise iterative methodology using living anionic polymerization. *Prog. Polym. Sci.* **2011**, *36*, 323–375. [CrossRef]
22. Morton, M.; Helminiak, T.E.; Gadkary, S.D.; Bueche, F. Preparation and properties of monodisperse branched polystyrene. *J. Polym. Sci.* **1962**, *57*, 471–482. [CrossRef]
23. Orofino, T.; Wenger, F. Dilute Solution Properties of Branched Polymers. Polystyrene Trifunctional Star Molecules 1. *J. Phys. Chem.* **1963**, *67*, 566–575. [CrossRef]
24. Mayer, R. Organized structures in amorphous styrene/cis-1,4-isoprene block copolymers: Low angle X-ray scattering and electron microscopy. *Polymer* **1974**, *15*, 137–145. [CrossRef]
25. Strazielle, C.; Herz, J. Synthese et etude physicochimique de polystyrenes ramifies en "etoile" et en "peigne". *Eur. Polym. J.* **1977**, *13*, 223. [CrossRef]

26. Gervasi, J.A.; Gosnell, A.B. Synthesis and characterization of branched polystyrene. Part I. Synthesis of four- and six-branch star polystyrene. *J. Polym. Sci. Part A Polym. Chem.* **1966**, *4*, 1391. [CrossRef]
27. Hsieh, H. Synthesis of radial thermoplastic elastomers. *Rubber Chem. Technol.* **1976**, *49*, 1305–1310. [CrossRef]
28. Herz, J.; Hert, M.; Straziel, C. Synthesis and characterization of threefold branched star-shaped polystyrenes. *Makromol. Chem.* **1972**, *160*, 213. [CrossRef]
29. Uraneck, C.A.; Short, J.N. Solution-polymerized rubbers with superior breakdown properties. *J. Appl. Polym. Sci.* **1970**, *14*, 1421–1432. [CrossRef]
30. Nicolas, J.; Guillaneuf, Y.; Lefay, C.; Bertin, D.; Gigmes, D.; Charleux, B. Nitroxide-mediated polymerization. *Prog. Polym. Sci.* **2013**, *38*, 63. [CrossRef]
31. El Yousfi, R.; Brahmi, M.; Dalli, M.; Achalhi, N.; Azougagh, O.; Tahani, A.; Touzani, R.; El Idrissi, A. Recent Advances in Nanoparticle Development for Drug Delivery: A Comprehensive Review of Polycaprolactone-Based Multi-Arm Architectures. *Polymers* **2023**, *15*, 1835. [CrossRef]
32. Sisson, A.L.; Ekinci, D.; Lendlein, A. The contemporary role of ε-caprolactone chemistry to create advanced polymer architectures. *Polymer* **2013**, *54*, 4333. [CrossRef]
33. Lorenzo, A.T.; Müller, A.J.; Lin, M.-C.; Chen, H.-L.; Jeng, U.S.; Priftis, D.; Pitsikalis, M.; Hadjichristidis, N. Influence of Macromolecular Architecture on the Crystallization of (PCL)$_2$-b-(PS)$_2$ 4-Miktoarm Star Block Copolymers in Comparison to Linear PCL-b-PS Diblock Copolymer Analogues. *Macromolecules* **2009**, *42*, 8353. [CrossRef]
34. Heise, A.; Trollsås, M.; Magbitang, T.; Hedrick, J.L.; Frank, C.W.; Miller, R.D. Star Polymers with Alternating Arms from Miktofunctional μ-Initiators Using Consecutive Atom Transfer Radical Polymerization and Ring-Opening Polymerization. *Macromolecules* **2001**, *34*, 2798–2804. [CrossRef]
35. Yang, L.; Zhou, H.; Shi, G.; Wang, Y.; Pan, C.-Y. Synthesis of ABCD 4-miktoarm star polymers by combination of RAFT, ROP, and "Click Chemistry". *J. Polym. Sci. Part A Polym. Chem.* **2008**, *46*, 6641–6653. [CrossRef]
36. Sathesh, V.; Chen, J.-K.; Chang, C.-J.; Aimi, J.; Chen, Z.-C.; Hsu, Y.-C.; Huang, Y.-S.; Huang, C.-F. Synthesis of Poly(ε-caprolactone)-Based Miktoarm Star Copolymers through ROP, SA ATRC, and ATRP. *Polymers* **2018**, *10*, 858. [CrossRef]
37. Gordin, C.; Delaite, C.; Medlej, H.; Josien-Lefebvre, D.; Hariri, K.; Rusu, M. Synthesis of ABC miktoarm star block copolymers from a new heterotrifunctional initiator by combination of ATRP and ROP. *Polym. Bull.* **2009**, *63*, 789–801. [CrossRef]
38. Nuyken, O.; Pask, S.D. Ring-Opening Polymerization—An Introductory Review. *Polymers* **2013**, *5*, 361–403. [CrossRef]
39. Nakamura, Y.; Arima, T.; Yamago, S. Modular Synthesis of Mid-Chain-Functionalized Polymers by Photoinduced Diene- and Styrene-Assisted Radical Coupling Reaction of Polymer-End Radicals. *Macromolecules* **2014**, *47*, 582–588. [CrossRef]
40. Fan, W.J.; Nakamura, Y.; Yamago, S. Synthesis of Multivalent Organotellurium Chain-Transfer Agents by Post-modification and Their Applications in Living Radical Polymerization. *Chem. Eur. J.* **2016**, *22*, 17006–17010. [CrossRef]
41. Huang, C.-F.; Huang, Y.-S.; Lai, K.-Y. Synthesis and self-assembly of Poly(N-octyl benzamide)-μ-poly(ε-caprolactone) miktoarm star copolymers displaying uniform nanofibril morphology. *Polymer* **2019**, *178*, 121582. [CrossRef]
42. Lu, Y.C.; Chou, L.C.; Huang, C.F. Iron-Catalysed Atom Transfer Radical Polyaddition for the Synthesis and Modification of Novel Aliphatic Polyesters Displaying Lower Critical Solution Temperature and pH-Dependent Release Behaviors. *Polym. Chem.* **2019**, *10*, 3912. [CrossRef]
43. Huang, Y.-S.; Chen, J.-K.; Chen, T.; Huang, C.-F. Synthesis of PNVP-based Copolymers with Tunable Thermosensitivity by Sequential Reversible Addition–Fragmentation Chain Transfer Copolymerization and Ring-Opening Polymerization. *Polymers* **2017**, *9*, 231. [CrossRef] [PubMed]
44. Huang, C.-F.; Kuo, S.-W.; Moravčíková, D.; Liao, J.-C.; Han, Y.-M.; Lee, T.-H.; Wang, P.-H.; Lee, R.-H.; Tsiang, R.C.-C.; Mosnáček, J. Effect of Variations of Cu^{II}X$_2$/L, Surface Area of Cu^0, Solvent, and Temperature on Atom Transfer Radical Polyaddition of 4-Vinylbenzyl 2-Bromo-2-isobutyrate Inimers. *RSC Adv.* **2016**, *6*, 51816. [CrossRef]
45. Han, Y.-M.; Chen, H.-H.; Huang, C.-F. Polymerization and Degradation of Aliphatic Polyesters Synthesized by Atom Transfer Radical Polyaddition. *Polym. Chem.* **2015**, *6*, 4565–4574. [CrossRef]
46. Aimi, J.; Yasuda, T.; Huang, C.-F.; Yoshio, M.; Chen, W.-C. Fabrication of Solution-Processable OFET Memory Using a Nano-Floating Gate Based on a Phthalocyanine-Cored Star-Shaped Polymer. *Mater. Adv.* **2022**, *3*, 3128–3134. [CrossRef]
47. Huang, Y.S.; Huang, C.F. Synthesis of Well-Defined PMMA-b-PDMS-b-PMMA Triblock Copolymer and Study of Its Self-Assembly Behaviors In Epoxy Resin. *Eur. Polym. J.* **2021**, *160*, 110787. [CrossRef]
48. Huang, Y.-S.; Hsueh, H.-Y.; Aimi, J.; Chou, L.-C.; Lu, Y.-C.; Kuo, S.-W.; Wang, C.-C.; Chen, K.-Y.; Huang, C.-F. Effects of Various Cu(0), Fe(0), and Proanthocyanidin Reducing Agents on Fe(III)-catalysed ATRP for the Synthesis of PMMA Block Copolymers and their Self-assembly Behaviours. *Polym. Chem.* **2020**, *11*, 5147–5155. [CrossRef]
49. Huang, C.-F.; Ohta, Y.; Yokoyama, A.; Yokozawa, T. Efficient Low-Temperature Atom Transfer Radical Coupling and Its Application to Synthesis of Well-Defined Symmetrical Polybenzamides. *Macromolecules* **2011**, *44*, 4140. [CrossRef]
50. Lai, K.-Y.; Huang, Y.-S.; Chu, C.-Y.; Huang, C.-F. Synthesis of Poly(N-H benzamide)-b-Poly(lauryl methacrylate)-b-Poly(N-H benzamide) Symmetrical Triblock Copolymers by Combinations of CGCP, SARA ATRP, and SA ATRC. *Polymer* **2018**, *137*, 385–394. [CrossRef]
51. Bunha, A.K.; Mangadlao, J.; Felipe, M.J.; Pangilinan, K.; Advincula, R. Catenated PS–PMMA Block Copolymers via Supramolecularly Templated ATRP Initiator Approach. *Macromol. Rapid Commun.* **2012**, *33*, 1214. [CrossRef]

52. Arce, M.M.; Pan, C.W.; Thursby, M.M.; Wu, J.P.; Carnicom, E.M.; Tillman, E.S. Influence of Solvent on Radical Trap-Assisted Dimerization and Cyclization of Polystyrene Radicals. *Macromolecules* **2016**, *49*, 7804–7813. [CrossRef]
53. Bevington, J.C.; Melville, M.H.; Taylor, R.P. The termination reaction in radical polymerizations. II. Polymerizations of styrene at 60° and of methyl methacrylate at 0 and 60°, and the copolymerization of these monomers at 60°. *J. Polym. Sci.* **1954**, *14*, 463. [CrossRef]
54. Zammit, M.D.; Davis, T.P.; Haddleton, D.M.; Suddaby, K.G. Evaluation of the mode of termination for a thermally initiated free-radical polymerization via matrix-assisted laser desorption ionization time-of-flight mass spectrometry. *Macromolecules* **1997**, *30*, 1915–1920. [CrossRef]
55. Tang, W.; Tsarevsky, N.V.; Matyjaszewski, K. Determination of equilibrium constants for atom transfer radical polymerization. *J. Am. Chem. Soc.* **2006**, *128*, 1598–1604. [CrossRef] [PubMed]
56. Tang, W.; Kwak, Y.; Braunecker, W.; Tsarevsky, N.V.; Coote, M.L.; Matyjaszewski, K. Understanding atom transfer radical polymerization: Effect of ligand and initiator structures on the equilibrium constants. *J. Am. Chem. Soc.* **2008**, *130*, 10702–10713. [CrossRef]
57. Tang, W.; Matyjaszewski, K. Kinetic Modeling of Normal ATRP, Normal ATRP with $[Cu^{II}]_0$, Reverse ATRP and SR&NI ATRP. *Macromol. Theory Simul.* **2008**, *17*, 359.
58. Fischer, H.; Radom, L. Factors controlling the addition of carbon-centered radicals to alkenes-an experimental and theoretical perspective. *Angew. Chem. Int. Ed.* **2001**, *40*, 1340–1371. [CrossRef]
59. Yoshikawa, C.; Goto, A.; Fukuda, T. Reactions of polystyrene radicals in a monomer-free atom transfer radical polymerization system. *e-Polymers* **2002**, *2*, 172–183. [CrossRef]
60. Moad, G.; Solomon, D.H.; Johns, S.R.; Willing, R.I. Fate of the initiator in the azobisisobutyronitrile-initiated polymerization of styrene. *Macromolecules* **1984**, *17*, 1094–1099. [CrossRef]
61. Barth, J.; Buback, M. SP-PLP-EPR Investigations into the Chain-Length-Dependent Termination of Methyl Methacrylate Bulk Polymerization. *Macromol. Rapid Commun.* **2009**, *30*, 1805–1811. [CrossRef]
62. Jang, S.; Lee, K.; Moon, H.C.; Kwak, J.; Park, J.; Jeon, G.; Lee, W.B.; Kim, J.K. Vertical Orientation of Nanodomains on Versatile Substrates through Self-Neutralization Induced by Star-Shaped Block Copolymers. *Adv. Funct. Mater.* **2015**, *25*, 5414. [CrossRef]
63. Gu, X.D.; Gunkel, I.; Russell, T.P. Pattern transfer using block copolymers. *Philos. Trans. R. Soc. A* **2013**, *371*, 20120306. [CrossRef] [PubMed]
64. Lo, T.Y.; Dehghan, A.; Georgopanos, P.; Avgeropoulos, A.; Shi, A.C.; Ho, R.M. Orienting Block Copolymer Thin Films via Entropy. *Macromolecules* **2016**, *49*, 624. [CrossRef]
65. Isono, T.; Otsuka, I.; Kondo, Y.; Halila, S.; Fort, S.; Rochas, C.; Satoh, T.; Borsali, R.; Kakuchi, T. Sub-10 nm Nano-Organization in AB_2- and AB_3-Type Miktoarm Star Copolymers Consisting of Maltoheptaose and Polycaprolactone. *Macromolecules* **2013**, *46*, 1461–1469. [CrossRef]
66. Isono, T.; Otsuka, I.; Suemasa, D.; Rochas, C.; Satoh, T.; Borsali, R.; Kakuchi, T. Synthesis, Self-Assembly, and Thermal Caramelization of Maltoheptaose-Conjugated Polycaprolactones Leading to Spherical, Cylindrical, and Lamellar Morphologies. *Macromolecules* **2013**, *46*, 8932. [CrossRef]

Disclaimer/Publisher's Note: The statements, opinions and data contained in all publications are solely those of the individual author(s) and contributor(s) and not of MDPI and/or the editor(s). MDPI and/or the editor(s) disclaim responsibility for any injury to people or property resulting from any ideas, methods, instructions or products referred to in the content.

Article

Doxorubicin-Loaded Polyelectrolyte Multilayer Capsules Modified with Antitumor DR5-Specific TRAIL Variant for Targeted Drug Delivery to Tumor Cells

Anastasia Gileva [1,*,†], Daria Trushina [2,†], Anne Yagolovich [1,3,†], Marine Gasparian [1], Leyli Kurbanova [1], Ivan Smirnov [1], Sergey Burov [4] and Elena Markvicheva [1]

1. Shemyakin-Ovchinnikov Institute of Bioorganic Chemistry RAS, 117997 Moscow, Russia
2. Laboratory of Bioorganic Structures, Shubnikov Institute of Crystallography of Federal Scientific Research Centre "Crystallography and Photonics" of Russian Academy of Sciences, 119333 Moscow, Russia
3. Faculty of Biology, Lomonosov Moscow State University, 119192 Moscow, Russia
4. Cytomed JSC, Orlovo-Denisovsky pr. 14, 197375 St. Petersburg, Russia
* Correspondence: sumina.anastasia@mail.ru; Tel.: +7-9164517818
† These authors contributed equally to this work.

Citation: Gileva, A.; Trushina, D.; Yagolovich, A.; Gasparian, M.; Kurbanova, L.; Smirnov, I.; Burov, S.; Markvicheva, E. Doxorubicin-Loaded Polyelectrolyte Multilayer Capsules Modified with Antitumor DR5-Specific TRAIL Variant for Targeted Drug Delivery to Tumor Cells. *Nanomaterials* 2023, 13, 902. https://doi.org/10.3390/nano13050902

Academic Editor: Daniela Iannazzo

Received: 19 January 2023
Revised: 22 February 2023
Accepted: 24 February 2023
Published: 27 February 2023

Copyright: © 2023 by the authors. Licensee MDPI, Basel, Switzerland. This article is an open access article distributed under the terms and conditions of the Creative Commons Attribution (CC BY) license (https://creativecommons.org/licenses/by/4.0/).

Abstract: Recently, biodegradable polyelectrolyte multilayer capsules (PMC) have been proposed for anticancer drug delivery. In many cases, microencapsulation allows to concentrate the substance locally and prolong its flow to the cells. To reduce systemic toxicity when delivering highly toxic drugs, such as doxorubicin (DOX), the development of a combined delivery system is of paramount importance. Many efforts have been made to exploit the DR5-dependent apoptosis induction for cancer treatment. However, despite having a high antitumor efficacy of the targeted tumor-specific DR5-B ligand, a DR5-specific TRAIL variant, its fast elimination from a body limits its potential use in a clinic. A combination of an antitumor effect of the DR5-B protein with DOX loaded in the capsules could allow to design a novel targeted drug delivery system. The aim of the study was to fabricate PMC loaded with a subtoxic concentration of DOX and functionalized with the DR5-B ligand and to evaluate a combined antitumor effect of this targeted drug delivery system in vitro. In this study, the effects of PMC surface modification with the DR5-B ligand on cell uptake both in 2D (monolayer culture) and 3D (tumor spheroids) were studied by confocal microscopy, flow cytometry and fluorimetry. Cytotoxicity of the capsules was evaluated using an MTT test. The capsules loaded with DOX and modified with DR5-B demonstrated synergistically enhanced cytotoxicity in both in vitro models. Thus, the use of the DR5-B-modified capsules loaded with DOX at a subtoxic concentration could provide both targeted drug delivery and a synergistic antitumor effect.

Keywords: polyelectrolyte multilayer capsules; antitumor protein DR5-B; targeted drug delivery; codelivery system; tumor spheroids; HCT-116 cells

1. Introduction

Biodegradable polyelectrolyte multilayer capsules are promising for the entrapment of a wide range of various biologically active molecules due to the versatility and simplicity of the fabrication technique [1,2]. This technique allows to fabricate capsules under mild conditions avoiding any organic solvents and using a rather wide selection of shell materials, including biocompatible and biodegradable polymers [3–7]. Moreover, all interactions, including electrostatics, hydrophobic, hydrogen binding, covalent binding, DNA hybridization, stereocomplexation, specific recognition, etc., can be used to provide driving forces for multilayer shell assembly [8]. Intracellular delivery of bioactive species has great potential for clinical applications [1,3,4]. Furthermore, an LbL approach can be proposed for the fabrication of high-performance nanosystems, in order to combine several drugs in PMC. Two up-to-date reviews comprehensively summarize current strategies that enable advanced functionalization of polyelectrolyte capsules, their potential biomedical applications, challenges in manufacturing as well as further safe translation into the clinical [2,5]. Despite PMC usage for targeted drug delivery, most of the researchers have been focused on magnetic drug targeting [6,7,9–12]. There are only a few studies, where magnetic antibody-modified capsules have been proposed for targeted drug delivery [13,14].

As well known, the surface design of the microcapsules with antibodies can provide highly specific antibody–antigen interactions with simultaneous suppression of any unspecific binding [15,16]. For example, antibody-functionalized capsules assembled from poly(allylamine hydrochloride) and poly(acrylic acid) on 6 μm CaCO$_3$ templates targeted major histocompatibility complex class I receptors into living cells [17]. The authors revealed that protein A functionalized polyelectrolyte microcapsules enhanced targeting efficiency by 40–50% over capsules with randomly attached antibodies via direct covalent coupling to the surface. Cortez et al. reported that non-degradable PMC from poly(sodium 4-styrenesulfonate) and polyallylamine hydrochloride were functionalized with the humanized A33 (HuA33) antibody in order to provide their binding to appropriate receptors overexpressed in the colorectal cancer cell line [18,19].

Other alternatives for the functionalization of the outer layer and the subsequent improvement of their targeting rely on the use of polysaccharides, in particular hyaluronic acid. Thus, some studies have been focused on the fabrication of capsules with an outer layer of hyaluronic acid, which allowed to enhance both the targeting and the internalization of the capsules by tumor cells [14,20]. The verification of cancer cell targeting and bio-adhesion abilities of hyaluronic acid was performed using shells of a combination of alginate, chitosan and hyaluronic acid assembled on 2 μm monodisperse spherical melamine formaldehyde resin particles [14] and poly-L-arginine, siRNA and hyaluronic acid polyelectrolytes adsorbed on 100 nm poly(lactic-co-glycolic acid) nanoparticles [20].

Tumor necrosis factor-related apoptosis-inducing ligand (TRAIL) is an antitumor cytokine, selectively targeting transformed cells, but not normal cells, upon binding to the DR4 and DR5 death receptors. However, a soluble wild type TRAIL did not demonstrate a sufficient antitumor activity in previous clinical trials, and therefore a list of modifications, including TRAIL immobilization on various nanocarriers, has been reported [21].

Recently, we have generated the DR5 receptor-selective TRAIL variant DR5-B, which can bind only to the DR5 death receptor, and, as a result, it can overcome a receptor-dependent resistance of tumor cells to TRAIL [22]. DR5 is considered the most important receptor for TRAIL apoptotic signaling [23]. Therefore, DR5-B can serve as a more specific ligand than TRAIL for tumor targeting.

Doxorubicin is widely employed in anticancer therapy due to its rather broad spectrum of antitumor activities. Presently, DOX is considered as a gold standard to treat several types of sarcomas, carcinomas, lymphomas, etc. However, like other anthracyclines, DOX is known to demonstrate irreversible cardiotoxicity which is related to the total cumulative dose to which the patient has been exposed to [24]. Nevertheless, due to its universal anticancer activity, DOX is still widely used in clinics. Therefore, the development of DOX-loaded capsules is of particular interest.

The aim of the current study was to fabricate biodegradable polyelectrolyte multilayer capsules loaded with doxorubicin and functionalized with the tumor-targeting DR5-B ligand, and to evaluate a combined antitumor effect of this targeted drug delivery system in vitro.

Here, for the first time, an influence of size and surface modifications of the nano-sized DOX-loaded capsules fabricated by the LbL technique on cell uptake as well as on in vitro cytotoxicity was studied. The combination of antitumor activity of the DR5-B ligand with DOX (at a subtoxic concentration) loaded into the capsules allowed us to provide both targeted drug delivery and a synergistic antitumor effect of the DR5-B ligand and doxorubicin.

2. Materials and Methods

2.1. Materials

All chemicals were of analytical grade and used as received without further purification. Calcium chloride dihydrate ($CaCl_2 \cdot 2H_2O$), anhydrous sodium carbonate (Na_2CO_3), glycerol, ethylenediaminetetraacetic acid (EDTA), dextran sulfate sodium (DS) salt (MW 50 kDa), poly-L-arginine hydrochloride (Parg) with MW 15–70 kDa, sodium chloride (NaCl), phosphate buffered saline (PBS), isopropyl-β-D-1-thiogalactopyranoside, Rhodamine 6G (MW 479), Fluorescein isothiocyanate (FITC-) dextran (MW 40 kDa), doxorubicin hydrochloride (MW 580) and Hoechst 33258 were purchased from Sigma-Aldrich; DMSO (dimethyl sulfoxide, 99.5%), Trypan Blue solution, phosphate-buffered saline (PBS, pH 7.4), 0.25% (v/v) trypsin-EDTA solution, Versene solution, Dulbecco's modified Eagle's medium with Phenol Red (DMEM) and MTT (Thiazolyl Blue Tetrazolium Bromide, 98%) were purchased from PanEko (Moscow, Russia). Fetal bovine serum (FBS) was obtained from PAA Laboratories (Pasching, Austria). Fluorescent dyes sulfo-Cyanine3 maleimide and BDP FL maleimide were obtained from Lumiprobe (Moscow, Russia). Deionized water from a three-stage Milli-Q Plus purification system was used in the experiments.

2.2. Expression and Purification of the Recombinant DR5-B Protein

The recombinant DR5-B protein was expressed and purified as previously described [25]. Briefly, the *E. coli* strain SHuffle B T7 was transformed with the pET32a/dr5-b plasmid vector, inoculated into the Terrific Broth (TB) medium containing ampicillin (100 μg/mL) and grown at 37 °C, 250 rpm for 5 h, followed by dilution (1:100) in the TB medium with ampicillin (100 μg/mL). At a cell optical density of 0.6, protein expression was induced by adding 0.05 mM isopropyl-β-D-1-thiogalactopyranoside solution, and the cells were grown overnight (250 rpm, 28 °C). Then, the cells were precipitated (5000 × g) at Beckman Coulter (USA) and stored at −80 °C. For protein purification, the cells were disrupted by a French press (Spectronic Instruments Inc., Irvine, CA, USA) under a pressure of 2000 psi, and DR5-B was purified from a soluble cytoplasmic fraction by metal-affinity chromatography on Ni-NTA agarose (Qiagen, Germantown, MD, USA), followed by ion exchange chromatography on SP Sepharose (GE Healthcare, Danderyd, Sweden). The purified protein was dialyzed against 5 mM Na_2HPO_4 (pH 7.0) and 150 mM of NaCl solutions and sterilized by filtration (Millipore, 0.22 μm). The protein concentration was determined by Bradford assay (Bio-Rad, Hercules, CA, USA). The expression level of the protein was analyzed using sodium dodecyl sulfate polyacrylamide gel electrophoresis (12% SDS-PAGE).

To obtain the fluorescently labeled protein DR5-B, the cysteine-modified protein DR5-B was obtained as described earlier [26]. Further, the cysteine-modified DR5-B protein was labeled by maleimide chemistry coupling either with fluorescent sulfo-Cyanine 3 maleimide or BDP-maleimide dye (Lumiprobe, Moscow, Russia) according to the manufacturer's protocol.

2.3. Preparation of Polyelectrolyte Multilayer Capsules Loaded with Doxorubicin and Functionalized with the DR5-B Protein

The polyelectrolyte multilayer capsules were prepared by LbL technique using colloidal $CaCO_3$ particles as a template. The $CaCO_3$ particles of a submicron size were synthesized as described earlier [27]. In brief, equal volumes of 0.1 M calcium chloride and sodium carbonate salts were mixed with glycerol in a ratio of 1:1:5 ($v/v/v$). After stirring (500 rpm, 60 min), a suspension was centrifuged (12,000× g, 5 min) and the $CaCO_3$ particles were washed three times with deionized water. Dextran sulfate and poly-L-arginine were used to obtain (Parg/DS)$_3$ capsules comprising 3 bilayers, which were assembled on the surface of $CaCO_3$ particles. DS and Parg were alternatively adsorbed from aqueous polymer solutions (2 mg/mL, 0.15 M NaCl), starting from the Parg layer. To dissolve $CaCO_3$ cores, the obtained particles were treated with an EDTA solution (0.5 M, 5 mL) for 15 min. The obtained hollow multilayer capsules (mean size of approx. 500 nm) were divided into two parts, and one part of them was subjected to size minimization via thermal treatment as described earlier [28]. For this purpose, the capsule suspension was incubated at 90 °C for 60 min.

To visualize the capsules by confocal laser scanning microscopy (CLSM), submicron capsules were post-loaded with Rhodamine 6G. For this purpose, the aliquoted PMC suspension (200 µL) samples were mixed with a dye solution (0.01 mg/mL, 200 µL) for 15 min. For doxorubicin loading, the pre-formed (Parg/DS)$_3$ capsules (in an amount of 10^{10} pieces) were re-suspended in a doxorubicin solution (1 mL, 0.02 mg/mL) and shaken for 60 min. Then the capsule samples were centrifuged (12,000× g, 5 min) to collect DOX-loaded capsules and then washed three times with deionized water.

To optimize the modification of the (Parg/DS)$_3$ capsules with the DR5-B protein, an aliquot of capsules was varied in a range of 10^8–3×10^{10} capsules in 250 µL, while a DR5-B amount was fixed (500 µL, 0.8 g/L). The aliquots of the capsule suspension were incubated with the aliquots of the DR5-B solution at stirring (500 rpm for 60 min). Then the capsules were separated by centrifugation (9000× g, 5 min) and washed twice with Milli Q water.

2.4. Characterization of the Polyelectrolyte Capsules

The doxorubicin loading efficiency was determined using UV-VIS PerkinElmer Lamdba C650 spectrophotometer. The absorbances of supernatant samples were detected at 485 nm wavelength (an absorption peak of doxorubicin). To calculate doxorubicin content in supernatants, a calibration curve was obtained based on doxorubicin solutions with previously known concentrations.

The concentrations of the remaining (non-adsorbed on the capsules) DR5-B protein in supernatants were determined by Bradford assay (Bio-Rad, Hercules, CA, USA) according to the manufacturer's protocol. A similar protocol was used to modify PMC samples with the fluorescently labeled DR5-B protein.

The amount of loaded doxorubicin and DR5-B was expressed in the encapsulation efficiency (EE%), which was calculated by the following equation:

$EE\% = \frac{M_{initial} - M_{sup}}{M_{initial}} * 100\%$, where $M_{initial}$ is the mass of doxorubicin/DR5-B in the solution before adding the capsule suspension, M_{sup}—is the mass of doxorubicin/DR5-B, which remained in the supernatant after encapsulation.

The drug loading capacity was calculated as the ratio of the weight of loaded DOX to the weight of CaCO$_3$ particles × 100%.

The shape and surface morphologies of capsule samples were analyzed using a Jeol 7401F field emission scanning electron microscope (SEM). A secondary electron image was taken with a 5 kV electron beam at a working distance of 8.0 mm. The microcapsule suspension was washed with water and loaded on an SEM specimen stub. After drying overnight, the sample was sputter-coated with gold prior to SEM observation.

A surface charge (ζ-potential) and a hydrodynamic capsule size were measured by a Zetasizer ZS (Malvern Instruments, UK). Each ζ-potential value was averaged from three subsequent measurements (each of 15 runs), while a mean size was calculated from five subsequent measurements (each of 20 runs).

Confocal laser scanning microscopy was used to visualize the Rhodamine 6G-modified PMC. Confocal microphotographs were taken on Leica TCS SPE (Leica microsystems, Wetzlar, Germany) equipped with a 100× oil immersion objective.

To quantify the DOX release from PMC, a suspension of DOX-loaded capsules (10 vol. %) was resuspended in PBS (2 mL, pH 7.2) and incubated for 24, 36, 48, 60 and 72 h, respectively. Then the samples were centrifuged (14000 g, 5 min) and the supernatant was replaced with fresh PBS (pH 7.2). The concentration of released DOX in filler fluids was calculated using the following formula:

$R = 100\% \times \frac{OD_{DOX}^R}{OD_{DOX}}$, where R is the drug release and %; OD_{DOX}^R is an optical density of the released DOX measured at certain time points; OD_{DOX} is an optical density of total DOX loaded into the capsules.

2.5. Cell Culture and Multicellular Tumor Spheroid Formation

Human colorectal carcinoma HCT-116 cells were cultivated in DMEM culture medium supplemented with 10% FBS in a humidified atmosphere (5% CO_2) at 37 °C. The cells were detached after treatment with a trypsin-EDTA solution (0.25% v/v), and the culture medium was replaced every 3–4 days. Multicellular tumor spheroids from HCT-116 cells were generated using a simple RGD-induced cell self-assembly technique, which has been recently developed by the authors of [29]. Briefly, cells (50,000 cells per 1 mL medium) were seeded in a 96-well culture plate (100 µL/well) and incubated at 37 °C (5% CO_2) for 2–3 h until the cells became attached to the plate bottom. Then the medium was replaced in each well with 100 µL of complete DMEM (10% FBS) containing cyclo-RGDfK(TPP) peptide (40 µM). Finally, the cells were transferred to a CO_2-incubator, and RGD-induced spheroid formation was observed in 72 h.

2.6. Confocal Microscopy

The cellular uptake of PMC and capsule localization were observed using confocal laser microscope (Nikon TE-2000, Japan). The cells or the tumor spheroids (40,000 cells per glass) were seeded on coverslips in complete DMEM (10% FBS) and transferred to the CO_2–incubator to stay overnight. To visualize nuclei both in 2D (monolayer cell culture) and 3D (the tumor spheroids) in vitro models, the medium was replaced with DMEM containing a 50 µM Hoechst 33258 solution, and the cells or the tumor spheroids were incubated for 15 and 30 min, respectively, followed by washing with PBS (pH 7.4). Then, in the case of the 2D in vitro model, the cells were incubated either with the free DR5-B protein or with the capsules modified with DR5-B (where DR5-B was labeled with red fluorescent dye sulfo-Cyanine3 maleimide) or with unmodified PMC loaded with R6G rhodamine. All capsule samples were previously re-suspended in serum free DMEM (0.5 µL of the capsule suspension per mL of DMEM) in a CO_2-incubator for 15 min and 1 h, respectively. In the case of the 3D in vitro model, the spheroids were incubated with unmodified PMC loaded with R6G rhodamine or with the DR5-B-modified capsules (where DR5-B was labeled with green fluorescent dye BDP FL maleimide) for 15 min and 1 h. Finally, the cells or the spheroids were washed three times with PBS (pH 7.4) and carefully transferred to the coverslips, fixed with a CC/Mount fluorophore protector and observed by confocal microscopy. The excitation/emission ($\lambda_{ex/em}$) wavelength values were 525/548 nm for R6G rhodamine, 555/569 for sulfo-Cyanine3, 503/509 for BDP FL and 352/454 nm for Hoechst 33258.

2.7. Flow Cytometry and Fluorimetry

2.7.1. Flow Cytometry

For flow cytometry analysis, a BD FACSCalibur fluorescent-activated flow cytometer with BD CellQuest software was used. The cells were seeded in a 24-well plate (50,000 cells/well) followed by overnight incubation (37 °C, 5% CO_2). Then the culture medium was removed, and PMC loaded with DOX (0.5 µL of the capsule aliquot per 1 mL of the medium) was added to each well. After treatment, the cells were washed with PBS (pH 7.4) to remove non-internalized PMC. Flow cytometry data were expressed as median fluorescent intensity divided by the background intensity of the control (non-treated cells) after 30 min, 1, 1.5, 3 and 7 h incubation.

2.7.2. Fluorimetry

To quantify the accumulation of free DOX, free DR5-B as well as PMC within the tumor spheroids, the samples previously re-suspended in DMEM were added to the spheroids in a 24-well plate (50,000 cells/well), and the plates were transferred to the CO_2-incubator (37 °C, 5% CO_2) for 1, 3 and 7 h, respectively. The samples were identified as DOX (free DOX), PMC_{300} + DOX (the 300 nm capsules loaded with DOX), PMC_{300}-DR5-B + DOX (the samples, loaded with DOX and modified with DR5-B) and DR5-B (free DR5-B protein). Then the spheroids were washed with PBS (pH 7.4) and treated with DMSO; fluorescence was measured using a Promega GloMax-Multi detection system (USA) at 317 nm for excitation and 375–410 nm for emission. Uptake data were expressed as a percentage of fluorescence associated with the cells versus fluorescence of the feed solution.

2.8. Cytotoxicity Study In Vitro (MTT-Test)

The cells or the spheroids were seeded in a 96-well plate (7500 cells/well) followed by overnight incubation in a CO_2-incubator (37 °C, 5% CO_2). The unmodified PMC, free DR5-B, free DOX as well as the capsules loaded with DOX and/or with the DR5-B protein at various dilutions (0.1, 1, 10, 100 and 1000 ng/mL) were added to each well. The cells or the spheroids were transferred to the CO_2-incubator for 24, 48 and 72 h, respectively. After treatment, the cells or the spheroids were incubated with an MTT-solution (0.05% w/v) in serum-free DMEM for 4 h. Then the medium was replaced with DMSO (100 µL/well), and an absorbance (570 nm) was measured using Multiskan FC reader (Thermo Scientific, USA). The half maximal inhibitory concentration (IC50) values were determined as a drug concentration, which resulted in 50% inhibition of cell growth.

2.9. Statistical Analysis

All data were normally distributed and were expressed as mean or mean ± SD. Statistical analysis of all results was performed using Student's *t*-test. All experiments were carried out with at least three repetitions. Collected data were processed using GraphPad Prism and were accepted as significantly different when $p < 0.05$.

3. Results and Discussion

3.1. Preparation and Characterization of Polyelectrolyte Multicellular Capsules Loaded with Doxorubicin and Modified with the DR5-B Protein

In order to develop a targeted anticancer drug delivery system, a series of biodegradable polyelectrolyte multilayer capsules were obtained. An electrostatic interaction is a dominant driving force for oppositely charged DS and Parg to assemble a multilayer structure on the surface of vaterite submicron particles. In the current study, $(Parg/DS)_3$ capsules comprising of three bilayers were obtained. Biodegradable PMCs were compacted according to the protocol of thermo-induced shrinking. After that, both intact and shrunken PMC were characterized in terms of their physico-chemical properties. Figure 1a,b reveals visible changes in the size and morphology of PMC after heat treatment, which follows from a rearrangement in the interpenetrating multilayers. The DLS-number distributions (Figure 1c,e) demonstrate an intense peak shift to the left as the result of heat treatment of the PMC suspension. As derived from the DLS-number data, the average hydrodynamic size of the heat-treated capsules was 320 ± 80 nm. The comparison of the DLS-volume data before and after heating (Figure 1d,f) shows that a significant amount of micron-sized objects remain in the sample after heating; aggregates probably remain as well. This can also be seen in the DLS-number distribution (Figure 1e) but in this case, the micron-sized peak is not very intense. It is worth mentioning that the number and volume distributions show a relative proportion of a number of differently sized objects as well as the volume occupied by them. As a result, a small amount of the aggregates or larger capsules can dominate in the DLS-volume distribution.

Successful Rhodamine 6G loading into the capsule cavity allowed us to visualize the intact capsules by confocal laser scanning microscopy (Figure 1g). Moreover, the capsules retain their aggregation stability when stored as an aqueous suspension at 4 °C for one month, which was confirmed by ζ-potential measurements. The ζ-potentials of the freshly prepared capsule samples and those ones after storage were −47 ± 6 and −42 ± 8 mV, respectively.

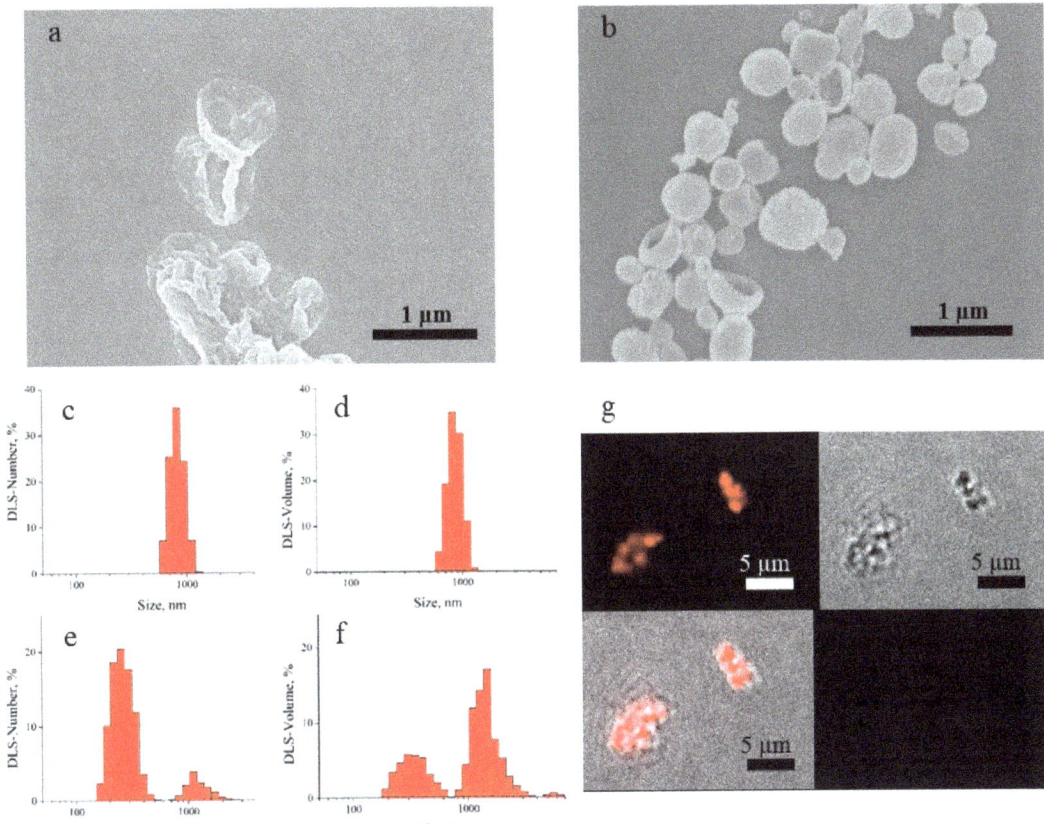

Figure 1. SEM images of the intact (**a**) and heat treated (**b**) (Parg/DS)$_3$ capsules; DLS of the capsule samples: DLS-Number (**c**) and DLS-Volume (**d**) distributions for the intact capsules, DLS-Number (**e**) and DLS-Volume (**f**) distributions for heat treated capsules; CLSM images of the Rhodamine 6G-loaded intact (Parg/DS)$_3$ capsules in fluorescent and transmission channels with their overlay (**g**).

To provide DOX loading in a sub-toxic concentration, DOX encapsulation procedure was optimized. The encapsulation efficiency was found to be 85 ± 7%, which corresponded to approximately 0.002 pg of DOX per capsule.

To optimize the surface modification of the capsules with the DR5-B protein, we carried out a series of experiments by loading a variable number of the capsules in a range of 1×10^8–3×10^{10} with a fixed DR5-B amount. The amount of DR5-B loaded into the capsules was found to depend on the number of capsules incubated in the DR5-B solution (Figure 2).

An increase of the capsule number from 1×10^8 to 2×10^{10} (in other words, decreasing the ratio of the DR5-B and the capsule number) resulted in an enhanced encapsulation efficiency from 12% to 81%. However, a further increase of the capsule number up to 3×10^{10} did not lead to a significant increase of the DR5-B protein loading. Therefore, we have optimized the conditions for efficient modification of the PMC surface with the DR5-B protein. Electrophoretic light scattering measurements showed that the modification of the PMC surface with DR5-B resulted in a change of the ζ-potential from a negative to a positive value (Supporting information, Figure S1a,b). Since the DR5-B molecules have regions that are charged differently (Supporting Information, Figure S1c), with an overall isoelectric point of about nine, the modification of the capsule surface by the protein is based on the electrostatic interaction of negatively charged sulfate groups and positively charged DR5-B units.

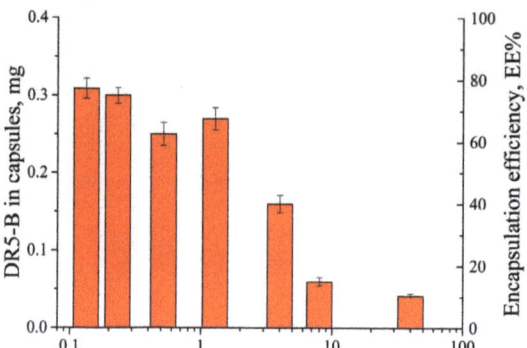

Figure 2. The DR5-B protein loading (mg) into the capsules and encapsulation efficiency (EE%) in the function of the DR5-B/a number of capsules ratio.

It has been shown that the amount of DR5-B protein released from the capsules at storage (4 °C, 4 weeks) was close to zero. Therefore, we could conclude that the DR5-B complex with the capsule shell was rather stable under these conditions.

Thus, it was of interest to compare two PMC types, namely untreated (here and further PMC_{500} according to the mean diameter) and heat-treated ones (here and further PMC_{300}). The DOX release efficiency from PMC_{500} and PMC_{300} is shown in Figure 3. It is clearly seen that the DOX release from the untreated PMC_{500} was significantly higher than that of thermally compressed PMC_{300}. Therefore, after 72 h, about 20% of the drug was released from the PMC_{500}, while only 6% was released from the PMC_{300}. For this reason, the PMC heat treatment could be used to reduce the drug release efficiency and, consequently, to ensure its sustained release.

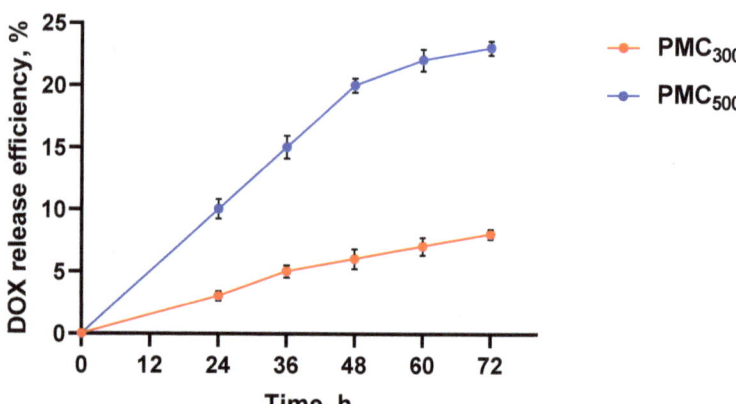

Figure 3. The DOX release efficiency from the PMC_{500} and PMC_{300}. Dialysis in vitro.

3.2. Modification of PMC with Fluorescently Labeled DR5-B

To study an entrapment of the DR5-B protein, we used the modified DR5-B with the amino acid substitution of valine 114 to cysteine (DR5-B/V114C). It was previously designed for the covalent conjugation with polymer-based micellar nanocarriers, as the sulfhydryl group of cysteine residue allows it to conjugate with maleimide groups of various compounds by click chemistry [26]. In this study, DR5-B/V114C was conjugated with fluorophores, namely either sulfo-Cyanine3 maleimide (DR5-B/V114C-Cy3) or BDP FL maleimide (DR5-B/V114C-BDP), to compare fluorophore accumulation efficiency and localization both within the cells alone or those entrapped in PMC. The introduced V114C substitution, as well as the fluorescent label, did not affect the DR5-B ability to induce apoptosis (see Supplementary materials, Figure S2). As seen in Figure S2, for a monolayer culture of HCT-116 cells the cytotoxic activity of free DR5-B, DR5-B/V114C and DR5-B/V114C-Cy3 were identical.

3.3. Study of Capsule Accumulation and Localization in the Cells

The capsule accumulation in the cells was evaluated in two in vitro models, namely, using the monolayer cell culture (2D in vitro model) and the multicellular tumor spheroids (3D in vitro model). For this purpose, confocal microscopy, flow cytometry and fluorimetry were used.

3.3.1. Study of Cellular Uptake of the Capsules by Confocal Microscopy

Qualitative analysis of the capsule accumulation and its intracellular localization was carried out by confocal laser microscopy (Figure 4). For this purpose, unmodified samples PMC_{500} and PMC_{300} loaded with R6G rhodamine dye, the PMC_{500} and PMC_{300} modified with the DR5-B protein and free DR5-B (as a control) were used. As seen from CLSM images, in the case of the monolayer cell culture (2D in vitro model) the free DR5-B protein was found to penetrate HCT-116 cells already after 15 min incubation, while in 1 h no significant fluorescence increase was observed.

The capsule samples PMC_{500} and PMC_{300}, which were not modified with the DR5-B protein but contained the R6G Rhodamine dye, were also uptaken by the HCT-116 cells after 15 min incubation. However, their fluorescence intensity levels were only slightly higher than those of the free DR5-B protein. As for PMC_{500} and PMC_{300} modified with DR5-B, after 15 min incubation their accumulation levels within the cells were obviously higher than those of the unmodified capsules. In addition, after 1 h, the fluorescence levels of the PMC_{500} and PMC_{300} samples were markedly enhanced, which could be explained by the gradual accumulation of the capsules within the cells. It should be noted that the capsules were mainly localized around the nuclei and in the cell cytoplasm. On the other hand, the capsules modified with DR5-B were adsorbed on the cytoplasmic membrane after 15 min and were mainly localized in the cytoplasm, but not in the nuclei. After 1 h, the fluorescence intensity levels of the PMC_{500} and PMC_{300} modified with DR5-B were more or less similar.

The monolayer cell culture (2D in vitro model) is widely used to study cytotoxicity effects as well as to estimate capsule accumulation within the cells. However, the monolayer culture suffers from the lack of cell–cell interactions as well as the cell–matrix ones, and, therefore, it could not mimic in vivo conditions well. Nowadays there are several techniques that can be used to generate the multicellular spheroids that are a 3D in vitro model. They are considered to be a more adequate model for in vitro studies than monolayer cell culture. We have recently developed a simple universal approach based on RGD-induced cell self-assembly technique. This technique was recently used by us successfully to study cytotoxic effects of various nanocarriers loaded with anticancer drugs, for instance diethylaminoethyl dextran/xanthan gum nanocontainers loaded with thymoquinon [30], cerasomes loaded with DOX [31] and liposomes, for gene delivery [32].

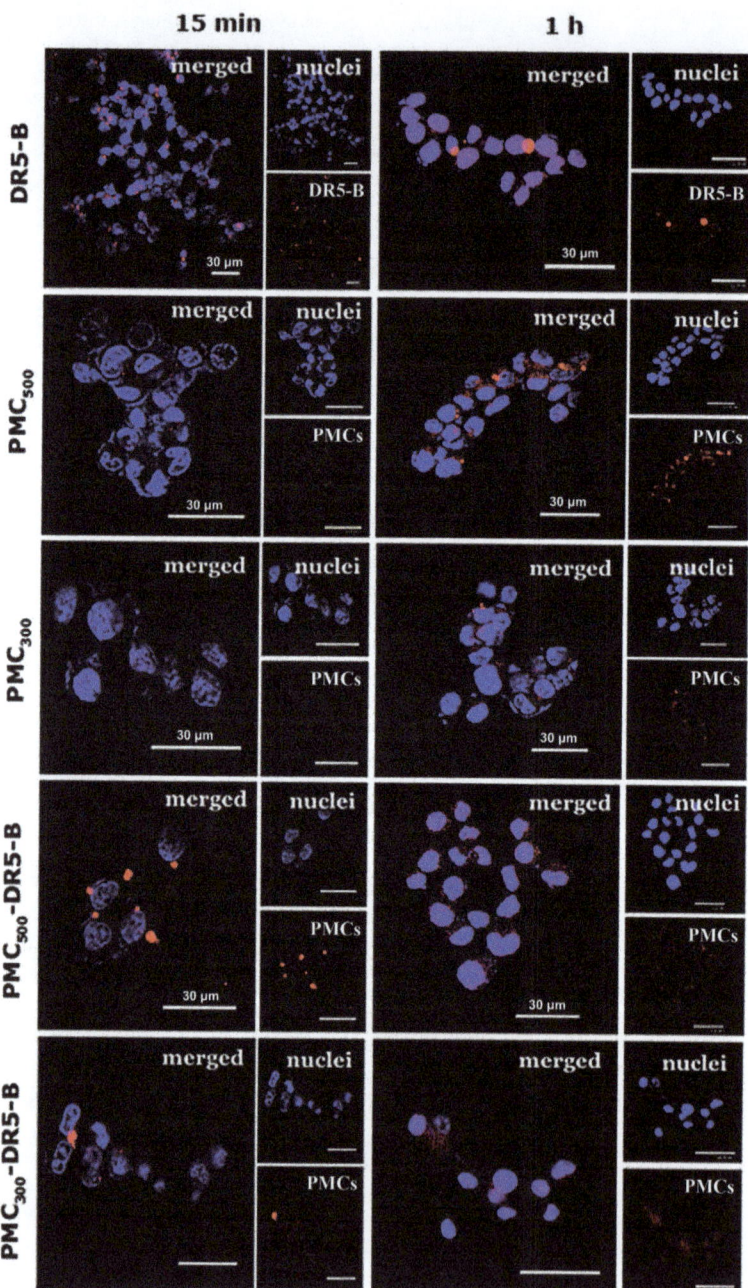

Figure 4. CLSM images of human colorectal carcinoma HCT-116 cells (monolayer cell culture) after 15 min and 1 h incubation with the free DR5-B protein, the unmodified capsule samples (PMC$_{500}$ and PMC$_{300}$ without DR5-B) as well as with the capsules modified with DR5-B. Free DR5-B (labeled with sulfo-Cyanine3 maleimide), the PMC samples modified with DR5-B (labeled with sulfo-Cyanine3 maleimide) and the unmodified PMC (loaded with R6G rhodamine) are in red. Cell nuclei stained with Hoechst 33258 are in blue. Confocal microscopy. The scale bar is 30 μm.

Thus, in the current study, a penetration and accumulation of PMC within the 3D multicellular tumor spheroids generated from HCT116 cells was also studied. As seen in Figure 5, after 15 min incubation unmodified PMC_{500} were observed within the spheroids, and in 1 h their fluorescence did not enhance. After 15 min incubation, the fluorescence intensity level of the unmodified PMC_{300} was a little bit lower than that of PMC_{500}. However, after 1 h incubation the fluorescence values of both PMC_{500} and PMC_{300} were similar. As for DR5-B-modified PMC, there was no significant difference after 15 min and 1 h incubation.

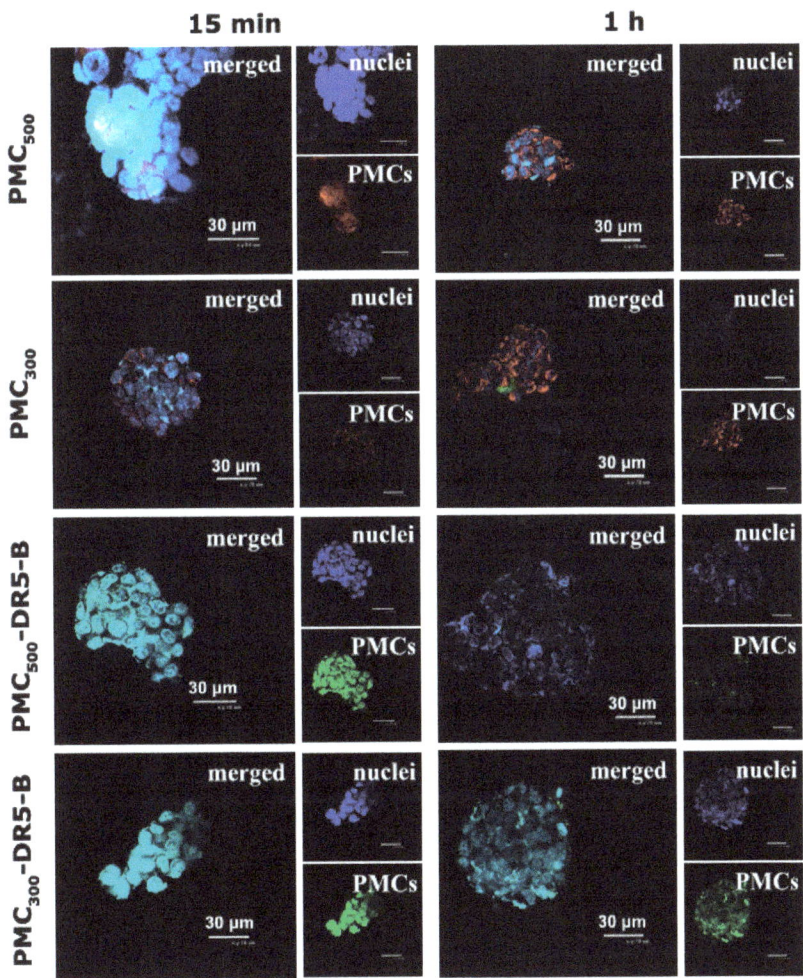

Figure 5. Spheroids from human colorectal carcinoma HCT116 cells (multicellular tumor spheroids) after 15 min and 1 h incubation with PMC_{500} and PMC_{300} samples. The unmodified PMC are stained with R6G rhodamine (in red) and the DR5-B is stained with a BDP FL maleimide dye (in green). Cell nuclei are stained with Hoechst 33258 (in blue). Confocal microscopy. The scale bar is 70 μm.

3.3.2. Study of Cellular Uptake of the Capsules by Flow Cytometry and Fluorimetry

Quantitative evaluation of accumulation efficiency was performed using flow cytometry for the 2D in vitro model and fluorimetry in the case of the 3D in vitro model. For this purpose, PMC_{300} and PMC_{500} loaded with doxorubicin (0.02 mg/mL, 85 ± 7% encapsulation efficiency) were used.

First of all, it was of interest to evaluate a difference in accumulation efficacy levels of DOX-loaded PMC_{300} and PMC_{500} for monolayer cell culture (Figure 6). The obtained results revealed that PMC_{300} accumulation efficiency in HCT116 cells was higher than that for PMC_{500} at all time points (0.5, 1.0, 1.5, 3 and 7 h). For both capsule samples, the accumulation levels were growing with time. The PMC_{300} accumulation levels were 19, 31, 58, 89 and 90% after 0.5, 1, 1.5, 3 and 7 h incubation, respectively, while for PMC_{500} the appropriate values were 17, 22, 40, 76 and 89%. Thus, one could conclude that the smaller capsules PMC_{300} accumulated in the cells faster than the bigger ones (PMC_{500}). Maximum accumulation of PMC_{300} was observed after 3 h incubation and it did not change in 7 h. As for PMC_{500}, their accumulation value continued to gradually increase after 3 h and reached its maximum only after 7 h.

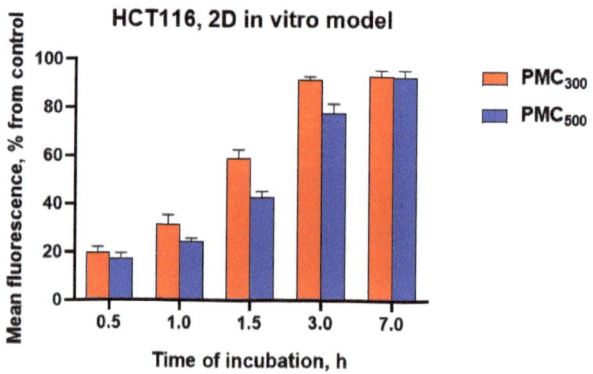

Figure 6. The accumulation levels of PMC_{300} and PMC_{500} loaded with DOX in HCT116 cells after 5, 30 and 60 min incubation. Monolayer cell culture. Flow cytometry.

Tumor vessels are known to be a chaotic mixture of abnormal hierarchically disorganized vessels with many structural and functional differences from healthy tissues. In particular, such differences include the increased pore diameter of tumor blood vessels, characterized by increased permeability to nanoparticles, whose size does not exceed 400 kDa or 500 nm [33]. Thus, a number of studies have revealed that an optimal size of nanoparticles for antitumor drug delivery is within a range of 50–450 nm [34]. Moreover, an accumulation of the particles within the cells was found to be in function of the nanoparticle mean size: the smaller the particles were the higher the accumulation that was revealed [35–37].

In the current study, the DOX-loaded PMC_{300} were modified with DR5-B, in order to provide targeted DOX delivery to tumor cells together with achieving the synergistic antitumor effect. The PMC_{300} sample was chosen since it has shown the higher accumulation efficiency (see Figure 6). To study the cellular uptake of the capsules, flow cytometry (in case of 2D in vitro model) and fluorimetry (for 3D in vitro model) were used (Figure 7). Free DR5-B and PMC_{300}-DR5-B samples were labeled with sulfo-Cyanine3 maleimide. The free DR5-B protein and free DOX were used as controls.

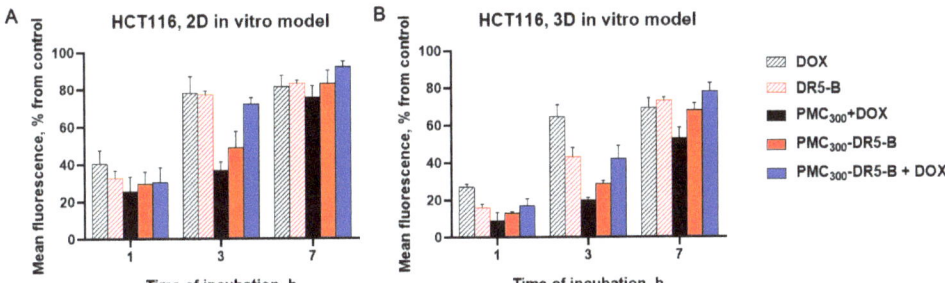

Figure 7. The accumulation efficiency levels of the capsules loaded with DOX (PMC$_{300}$ + DOX), modified with DR5-B (PMC$_{300}$-DR5-B), and the capsules loaded with DOX and modified with DR5-B (PMC$_{300}$-DR5-B + DOX) in monolayer cell culture (**A**) and the tumor spheroids from HCT-116 cells (**B**) after 1, 3 and 7 h incubation. Free DOX and free DR5-B were used as controls. The results of flow cytometry (**A**) and fluorimetry (**B**).

As seen in Figure 7, the maximum fluorescence levels in both 2D and 3D models were revealed for the capsules, which were loaded with DOX and modified with DR5-B. On the other hand, as was expected, the capsules accumulated in the spheroids more slowly than in the 2D in vitro model. This could be explained by the capsule penetration through several cell layers of the tumor spheroids. It should be noted that there was no significant difference in the accumulation of free DOX in 2D and 3D models, which could be related to rather low DOX molecular weight. A similar accumulation tendency for all samples was observed in both 2D and 3D in vitro models. Thus, in the case of the 2D model, for free DOX and free DR5-B samples the highest accumulation levels of 40 and 34%, respectively, were revealed after 1h. These values enhanced up to 78 and 77%, relatively, after 3 h. As for the spheroids, accumulations of both DOX and free DR5-B was slower in them than in the 2D model, namely 26 and 18% (in 1 h) and 63 and 43% (in 3h), respectively. The maximum accumulation levels of free DOX and the free DR5-B protein were revealed after 3 h incubation, in monolayer cell culture, while in tumor spheroids it took 7 h. This is in a good agreement with our previous results [30,31]. As for the capsule samples, the capsules loaded with DOX (PMC$_{300}$+DOX) demonstrated the lowest accumulation efficiencies in all-time points both in 2D and 3D in vitro models. These values were 24, 38 and 77% (2D in vitro model) and 9, 19 and 53% (3D in vitro model) after 1, 3 and 7 h incubation, respectively. As for the capsule samples loaded with DOX and modified with the DR5-B protein (PMC$_{300}$-DR5-B+DOX), their accumulation levels reached maximum values after 7 h in both 2D and 3D models. Therefore, as we expected, the modification of the DOX-loaded capsules with the DR5-B ligand provided their enhanced cellular uptake by HCT-116 cells.

3.4. Study of In Vitro Cytotoxicity Effects of the Capsules

Cytotoxicity effects of the obtained DR5-B-modified PMC were evaluated by MTT-test (Figure 8, Table 1). In monolayer cell culture, the free DR5-B protein was the most cytotoxic in all time points, and IC50 values were 31 ± 2.6, 27 ± 3.2 and 10 ± 0.7 ng/mL after 24, 48 and 72 h, respectively. In contrast, PMC$_{300}$ and PMC$_{500}$ (without DR5-B) were completely non-toxic for the cells even after 72 h incubation with the highest sample concentration (IC50 > 1000 ng/mL). The DR5-B-modified capsules (PMC$_{300}$-DR5-B) were more toxic than PMC$_{500}$-DR5-B, which could be explained by their higher accumulation efficiency (see Figure 6). Thus, IC50 values were 72 ± 8.3 and 519 ± 38.5 ng/mL after 24 h, while after 72 h incubation appropriate values were 10 ± 3.4 and 24 ± 3.1 ng/mL for PMC$_{300}$-DR5-B and PMC$_{500}$-DR5-B, respectively.

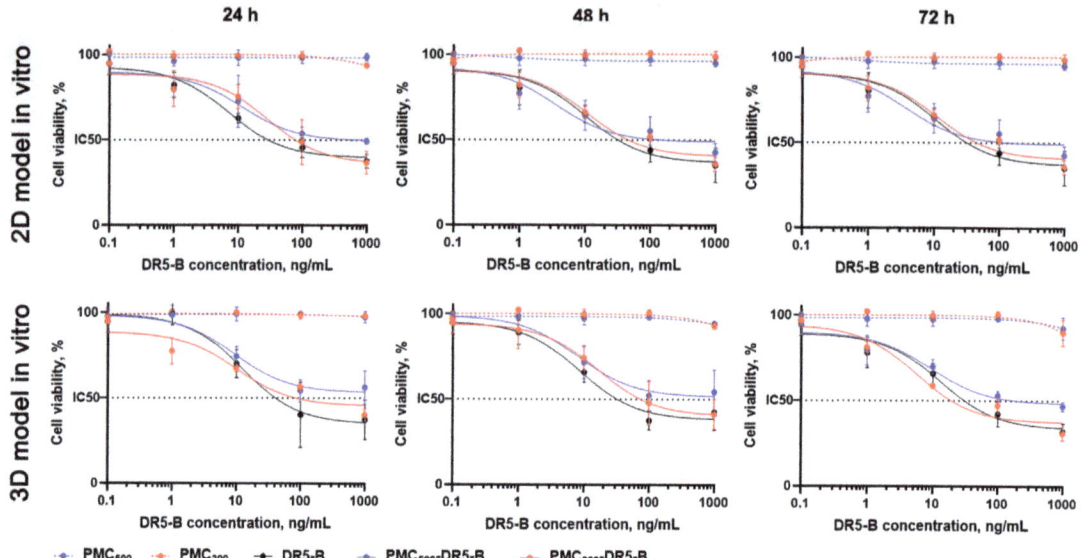

Figure 8. The cytotoxicity effects of the free capsules (PMC_{300} and PMC_{500}) and the capsules modified with DR5-B (PMC_{300}-DR5-B and PMC_{500}-DR5-B) after incubation with human colorectal carcinoma HCT-116 cells after 24, 48 and 72 h incubation. Monolayer cell culture (a top panel) and the tumor spheroids (a lower panel). Free DR5-B protein was used as a control. MTT-test. $p < 0.05$ indicates significant difference according to Student's t-test.

Table 1. The IC50 values of the capsules after incubation with human colorectal carcinoma HCT-116 cells. Monolayer cell culture (2D in vitro model) and tumor spheroids (3D in vitro model). Free DR5-B protein was taken as a control. MTT-test.

Samples	IC50 * Values, ng/mL					
	2D In Vitro Model			3D In Vitro Model		
	24 h	48 h	72 h	24 h	48 h	72 h
PMC_{300}	>1000					
PMC_{500}						
DR5-B	31 ± 2.6	27 ± 3.2	10 ± 0.7	39 ± 4.0	33 ± 2.4	28 ± 5.1
PMC_{300}-DR5-B	72 ± 8.3	42 ± 3.8	10 ± 3.4	96 ± 7.7	77 ± 5.2	19 ± 1.6
PMC_{500}-DR5-B	519 ± 38.5	111 ± 10.9	24 ± 3.1	>1000		109 ± 18.2

* IC50 is half maximal inhibitory concentration.

As we expected, for all spheroids cytotoxicity values were lower than those for the monolayer cell culture. This could be explained by lower capsule penetration and accumulation within the cells in the 3D in vitro model due to a multilayer structure of the spheroids. These data are in good agreement with our previous results [29,32]. The free DR5-B protein was used as a control and demonstrated maximal cytotoxicity effects (see Table 1). The unmodified PMC_{300} and PMC_{500} were non-toxic (>1000 ng/mL). As seen in Table 1, the smaller capsules (PMC_{300}-DR5-B) were more cytotoxic than the bigger ones (PMC_{500}-DR5-B). For instance, after 72 h the IC50 values were 19 ± 1.6 and 109 ± 18.2 ng/mL for PMC$_{300}$-DR5-B and PMC$_{500}$-DR5-B samples, respectively. These results are in good agreement with our previously reported data obtained from the MCF-7 cells [38].

Thus, we have demonstrated that the DR5-B protein entrapped by PMC kept its cytotoxicity in tumor cells. Next, we aimed to investigate the cytotoxic effects of the capsules loaded with DOX and modified with DR5-B. Since the PMC$_{300}$-DR5-B sample exhibited a higher antitumor effect compared to PMC$_{500}$-DR5-B, it was selected for further experiments. The toxicity of the PMC$_{300}$-DR5-B + DOX capsules was studied in both 2D and 3D in vitro models by MTT-test. The results are shown in Figure 9 and Table 2. Free DOX, free DR5-B, a combination of free DR5-B with free DOX (DR5-B+DOX) as well as hollow PMC$_{300}$ capsules were used as controls.

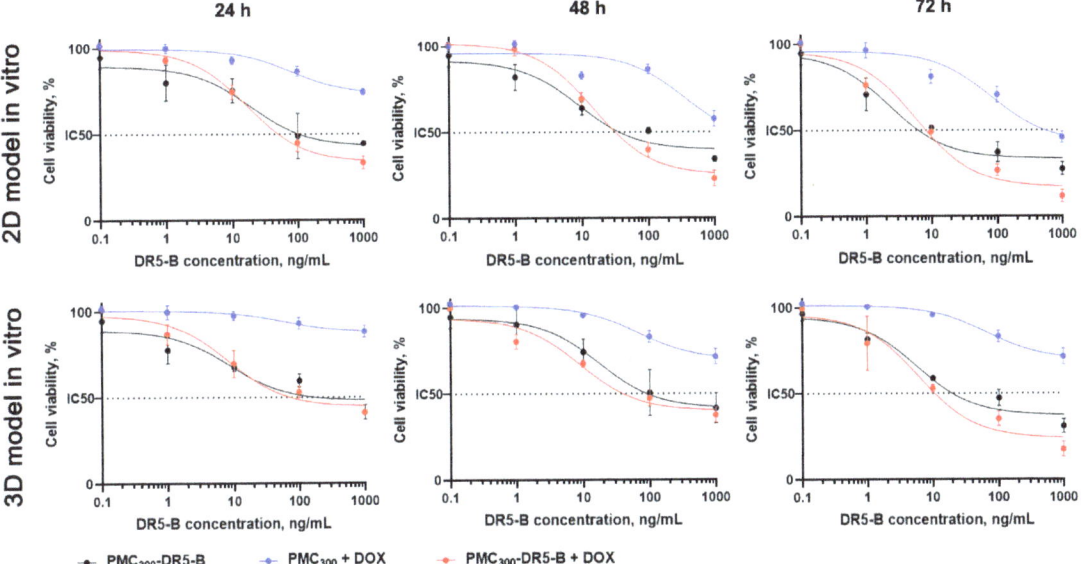

Figure 9. The cytotoxicity effects of the capsules modified with DR5-B (PMC$_{300}$-DR5-B), the DOX-loaded capsules (PMC$_{300}$ + DOX) and the DOX-loaded capsules modified with DR5-B (PMC$_{300}$-DR5-B + DOX) on human colorectal carcinoma HCT-116 cells after 24, 48 and 72 h incubation. Monolayer cell culture (a top panel) and the tumor spheroids (a lower panel). MTT-test. $p < 0.05$ indicates significant difference according to Student's t-test.

Table 2. The IC50 values of the capsules on human colorectal carcinoma HCT-116 cells. Monolayer cell culture (2D in vitro model) and tumor spheroids (3D in vitro model). Free DR5-B protein, free DOX and the combination of free DR5-B with free DOX were taken as controls. MTT-test.

Samples	IC50 * Values, ng/mL					
	2D In Vitro Model			3D In Vitro Model		
	24 h	48 h	72 h	24 h	48 h	72 h
PMC$_{300}$	>1000					
DR5-B	40 ± 5.8	31 ± 4.2	12 ± 2.2	56 ± 8.0	40 ± 8.8	31 ± 5.1
DOX + DR5-B	19 ± 2.4	7 ± 3.0	3 ± 0.9	53 ± 7.6	31 ± 4.3	10 ± 2.0
DOX	>1000	809 ± 31.3	665 ± 27.3	>1000		847 ± 65.3
PMC$_{300}$ + DOX	>1000		593 ± 43.8	>1000		

Table 2. Cont.

Samples	IC50 * Values, ng/mL					
	2D In Vitro Model			3D In Vitro Model		
	24 h	48 h	72 h	24 h	48 h	72 h
PMC$_{300}$-DR5-B	101 ± 9.6	38 ± 6.2	6 ± 1.2	107 ± 25.5	84 ± 10.4	19 ± 6.0
PMC$_{300}$-DR5-B+DOX	52 ± 3.9	30 ± 3.7	7 ± 0.8	78 ± 8.1	47 ± 4.5	11 ± 3.5

* IC50 is half maximal inhibitory concentration.

As seen in Table 2, the IC50 values of PMC$_{300}$-DR5-B+DOX were markedly lower than those of the PMC$_{300}$-DR5-B sample. A similar effect was observed for both controls, namely free DR5-B and the combination of free DR5-B with free DOX. Thus, the capsules, which were loaded with DOX and modified with DR5-B (PMC$_{300}$-DR5-B+DOX), demonstrated synergistically enhanced cytotoxicity in both 2D and 3D in vitro models.

Previously, attempts have been made to modify a variety of nanosystems fabricated by various techniques with TRAIL for tumor targeting [21]. For instance, biodegradable nanoparticles based on poly(epsilon-caprolactone)-poly(ethylene glycol)-poly(epsilon-caprolactone) (PCL-PEG-PCL) loaded with curcumin were prepared by solvent emulsion evaporation technique and then coated with TRAIL protein via electrostatic interactions [39]. Another example is the gel-like PEGylated coacervate microdroplets (mPEG-Coa) loaded with TRAIL for the treatment of colon cancer. These microdroplets were fabricated from poly(ethylene arginyl-aspartate diglyceride) (PEAD) (i.e., polycation), which interacted with heparin (i.e., polyanionic counterpart) by electrostatic interaction in aqueous conditions, forming a gel-like Coa structure. TRAIL was simultaneously entrapped into the gel-like mPEG-Coa droplets [40]. In our study, we have used a simple and universal LbL technique for the fabrication of biodegradable multilayer capsules loaded with DOX and modified with the DR5-B, which is a DR5-specific TRAIL variant. The LbL technique has advantages over both the solvent emulsion evaporation method and the coacervation approach mentioned above. It allowed us to fabricate the hollow capsules (with mean sizes of 300 and 500 nm) with higher loading capacity compared to that of any polymer particles. Moreover, these capsules have potentially reduced immune response due to a variety of polymers for capsule preparation. Finally, the capsules modified with the DR5-selective TRAIL variant could be more advantageous over nanosystems containing a wild-type TRAIL due to DR5-B specific tumor targeting and the enhanced antitumor effect.

4. Conclusions

In this study, the biodegradable (Parg/DS)$_3$ capsules modified with the DR5-B protein and loaded with DOX at subtoxic concentrations were prepared by the LbL technique. The capsules were characterized in terms of their mean size (500 and 300 nm after temperature treatment), ζ-potentials, DR5-B protein loading ability and DOX encapsulation efficacy. Cellular uptake of the capsules as a function of their size and surface modification with the DR5-B protein was evaluated by confocal microscopy, flow cytometry and fluorimetry using human colorectal carcinoma HCT-116 cells in both 2D and 3D in vitro models. The smaller capsules (PMC$_{300}$-DR5-B) were found to be more cytotoxic than the bigger ones (PMC$_{500}$-DR5-B) which was confirmed by MTT-test. As we expected, IC50 values in case of the spheroids (3D in vitro model) were higher than those for the monolayer cell culture (2D model). This could be explained by lower capsule penetration and accumulation within the cells due to a spheroid multilayer structure. The smaller capsules, which were loaded with DOX and modified with DR5-B (PMC$_{300}$-DR5-B + DOX), demonstrated a 2–3-fold cytotoxicity enhancement compared to the capsules without DOX (PMC$_{300}$-DR5-B) in both 2D and 3D in vitro models.

Thus, the developed DOX-loaded capsules modified with the DR5-B protein could be proposed as a novel targeted codelivery system, which was found to show an enhanced antitumor activity due to a rather high tumor targeting and synergistic effect of DR5-B and DOX. The inhibitory effect provided by the capsules containing subtoxic concentrations of DOX holds promise for reducing systemic toxicity while maintaining the therapeutic potential. This approach could be of great interest for the development of various nanodrugs for targeted anticancer therapy.

Supplementary Materials: The following supporting information can be downloaded at: https://www.mdpi.com/xxx/s1, Figure S1: ζ-potential distributions for intact (Parg/DS)$_3$ capsules (a), (Parg/DS)$_3$ capsules after modification with targeted protein DR5-B (b), and for DR5-B protein molecules (c); Figure S2: Cytotoxicity of free DR5-B, DR5-B/V114C and DR5-B/V114C-Cy3 for colorectal carcinoma HCT-116 monolayer cell culture. MTT-test.

Author Contributions: Conceptualization, A.Y.; methodology, D.T. and A.Y.; software, A.G.; investigation, D.T., A.Y., M.G., L.K., I.S. and A.G.; resources, D.T., S.B. and M.G.; data curation, A.Y., A.G. and E.M.; writing—original draft preparation, D.T., A.Y. and A.G.; writing—review and editing, E.M. and S.B.; visualization, D.T. and A.G.; supervision, E.M.; project administration, E.M.; funding acquisition, D.T., A.Y. All authors have read and agreed to the published version of the manuscript.

Funding: This work was done using equipment of the Shared Research Center FSRC Crystallography and Photonics, RAS, and was partly supported by the Ministry of Science and Higher Education of the Russian Federation within the state assignment FSRC Crystallography and Photonics RAS in the part of LbL capsules assembly, characterization and drug loading. The study was partly supported (DR5-B protein expression, purification and fluorescent labeling) by the Russian Science Foundation grant No. 21-14-00224, https://rscf.ru/project/21-14-00224/ (accessed on 12 January 2023).

Data Availability Statement: Not applicable.

Acknowledgments: The authors are grateful to acknowledge V.V. Artemov for obtaining SEM images of the samples.

Conflicts of Interest: The authors declare no conflict of interest.

List of Acronyms

DOX	doxorubicin
DR4	death receptor 4
DR5	death receptor 5
DS	dextran sulfate sodium salt (MW 50 kDa)
FBS	fetal bovine serum
HCT-116	human colorectal carcinoma cell line
LbL	layer-by-layer
MTT	thiazolyl blue tetrazolium bromide, 98%
Parg	poly-L-arginine hydrochloride (MW 15–70 kDa)
PMC	polyelectrolyte multilayer capsules
TB medium	Terrific Broth medium
TRAIL	tumor necrosis factor-related apoptosis inducing ligand

References

1. Mateos-Maroto, A.; Fernández-Peña, L.; Abelenda-Núñez, I.; Ortega, F.; Rubio, R.G.; Guzmán, E. Polyelectrolyte Multilayered Capsules as Biomedical Tools. *Polymers* **2022**, *14*, 479. [CrossRef] [PubMed]
2. Marin, E.; Tapeinos, C.; Sarasua, J.R.; Larrañaga, A. Exploiting the layer-by-layer nanoarchitectonics for the fabrication of polymer capsules: A toolbox to provide multifunctional properties to target complex pathologies. *Adv. Colloid Interface Sci.* **2022**, *304*, 102680. [CrossRef]
3. Cui, J.; Van Koeverden, M.P.; Müllner, M.; Kempe, K.; Caruso, F. Emerging methods for the fabrication of polymer capsules. *Adv. Colloid Interface Sci.* **2014**, *207*, 14–31. [CrossRef] [PubMed]
4. Linnik, D.S.; Tarakanchikova, Y.V.; Zyuzin, M.V.; Lepik, K.V.; Aerts, J.L.; Sukhorukov, G.; Timin, A.S. Layer-by-Layer technique as a versatile tool for gene delivery applications. *Expert Opin. Drug Deliv.* **2021**, *18*, 1047–1066. [CrossRef] [PubMed]
5. del Mercato, L.L.; Rivera-Gil, P.; Abbasi, A.Z.; Ochs, M.; Ganas, C.; Zins, I.; Sönnichsen, C.; Parak, W.J. LbL multilayer capsules: Recent progress and future outlook for their use in life sciences. *Nanoscale* **2010**, *2*, 458–467. [CrossRef]

6. Meng, Q.; Zhong, S.; Gao, Y.; Cui, X. Advances in polysaccharide-based nano/microcapsules for biomedical applications: A review. *Int. J. Biol. Macromol.* **2022**, *220*, 878–891. [CrossRef]
7. Voronin, D.V.; Sindeeva, O.A.; Kurochkin, M.A.; Mayorova, O.; Fedosov, I.V.; Semyachkina-Glushkovskaya, O.; Gorin, D.A.; Tuchin, V.V.; Sukhorukov, G.B. In Vitro and in Vivo Visualization and Trapping of Fluorescent Magnetic Microcapsules in a Bloodstream. *ACS Appl. Mater. Interfaces* **2017**, *9*, 6885–6893. [CrossRef]
8. Novoselova, M.V.; German, S.V.; Sindeeva, O.A.; Kulikov, O.A.; Minaeva, O.V.; Brodovskaya, E.P.; Ageev, V.P.; Zharkov, M.N.; Pyataev, N.A.; Sukhorukov, G.B.; et al. Submicron-Sized Nanocomposite Magnetic-Sensitive Carriers: Controllable Organ Distribution and Biological Effects. *Polymers* **2019**, *11*, 1082. [CrossRef]
9. Verkhovskii, R.; Ermakov, A.; Sindeeva, O.; Prikhozhdenko, E.; Kozlova, A.; Grishin, O.; Makarkin, M.; Gorin, D.; Bratashov, D. Effect of Size on Magnetic Polyelectrolyte Microcapsules Behavior: Biodistribution, Circulation Time, Interactions with Blood Cells and Immune System. *Pharmaceutics* **2021**, *13*, 2147. [CrossRef]
10. Pavlov, A.M.; Gabriel, S.A.; Sukhorukov, G.B.; Gould, D.J. Improved and targeted delivery of bioactive molecules to cells with magnetic layer-by-layer assembled microcapsules. *Nanoscale* **2015**, *7*, 9686–9693. [CrossRef]
11. Svenskaya, Y.; Garello, F.; Lengert, E.; Kozlova, A.; Verkhovskii, R.; Bitonto, V.; Ruggiero, M.R.; German, S.; Gorin, D.; Terreno, E. Biodegradable polyelectrolyte/magnetite capsules for MR imaging and magnetic targeting of tumors. *Nanotheranostics* **2021**, *5*, 362–377. [CrossRef] [PubMed]
12. Zebli, B.; Susha, A.S.; Sukhorukov, G.B.; Rogach, A.L.; Parak, W.J. Magnetic Targeting and Cellular Uptake of Polymer Microcapsules Simultaneously Functionalized with Magnetic and Luminescent Nanocrystals. *Langmuir* **2005**, *21*, 4262–4265. [CrossRef] [PubMed]
13. Valdepérez, D.; del Pino, P.; Sánchez, L.; Parak, W.J.; Pelaz, B. Highly active antibody-modified magnetic polyelectrolyte capsules. *J. Colloid Interface Sci.* **2016**, *474*, 1–8. [CrossRef] [PubMed]
14. Deng, L.; Li, Q.; Al-Rehili, S.; Omar, H.; Almalik, A.; Alshamsan, A.; Zhang, J.; Khashab, N.M. Hybrid Iron Oxide-Graphene Oxide-Polysaccharides Microcapsule: A Micro-Matryoshka for On-Demand Drug Release and Antitumor Therapy in Vivo. *ACS Appl. Mater. Interfaces* **2016**, *8*, 6859–6868. [CrossRef]
15. Deo, D.I.; Gautrot, J.E.; Sukhorukov, G.B.; Wang, W. Biofunctionalization of PEGylated Microcapsules for Exclusive Binding to Protein Substrates. *Biomacromolecules* **2014**, *15*, 2555–2562. [CrossRef]
16. Nifontova, G.; Kalenichenko, D.; Baryshnikova, M.; Ramos Gomes, F.; Alves, F.; Karaulov, A.; Nabiev, I.; Sukhanova, A. Biofunctionalized Polyelectrolyte Microcapsules Encoded with Fluorescent Semiconductor Nanocrystals for Highly Specific Targeting and Imaging of Cancer Cells. *Photonics* **2019**, *6*, 117. [CrossRef]
17. Kolesnikova, T.A.; Kiragosyan, G.; Le, T.H.N.; Springer, S.; Winterhalter, M. Protein A Functionalized Polyelectrolyte Microcapsules as a Universal Platform for Enhanced Targeting of Cell Surface Receptors. *ACS Appl. Mater. Interfaces* **2017**, *9*, 11506–11517. [CrossRef]
18. Cortez, C.; Tomaskovic-Crook, E.; Johnston, A.P.R.; Radt, B.; Cody, S.H.; Scott, A.M.; Nice, E.C.; Heath, J.K.; Caruso, F. Targeting and Uptake of Multilayered Particles to Colorectal Cancer Cells. *Adv. Mater.* **2006**, *18*, 1998–2003. [CrossRef]
19. Cortez, C.; Tomaskovic-Crook, E.; Johnston, A.P.R.; Scott, A.M.; Nice, E.C.; Heath, J.K.; Caruso, F. Influence of Size, Surface, Cell Line, and Kinetic Properties on the Specific Binding of A33 Antigen-Targeted Multilayered Particles and Capsules to Colorectal Cancer Cells. *ACS Nano* **2007**, *1*, 93–102. [CrossRef]
20. Choi, K.Y.; Correa, S.; Min, J.; Li, J.; Roy, S.; Laccetti, K.H.; Dreaden, E.; Kong, S.; Heo, R.; Roh, Y.H.; et al. Binary Targeting of siRNA to Hematologic Cancer Cells In Vivo Using Layer-by-Layer Nanoparticles. *Adv. Funct. Mater.* **2019**, *29*, 1900018. [CrossRef]
21. Dianat-Moghadam, H.; Heidarifard, M.; Mahari, A.; Shahgolzari, M.; Keshavarz, M.; Nouri, M.; Amoozgar, Z. TRAIL in oncology: From recombinant TRAIL to nano- and self-targeted TRAIL-based therapies. *Pharmacol. Res.* **2020**, *155*, 104716. [CrossRef] [PubMed]
22. Gasparian, M.E.; Chernyak, B.V.; Dolgikh, D.A.; Yagolovich, A.V.; Popova, E.N.; Sycheva, A.M.; Moshkovskii, S.A.; Kirpichnikov, M.P. Generation of new TRAIL mutants DR5-A and DR5-B with improved selectivity to death receptor 5. *Apoptosis* **2009**, *14*, 778–787. [CrossRef] [PubMed]
23. McCarthy, M.M.; Sznol, M.; DiVito, K.A.; Camp, R.L.; Rimm, D.L.; Kluger, H.M. Evaluating the Expression and Prognostic Value of TRAIL-R1 and TRAIL-R2 in Breast Cancer. *Clin. Cancer Res.* **2005**, *11*, 5188–5194. [CrossRef] [PubMed]
24. Nolan, M.T.; Lowenthal, R.M.; Venn, A.; Marwick, T.H. Chemotherapy-related cardiomyopathy: A neglected aspect of cancer survivorship. *Intern. Med. J.* **2014**, *44*, 939–950. [CrossRef]
25. Yagolovich, A.V.; Artykov, A.A.; Dolgikh, D.A.; Kirpichnikov, M.P.; Gasparian, M.E. A New Efficient Method for Production of Recombinant Antitumor Cytokine TRAIL and Its Receptor-Selective Variant DR5-B. *Biochem.* **2019**, *84*, 627–636. [CrossRef]
26. Yagolovich, A.V.; Kuskov, A.; Kulikov, P.; Kurbanova, L.; Bagrov, D.; Artykov, A.; Gasparian, M.; Sizova, S.; Oleinikov, V.; Gileva, A.; et al. Amphiphilic Poly(N-vinylpyrrolidone) Nanoparticles Conjugated with DR5-Specific Antitumor Cytokine DR5-B for Targeted Delivery to Cancer Cells. *Pharmaceutics* **2021**, *13*, 1413. [CrossRef]
27. Trushina, D.B.; Bukreeva, T.V.; Antipina, M.N. Size-Controlled Synthesis of Vaterite Calcium Carbonate by the Mixing Method: Aiming for Nanosized Particles. *Cryst. Growth Des.* **2016**, *16*, 1311–1319. [CrossRef]
28. Trushina, D.B.; Bukreeva, T.V.; Borodina, T.N.; Belova, D.D.; Belyakov, S.; Antipina, M.N. Heat-driven size reduction of biodegradable polyelectrolyte multilayer hollow capsules assembled on CaCO3 template. *Colloids Surfaces B Biointerfaces* **2018**, *170*, 312–321. [CrossRef]

29. Akasov, R.; Zaytseva-Zotova, D.; Burov, S.; Leko, M.; Dontenwill, M.; Chiper, M.; Vandamme, T.; Markvicheva, E. Formation of multicellular tumor spheroids induced by cyclic RGD-peptides and use for anticancer drug testing in vitro. *Int. J. Pharm.* **2016**, *506*, 148–157. [CrossRef] [PubMed]
30. Borodina, T.; Gileva, A.; Akasov, R.; Trushina, D.; Burov, S.; Klyachko, N.; González-Alfaro, Y.; Bukreeva, T.; Markvicheva, E. Fabrication and evaluation of nanocontainers for lipophilic anticancer drug delivery in 3D in vitro model. *J. Biomed. Mater. Res. Part B Appl. Biomater.* **2021**, *109*, 527–537. [CrossRef]
31. Gileva, A.; Sarychev, G.; Kondrya, U.; Mironova, M.; Sapach, A.; Selina, O.; Budanova, U.; Burov, S.; Sebyakin, Y.; Markvicheva, E. Lipoamino acid-based cerasomes for doxorubicin delivery: Preparation and in vitro evaluation. *Mater. Sci. Eng. C* **2019**, *100*, 724–734. [CrossRef]
32. Koloskova, O.O.; Gileva, A.M.; Drozdova, M.G.; Grechihina, M.V.; Suzina, N.E.; Budanova, U.A.; Sebyakin, Y.L.; Kudlay, D.A.; Shilovskiy, I.P.; Sapozhnikov, A.M.; et al. Effect of lipopeptide structure on gene delivery system properties: Evaluation in 2D and 3D in vitro models. *Colloids Surfaces B Biointerfaces* **2018**, *167*, 328–336. [CrossRef]
33. Jain, R.K. Transport of Molecules, Particles, and Cells in Solid Tumors. *Annu. Rev. Biomed. Eng.* **1999**, *1*, 241–263. [CrossRef]
34. Euliss, L.E.; DuPont, J.A.; Gratton, S.; DeSimone, J. Imparting size, shape, and composition control of materials for nanomedicine. *Chem. Soc. Rev.* **2006**, *35*, 1095. [CrossRef]
35. Yuan, X.; Zhao, X.; Lin, Y.; Su, Z. Polydopamine-Based Nanoparticles for an Antibiofilm Platform: Influence of Size and Surface Charge on Their Penetration and Accumulation in *S. aureus* Biofilms. *Langmuir* **2022**, *38*, 10662–10671. [CrossRef] [PubMed]
36. Jiang, Q.; Liu, Y.; Guo, R.; Yao, X.; Sung, S.; Pang, Z.; Yang, W. Erythrocyte-cancer hybrid membrane-camouflaged melanin nanoparticles for enhancing photothermal therapy efficacy in tumors. *Biomaterials* **2019**, *192*, 292–308. [CrossRef] [PubMed]
37. Schädlich, A.; Caysa, H.; Mueller, T.; Tenambergen, F.; Rose, C.; Göpferich, A.; Kuntsche, J.; Mäder, K. Tumor Accumulation of NIR Fluorescent PEG–PLA Nanoparticles: Impact of Particle Size and Human Xenograft Tumor Model. *ACS Nano* **2011**, *5*, 8710–8720. [CrossRef] [PubMed]
38. Trushina, D.B.; Akasov, R.A.; Khovankina, A.V.; Borodina, T.N.; Bukreeva, T.V.; Markvicheva, E.A. Doxorubicin-loaded biodegradable capsules: Temperature induced shrinking and study of cytotoxicity in vitro. *J. Mol. Liq.* **2019**, *284*, 215–224. [CrossRef]
39. Yang, X.; Li, Z.; Wu, Q.; Chen, S.; Yi, C.; Gong, C. TRAIL and curcumin codelivery nanoparticles enhance TRAIL-induced apoptosis through upregulation of death receptors. *Drug Deliv.* **2017**, *24*, 1526–1536. [CrossRef]
40. Kim, S.; Jwa, Y.; Hong, J.; Kim, K. Inhibition of Colon Cancer Recurrence via Exogenous TRAIL Delivery Using Gel-like Coacervate Microdroplets. *Gels* **2022**, *8*, 427. [CrossRef]

Disclaimer/Publisher's Note: The statements, opinions and data contained in all publications are solely those of the individual author(s) and contributor(s) and not of MDPI and/or the editor(s). MDPI and/or the editor(s) disclaim responsibility for any injury to people or property resulting from any ideas, methods, instructions or products referred to in the content.

MDPI AG
Grosspeteranlage 5
4052 Basel
Switzerland
Tel.: +41 61 683 77 34

Nanomaterials Editorial Office
E-mail: nanomaterials@mdpi.com
www.mdpi.com/journal/nanomaterials

Disclaimer/Publisher's Note: The statements, opinions and data contained in all publications are solely those of the individual author(s) and contributor(s) and not of MDPI and/or the editor(s). MDPI and/or the editor(s) disclaim responsibility for any injury to people or property resulting from any ideas, methods, instructions or products referred to in the content.

www.ingramcontent.com/pod-product-compliance
Lightning Source LLC
LaVergne TN
LVHW070725100526
838202LV00013B/1173